Lecture Notes in Economics and Mathematical Systems

686

Founding Editors:

M. Beckmann
H.P. Künzi

Managing Editors:

Prof. Dr. G. Fandel
Fachbereich Wirtschaftswissenschaften
Fernuniversität Hagen
Hagen, Germany

Prof. Dr. W. Trockel
Murat Sertel Institute for Advanced Economic Research
Istanbul Bilgi University
Istanbul, Turkey

and

Institut für MathematischeWirtschaftsforschung (IMW)
Universität Bielefeld
Bielefeld, Germany

Editorial Board:

H. Dawid, D. Dimitrov, A. Gerber, C.-J. Haake, C. Hofmann, T. Pfeiffer,
R. Slowiński, W.H.M. Zijm

More information about this series at http://www.springer.com/series/300

Torben Kuschel

Capacitated Planned Maintenance

Models, Optimization Algorithms, Combinatorial and Polyhedral Properties

Torben Kuschel
WINFOR–Business Computing
  and Operations Research
University of Wuppertal
Wuppertal, Germany

Accepted dissertation at Bergische Universität Wuppertal, Germany

ISSN 0075-8442　　　　　　　ISSN 2196-9957　(electronic)
Lecture Notes in Economics and Mathematical Systems
ISBN 978-3-319-40288-8　　　ISBN 978-3-319-40289-5　(eBook)
DOI 10.1007/978-3-319-40289-5

Library of Congress Control Number: 2016957870

© Springer International Publishing Switzerland 2017
This work is subject to copyright. All rights are reserved by the Publisher, whether the whole or part of
the material is concerned, specifically the rights of translation, reprinting, reuse of illustrations, recitation,
broadcasting, reproduction on microfilms or in any other physical way, and transmission or information
storage and retrieval, electronic adaptation, computer software, or by similar or dissimilar methodology
now known or hereafter developed.
The use of general descriptive names, registered names, trademarks, service marks, etc. in this publication
does not imply, even in the absence of a specific statement, that such names are exempt from the relevant
protective laws and regulations and therefore free for general use.
The publisher, the authors and the editors are safe to assume that the advice and information in this book
are believed to be true and accurate at the date of publication. Neither the publisher nor the authors or
the editors give a warranty, express or implied, with respect to the material contained herein or for any
errors or omissions that may have been made.

Printed on acid-free paper

This Springer imprint is published by Springer Nature
The registered company is Springer International Publishing AG Switzerland

Meinen lieben Eltern.

*Dank Euch konnte ich studieren
und dieses Buch schreiben.*

*For in O.R. we work for management:
ours is essentially empirical science, and
its subject matter is decision-taking,
policy-making and control.*

S. Beer, "An operational research approach to the nature of conflict,"
Political Studies, vol. XIV (2), pp. 119–132, 1966.

Preface

While I wrote my bachelors thesis in chemistry, I attended my first two lectures in business and economics. In the first lesson of operations research, Prof. Dr. Andreas Drexl from the University of Kiel introduced the set packing problem and its application to auctions. I was completely taken away. Thereupon, operations research became my academic passion, which culminates in this work. In the thesis, the intrinsic beauty of the polyhedral Theorem 4.1, the enumeration in Lemma 5.15, and the Primal-Dual Algorithm 6.5 is most appealing.

Because a research project is made possible with the support of several colleagues and friends, I would like to acknowledge all that directly or indirectly contributed.

It is truly a great honor to express my gratitude and indebtedness to my research supervisor *Prof. Dr. Stefan Bock*. I was employed at his chair from October 2006 to March 2012 at the University of Wuppertal, where I worked this dissertation out. Thank you very much for your scientific guidance, the valuable discussions, and the freedom within my research. I would also like to thank *Prof. Dr. Dirk Briskorn* for refereeing this thesis.

My thanks go to *Dipl.-Math. Christian Rathjen* for the collaboration on the scheduling problem, *Dr. Simon Görtz* for the collaboration on the lot-sizing problem, *Dr. Volker Arendt* and *Dipl.-Wirt.-Inf. Rudolf Bauer* for computer systems support, *Dipl.-Wirtschaftschem. Andreas Hofmann* because your project inspired me to the maintenance problem, my colleagues Christian T., Francesco, Iwona, Melanie, Mila, Paul, and Sarah, the students (teaching was always fun), Christoph, Benjamin and Lennart, and Ana for the support.

My innermost thanks and my heart belong to my family and to my friends.

Düsseldorf, Germany
July 2015

Torben Kuschel

Contents

1	**Introduction**	1
	References	5
2	**The Capacitated Planned Maintenance Problem**	7
	2.1 Problem Definition and Motivation	7
	2.2 Known Maintenance Problems	10
	2.2.1 Periodic Maintenance	10
	2.2.2 Machine Scheduling with Periodic Maintenance	11
	2.2.3 Periodic Maintenance Inspection	12
	2.2.4 Aircraft Maintenance	13
	2.2.5 Other Maintenance Approaches	13
	2.3 The Mathematical Formulation	15
	2.3.1 Assumptions and Terminology	17
	2.3.2 Data Structure	19
	References	21
3	**Known Concepts and Solution Techniques**	25
	3.1 Computational Complexity	25
	3.2 Linear and Integer Programming	27
	3.2.1 The Simplex Algorithm	34
	3.2.2 The Primal-Dual Simplex Algorithm	34
	3.3 Dual Decomposition: Lagrangean Relaxation	35
	3.3.1 Column Generation	41
	3.3.2 Subgradient Optimization	43
	3.4 Primal Decomposition: Benders' Reformulation	49
	3.5 Local Search and Tabu Search	52
	3.5.1 Local Search	52
	3.5.2 Tabu Search	53
	3.6 The Knapsack Problem	55
	3.6.1 Valid Inequalities	56
	3.6.2 Lower and Upper Bounds	59
	3.6.3 An Exact Algorithm	62
	References	65

4 The Weighted Uncapacitated Planned Maintenance Problem ... 71
- 4.1 The Mathematical Formulation ... 71
- 4.2 Polyhedral Properties ... 74
- 4.3 Computational Complexity ... 82
- 4.4 The Single Weighted Uncapacitated Planned Maintenance Problem ... 89
  - 4.4.1 An Optimal Solution to the Primal Problem ... 89
  - 4.4.2 An Optimal Solution to the Corresponding Dual Problem of the LP Relaxation ... 91
- 4.5 The Uncapacitated Planned Maintenance Problem ... 94
- References ... 101

5 Analyzing the Solvability of the Capacitated Planned Maintenance Problem ... 103
- 5.1 Valid Inequalities and Polyhedral Properties ... 103
- 5.2 Computational Complexity ... 112
- 5.3 Lower Bounds ... 127
  - 5.3.1 Considered Lower Bounds ... 127
  - 5.3.2 Relative Strengths and Computational Complexity ... 133
  - 5.3.3 Transformations of the Lower Bounds ... 144
    - 5.3.3.1 Network Flow Problems ... 145
    - 5.3.3.2 Facility Location Problems ... 147
  - 5.3.4 Mathematical Formulations of the Lagrangean Duals ... 150
    - 5.3.4.1 Lagrangean Duals $Z_{(R)}^{(P)+(Q)}$ and $Z_{(R)}^{(P)(X)+(Q)}$ ... 150
    - 5.3.4.2 Lagrangean Dual $Z_{(P)}$ ... 154
    - 5.3.4.3 Lagrangean Dual $Z_{(C)}$ ... 156
    - 5.3.4.4 Lagrangean Dual $Z_{(P)/(C)}$ ... 157
    - 5.3.4.5 Lagrangean Dual $Z_{(V)}^{(Y)}$ ... 161
- References ... 162

6 Algorithms for the Capacitated Planned Maintenance Problem ... 165
- 6.1 Three Construction Heuristics ... 165
  - 6.1.1 The First Fit Heuristic ... 166
  - 6.1.2 The Overlap Heuristic ... 167
  - 6.1.3 The Iterated Best-of-Three Heuristic ... 170
- 6.2 Two Lagrangean Heuristics ... 170
  - 6.2.1 The Lagrangean Relaxation of the Capacity Constraint ... 171
    - 6.2.1.1 The LP Lower Bound ... 172
    - 6.2.1.2 The Dual Priority Rule Lower Bound ... 174
    - 6.2.1.3 The Combined Lower Bound ... 176
    - 6.2.1.4 The Primal-Dual Lower Bound ... 176
    - 6.2.1.5 The Shortest Path Lower Bound ... 183
    - 6.2.1.6 The Upper Bound Heuristic ... 183

		6.2.2	The Lagrangean Relaxation of the Period Covering Constraint	186

- 6.2.2 The Lagrangean Relaxation of the Period Covering Constraint 186
- 6.2.3 The Lagrangean Heuristic for Both Relaxations 187
  - 6.2.3.1 Initial Lagrangean Multiplier 189
  - 6.2.3.2 A Lagrangean Heuristic............................ 191
- 6.3 A More Sophisticated Lagrangean Hybrid Heuristic 195
  - 6.3.1 A General Approach to Link Two Lagrangean Relaxations 195
  - 6.3.2 The Lagrangean Hybrid Heuristic 201
    - 6.3.2.1 Initial Extended Cover Inequalities 202
    - 6.3.2.2 A Hybrid Heuristic 202
  - 6.3.3 Mathematical Formulations of the Auxiliary LPs........... 207
    - 6.3.3.1 The Auxiliary LP $DL_{(C)}$ 207
    - 6.3.3.2 The Auxiliary LP $DL_{(P)}$ 209
- 6.4 A Tabu Search Heuristic 211
  - 6.4.1 The Tabu Search Heuristic 211
  - 6.4.2 Bitshifting in the Calculation of the Tabu Search Objective Function 218
- References 221

7 Computations for the Capacitated Planned Maintenance Problem 223
- 7.1 Instance Generation and Test-Sets..................................... 223
  - 7.1.1 A Value for the Mean $\bar{r}^{\varnothing}$ 224
  - 7.1.2 Test-Sets 225
- 7.2 Absolute Strength of the Lower Bounds 225
- 7.3 Performance of the Heuristics 230
  - 7.3.1 Heuristics to the Lagrangean Relaxations and Pseudo-Subgradient Optimization 231
    - 7.3.1.1 The Performance of the Heuristics 231
    - 7.3.1.2 Applying Different Heuristics in the Lagrangean Heuristic 233
    - 7.3.1.3 The Approximation of the Lagrangean Duals 237
  - 7.3.2 Heuristics to the Planned Maintenance Problem 239
    - 7.3.2.1 The Construction Heuristics 239
    - 7.3.2.2 Variants of the Lagrangean Hybrid Heuristic 241
    - 7.3.2.3 The Metaheuristics 254
- Reference 265

8 Final Remarks and Future Perspectives................................... 267

A Additional Material for the Capacitated Planned Maintenance Problem ... 273
A.1 Problem Extensions ... 273
A.1.1 Capacitated Weighted Planned Maintenance Problems ... 273
A.1.2 Multiple Machines ... 274
A.1.3 Multiple Repairmen ... 274
A.1.4 Maintenance Precedence Constraints ... 275
A.1.5 Minimal Time Between Maintenance Activities ... 275
A.1.6 Integration with Lot-Sizing ... 276
A.2 Mathematical Formulations ... 278
References ... 282

Index ... 283

List of Algorithms

3.1	A column generation algorithm	42
3.2	A Lagrangean subgradient heuristic	48
3.3	A local search heuristic	52
3.4	A tabu search heuristic	54
3.5	All extended cliques for the KP	59
3.6	The greedy solution for the KP	61
3.7	The Combo Algorithm for the KP	64
4.1	An optimal solution to the SWUPMP	91
4.2	An optimal solution to the corresponding dual problem of the SWUPMP LP relaxation	94
6.1	The First Fit Heuristic for the CPMP	166
6.2	The Overlap Heuristic for the CPMP	167
6.3	The Iterated Best-of-Three Heuristic for the CPMP	170
6.4	The Dual Priority Rule Lower Bound for $ZL_{(C)}$	174
6.5	The Primal-dual Lower Bound for $ZL_{(C)}$	182
6.6	The Upper Bound Heuristic for $ZL_{(C)}$	184
6.7	The Knapsack Heuristic for $ZL_{(P)}$	186
6.8	The Lagrangean Heuristic to the CPMP for $Z_{(C)}$ and $Z_{(P)}$	193
6.9	Initial extended covers for the CPMP	202
6.10	The Lagrangean Hybrid Heuristic for the CPMP starting with $ZL_{(C)}$	203
6.11	The Tabu Search Heuristic for the CPMP	216
7.1	A value for the mean $\bar{r}^{\varnothing}$	224

List of Figures

Fig. 1.1	Decision making grid for maintenance strategies	3
Fig. 2.1	The data structure representing a solution to the CPMP	20
Fig. 3.1	Concavity of the Lagrangean objective function	44
Fig. 3.2	A subgradient step	46
Fig. 4.1	Reformulation of (P) as a path finding problem	73
Fig. 5.1	Network-transformation: directed graph	146
Fig. 6.1	Algorithm 6.8 solves one instance	194
Fig. 6.2	Linking two Lagrangean relaxations	195
Fig. 6.3	The Lagrangean Hybrid Heuristic	201
Fig. 6.4	An example to the Lagrangean Hybrid Heuristic	206
Fig. 6.5	Algorithm 6.10 solves one instance	207
Fig. 6.6	Shifting the scope through the planning horizon in various states	214
Fig. 6.7	Algorithm 6.11 solves one instance	217
Fig. 7.1	Estimated evolution of the relative error of the upper bounds and of BestLB from Table 7.14	260
Fig. 7.2	Evolution of the relative error of the upper bounds of the CPMP from Table 7.14 to BestLB for a test-set from Table 7.20 with low capacity	264
Fig. 7.3	Evolution of the relative error of the upper bounds of the CPMP from Table 7.14 to BestLB for a test-set from Table 7.20 with medium capacity	264
Fig. 7.4	Evolution of the relative error of the upper bounds of the CPMP from Table 7.14 to BestLB for a test-set from Table 7.20 with high capacity	265

List of Tables

Table 2.1	Parameters and variables of the CPMP	16
Table 2.2	A feasible solution to the CPMP	16
Table 2.3	Variable fixation in an initial and a terminal planning horizon	18
Table 3.1	Dualization rules	29
Table 3.2	Combinations of solutions to the primal and the dual problem	30
Table 3.3	Comparison of primal and dual decomposition	51
Table 4.1	Coefficient matrix of (P) and its stairway structure	72
Table 4.2	The convex combination yields the fractional solution $\bar{x}$	75
Table 4.3	The transformation of the WUPMP to an extended UFLP	80
Table 4.4	Computational complexity of uncapacitated planned maintenance problems	83
Table 4.5	Coefficient matrix of (P) and (V) for $n = 2$ and $\pi_i = T \ \forall i$	86
Table 4.6	Coefficient matrix of (P) and (V) for $\pi_i = 2 \ \forall i$	87
Table 4.7	Example of an optimal solution to the UPMP	95
Table 4.8	Example that Lemma 4.17 does not minimize the total number of tasks	96
Table 4.9	The convex combination yields the fractional solution $(\bar{x}, \bar{y})$	98
Table 4.10	The convex combination yields the fractional solution $(\bar{x}, \bar{y})$	101
Table 4.11	Optimal solutions to Lemmata 4.20 and 4.12	101
Table 5.1	Coefficient matrix of (Q) and (R)	105
Table 5.2	Upper bounds on the number of tasks for (U)	108
Table 5.3	Coefficient matrix of (P) and (C) for $r_i = 1 \ \forall i \wedge \pi_i = T \ \forall i \wedge \bar{r}_t \in \{0, 1\} \ \forall t$	111
Table 5.4	The convex combination yields the fractional solution $(\bar{x}, \bar{y})$	112
Table 5.5	Computational complexity of capacitated planned maintenance problems	113
Table 5.6	The convex combination yields the fractional solution $(\bar{x}, \bar{y})$	131
Table 5.7	Computational complexity of the relevant lower bounds from Theorem 5.22	133

Table 5.8	Relative strength: instances	135
Table 5.9	Relative strength: optimal objective function values for Table 5.8	135
Table 5.10	Relative strength: comparison of Table 5.9	136
Table 5.11	Coefficient matrix of (Q) and (R) for Lemma 5.27	142
Table 5.12	Facility-transformation: transportation costs C_{jt}	149
Table 5.13	Coefficient matrix of Example 5.11	151
Table 6.1	Bitshift implementation of $a \cdot c$	218
Table 6.2	Bitshift implementations of $\left(\frac{a}{b}\right)^2 \cdot c$	221
Table 7.1	Values R for the mean $\bar{r}^\varnothing$ for all test-subsets obtained with Algorithm 7.1	226
Table 7.2	Relative error in % of the lower bounds from Theorem 5.22 to the optimum of the CPMP generated by CPLEX	227
Table 7.3	Relative error in % of the lower bounds from Theorem 5.22, which are rounded up to the next integer, to the optimum of the CPMP generated by CPLEX	228
Table 7.4	Computational time in seconds to obtain the values of Tables 7.2 and 7.3	229
Table 7.5	Lower and upper bounds used in Tables 7.6 and 7.7	232
Table 7.6	Performance of the heuristics to $ZL_{(P)}$ and $ZL_{(C)}$ from Table 7.5. Relative error in % of the lower bounds to the optimum of the Lagrangean relaxation	234
Table 7.7	Performance of the heuristics to $ZL_{(P)}$ and $ZL_{(C)}$ from Table 7.5. Relative error in % of the upper bounds to the optimum of the Lagrangean relaxation	236
Table 7.8	Lagrangean heuristics and lower bounds used in Tables 7.9, 7.10, 7.11, 7.12, and 7.13	238
Table 7.9	Performance of the Lagrangean heuristics and lower bounds from Table 7.8. Relative error in % of the lower bounds, which are rounded up to the next integer, to the optimum of the CPMP generated by CPLEX	240
Table 7.10	Performance of the Lagrangean heuristics from Table 7.8. Relative error in % of the upper bounds to the optimum of the CPMP generated by CPLEX	242
Table 7.11	Performance of the Lagrangean heuristics from Table 7.8. Number of iterations per second in s^{-1} to yield the results from the Tables 7.9 and 7.10	244
Table 7.12	Approximation of the Lagrangean dual by the Lagrangean heuristics and lower bounds from Table 7.8	244
Table 7.13	Performance of the Lagrangean heuristics from Table 7.8. Relative error in % of the lower bounds to the optimum of the CPMP generated by CPLEX	245

Table 7.14	Construction heuristics, metaheuristics, lower and upper bounds used in Tables 7.15, 7.16, 7.17, 7.18, 7.19, and 7.20	245
Table 7.15	Performance of the construction heuristics from Table 7.14. Relative error in % of the upper bounds to the optimum of the CPMP generated by CPLEX	246
Table 7.16	Relative error in % of the results of UChKh from Table 7.14 to the optimum of the CPMP generated by CPLEX for different values of α	248
Table 7.17	Variants of the Lagrangean hybrid heuristics from Table 7.14. Relative error in % of the results to the optimum of the CPMP generated by CPLEX	251
Table 7.18	Performance of the metaheuristics from Table 7.14. Relative error in % of the results to the optimum of the CPMP generated by CPLEX	255
Table 7.19	Performance of the metaheuristics from Table 7.14. Relative error in % of the lower bounds, which are rounded up to the next integer, to BestLB	258
Table 7.20	Performance of the metaheuristics from Table 7.14. Relative error in % of the upper bounds to BestLB	261

Abbreviations

$\ll$	The bitshift operator $a \ll b$ shifts the number a by b positions to the left in the CPU register	218
$\lfloor a \rfloor$	Rounding down a real a to the next integer (Gauss bracket)	74
$\lceil a \rceil$	Rounding up a real a to the next integer (Gauss bracket)	74
$\leq_p$	A decision problem P′ polynomially reduces to the decision problem under investigation P that is P′ $\leq_p$ P	26
$\Vert a \Vert$	Euclidian norm of a vector a	45
(A)	Aggregated covering constraint	105
A	Goal weight of penalization infeasibility of (C) in the tabu search objective function	212
A_t	Pointer to a list that contains all list elements associated with tasks scheduled in period t	19
α_{it}	Dual variable of the i, tth simple upper bound constraint (S^x)	172
B	Goal weight of the number of established maintenance slots in the tabu search objective function	212
B2e	Best bounds of Ue, Ke, each with halved computational time	239
B2h	Best bounds of UCh, Kh, each with halved computational time	239
B3	Best upper bound of FF, OF and OS	239
BestLB	Best lower bound of Ke, Kh, Ue, UCh, UChKh, B2e, B2h, $Z^{(X)(Y)}$ that are rounded up to the next integer	239
β_t	Dual variable of the tth simple upper bound constraint (S^y)	172
c	Capacity in the KP	55
$\bar{c}$	Residual capacity in the KP	59
(C)	Capacity constraint	16
C	Goal weight of the bonus per period in the tabu search objective function	212

xxi

C_{iab}	Flow costs of the arc $(a, b) \in E$ of maintenance activity i in the network-transformation	145
c_{it}	Variable costs of the WUPMP	71
C_{jt}	Transportation cost if customer j is served from facility t in the facility-transformation	148
C'_t	Index set of the extended cover inequalities (O) for period t	103
C''_t	Index set of the extended clique inequalities (L) for period t	103
CBM	Condition-Based Maintenance	2
cf.	confer, compare	3
CFLP	Capacitated Facility Location Problem	144
CM	Corrective Maintenance	1
$Conv(1)\ldots(K)$	Convex hull of the constraints $(1),\ldots,(K)$	31
CPMP	Capacitated Planned Maintenance Problem	4
CRC	Cyclic Redundancy Check	214
(C^y)	Minimal capacity constraint	109
$d(j, i, q)$	If customer j is the pth regular customer of maintenance activity i in the qth subproblem, then let $d(j, i, q)$ yield the pth period of $S_i \cap Q_q$ in the facility-transformation	148
$d^{(K)}$	Slack of a constraint (K)	180
d'_{it}	Slack of the i, tth constraint	173
d''_t	Slack of the tth constraint	175
(D)	Non-coupling capacity constraint	108
$DL_{(I)}$	Auxiliary LP to yield novel Lagrangean multiplier for $ZL_{(II)}$ when changing from $ZL_{(I)}$ to $ZL_{(II)}$	195
$DL_{(C)}$	Auxiliary LP (6.71)–(6.77) that yields novel Lagrangean multiplier υ for $ZL_{(P)}$ when changing from $ZL_{(C)}$ to $ZL_{(P)}$	202
$DL_{(P)}$	Auxiliary LP (6.79)–(6.85) to yield novel Lagrangean multiplier μ for $ZL_{(C)}$ when changing from $ZL_{(P)}$ to $ZL_{(C)}$	202
DOM	Design Out Maintenance	2
e.g.	exempli gratia, for example	1
E	Arc set of the network-transformation	145
E_i	Arc set of the ith maintenance activity	72
E_m	Identity matrix of size m	27
E'	Arc set that represents maintenance slots and constraint (C) in the network-transformation	145
E'_{tc}	Extension of the cover for the cth inequality (O) for period t	103
E''_i	Arc set for the reformulation of (P) of maintenance activity i in the network-transformation	145
E''_{tc}	Extension of the clique for the cth inequality (L) for period t	104

$\eta_{(II)}$	Best known lower bound to $Z_{(II)}$	196
$\eta_{(C)}$	Best known lower bound to $Z_{(C)}$	202
$\eta_{(P)}$	Best known lower bound to $Z_{(P)}$	203
f_i	Flexibility of maintenance activity i	168
$F_{(II)}$	Index set of some feasible solutions to $ZL_{(II)}$	196
F_{ab}	Fixed costs of the arc $(a,b) \in E$ in the network-transformation	145
$F_{(C)}$	Index set of some feasible solutions to $ZL_{(C)}$	204
$F_{(P)}$	Index set of some feasible solutions to $ZL_{(P)}$	204
f_t	Fixed costs of the WUPMP	71
F_t	Fixed costs of a facility t in the facility-transformation	148
FCNDP	Fixed-Charge Capacitated Network Design Problem	146
FF	First Fit Heuristic	239
$First_i$	Pointer to the list element that represents the maintenance activity i in the list A_0	19
g	Subgradient	43
$\hat{g}$	Pseudo-subgradient	187
$G(V,E)$	Directed graph with V and E	145
$G_i(V,E_i)$	Weighted, directed graph with V and E_i for maintenance activity i	72
$H(1)\ldots(K)$	Intersection of half spaces of the constraints $(1),\ldots,(K)$	31
i.e.	id est, that is	17
IB3	Iterated Best-of-three Heuristic	239
IP	Integer Program	30
ITER	Number of iterations the scope is explored	213
ITER^{final}	Number of iterations the scope is explored in the final state	213
J	Set of all customers in the facility-transformation	147
J	Set of items in the KP	55
J^x	Potential primal basic variable x_{it}	176
J^y	Potential primal basic variable y_t	176
J^z	Potential primal basic variable z_{it}	176
K	Knapsack Heuristic	231
Ke	Lagrangean Heuristic, $ZL_{(P)}$ solved to optimality	233
KeUe	Lagrangean Hybrid Heuristic, start with Ke and change to Ue	239
Kh	Heuristic K in the Lagrangean Heuristic	233
KhUCh	Lagrangean Hybrid Heuristic, start with Kh and change to UCh	239
KP	Knapsack Problem	4
L	Number of periods the scope of the next iteration overlaps with the scope of the previous iteration	213
(L)	Extended clique inequality	103
L^{bin}	List of unused list elements	19

L_i	Set of periods for the maintenance activity i where the complementary slackness criterion $d'_{it} \cdot x_{it} = 0$ with $t \in L_i$ holds	174
λ	Scalar for the update of $\left(\bar{v}^{iter+1}, \bar{u}^{iter+1}\right)$ in (6.33)	180
$Last_i$	Pointer to the list element that represents the maintenance activity i in the list A_{T+1}	19
LB	Lower bound	31
LB^*	Best lower bound	31
LB^B	Lower bound of the backward greedy solution	61
LB^F	Lower bound of the forward greedy solution	60
LB^G	Lower bound of the greedy solution	61
LB^{init}	Initial lower bound that is obtained after the first iteration of a Lagrangean heuristic	192
LB_t^{KP}	Knapsack lower bound to the tth KP	186
LB^S	Lower bound of the single item solution	61
$left_{it}$	Pointer of a list element in the referenced list A_t that references to the list element of the left neighbor of the maintenance activity i	19
$\log_b n$	Logarithm of a number n of some base b	25
LP	Linear Program	27
MAPE	Mean Absolute Percentage Error	220
MIP	Mixed Integer Program	30
MRO	Maintenance, Repair and Overhaul	2
μ_t	Dual variable, Lagrangean multiplier of tth constraint (C)	156
μ_t^{init}	Initial Lagrangean multiplier of the tth constraint (C)	189
$\mathbb{N}$	Set of all natural numbers including zero	15
n	Number of maintenance activities	16
$next_{it}$	Pointer of a list element that references the successor in the referenced list A_t	19
$\mathcal{NP}$	Set of decision problems that are solvable by a non-deterministic Turing machine in polynomial time	26
$O(f)$	Big O-notation of a function f	25
(O)	Extended cover inequality	103
OF	Overlap Heuristic with the priority rule flexibility	239
OS	Overlap Heuristic with the priority rule simplicity	239
p.	page	1
p_j	Profit of the jth item in the KP	55
(P)	Period covering constraint	15
$\mathcal{P}$	Set of decision problems that are solvable by a deterministic Turing machine in polynomial time	26
P	Set of potential integers	173
$P_{(I)}$	Index set of some Lagrangean multiplier to $ZL_{(I)}$	198
$P_{(C)}$	Index set of some Lagrangean multiplier to $ZL_{(C)}$	203
$P_{(P)}$	Index set of some Lagrangean multiplier to $ZL_{(P)}$	204

$Path_i$	Set of nodes in the shortest path of maintenance activity i	185
π_i	Coverage of the ith maintenance activity	16
$\bar{\pi}_i$	Integer multiple of π_1 with $\bar{\pi}_i \leq \pi_i$ for maintenance activity i	94
$\pi^{\varnothing}$	Mean of the coverage in the instance generation	223
$\pi^{\%}$	Scalar that determines a symmetric interval around the mean $\pi^{\varnothing}$	223
PLM	Planned Maintenance	2
PM	Preventive Maintenance	1
pp.	pages	1
$prev_{it}$	Pointer of a list element that references the predecessor in the referenced list A_t	19
(P^y)	Slot covering constraint	104
(Q)	Section covering constraint	104
Q_q^{max}	Largest period of a set Q_q	104
Q_q^{min}	Smallest period of a set Q_q	104
Q_q	Set of periods of the qth section	104
$\mathbb{R}$	Set of all real numbers	27
$\bar{r}^{max}$	Maximal capacity of the capacities of all periods	17
$\bar{r}^{sum}$	Sum of the capacities of all periods	17
$\bar{r}_t$	Capacity of the tth period	16
$\bar{r}^{\varnothing}$	Mean of the capacity in the instance generation	223
$\bar{r}^{\%}$	Scalar that determines a symmetric interval around the mean $\bar{r}^{\varnothing}$	223
r_i	Maintenance time of the ith maintenance activity	16
$r^{\varnothing}$	Mean of the maintenance time in the instance generation	223
$r^{\%}$	Scalar that determines a symmetric interval around the mean $r^{\varnothing}$	223
r^{min}	Smallest maintenance time among all maintenance activities	17
r_t^{min}	Smallest maintenance time among all maintenance activities i with $r_i \leq \bar{r}_t$ for period t	186
r^{sum}	Sum of the maintenance time of all maintenance activities	17
r_t^{sum}	Sum of the maintenance time of all maintenance activities i with $r_i \leq \bar{r}_t$ for period t	186
(R)	Remaining task horizons from (P) without (Q)	104
$\bar{R}_t$	Capacity of a facility t in the facility-transformation	148
R_j	Capacity of a customer j in the facility-transformation	148
RCM	Reliability Centered Maintenance	2
RDP	Reduced Dual Problem	35
$right_{it}$	Pointer of a list element in the referenced list A_t that references to the list element of the right neighbor of the maintenance activity i	19

Abbreviation	Description	Page
RMP	Restricted Master Problem	41
RP	Reduced Primal Problem	35
s	Critical item in the KP	59
s	second(s)	233
(S)	Simple upper bound constraint	16
(S^x)	Simple upper bound constraint for the variables x	16
(S^y)	Simple upper bound constraint for the variables y	16
S	Solution space	27
S^{bin}	Stack of pointers of empty lists	19
S_i	All first periods of the successive task horizons in all sections $Q_1, \ldots, Q_m$ for a maintenance activity i	104
SetPP	Set Partitioning Problem	32
σ	Step size	45
SLU	Skill Level Upgrade	2
SP_{ib}	Length of the shortest path from node 0 to a node b in the graph $G_i(V, E_i)$ for maintenance activity i	89
SPP	Shortest Path Problem	31
SSCFLP	Single Source Capacitated Facility Location Problem	133
SSP	Subset Sum Problem	55
STEPSIZE	Step size parameter	47
STEPSIZE^{init}	Initial step size parameter	191
SWUPMP	Single Weighted Uncapacitated Planned Maintenance Problem	89
t^{in}	A move schedules a maintenance activity in the period t^{in}	212
t^{out}	A move plans out a task from the period t^{out}	212
t'	Left neighbor of a period	18
t''	Right neighbor of a period	18
(T)	Task redundancy constraint	104
T	Number of periods in the planning horizon	16
Tabu	Tabu Search Heuristic	239
TBM	Time-Based Maintenance	3
TPM	Total Productive Maintenance	2
u_i^{max}	Largest value of u_{iq} among all sections $q = 1, \ldots, m$ of the maintenance activity i	147
u_{iq}	Upper bound on the number of tasks of maintenance activity i in section Q_q of an optimal solution to the CPMP	106
u_{it}	Dual variable, Lagrangean multiplier of i, tth constraint (V)	161
(U)	Task upper bound constraint	106
U_i	Set of all customers assigned to a maintenance activity i in the facility-transformation	147
UB	Upper bound	31
UB^*	Best upper bound	31

UB^D	Dantzig bound	59
UB_t^{KP}	Knapsack upper bound to the tth KP	186
UB^{MM}	Müller-Meerbach bound	60
UB^{MT}	Martello-Toth bound	60
UBL	Upper bound on a Lagrangean relaxation	185
UC	Combined Lower Bound; Upper Bound Heuristic: No potential integers, with complementary slackness	231
UCh	Heuristic UC in the Lagrangean Heuristic	233
UChKh	Lagrangean Hybrid Heuristic, start with UCh and change to Kh	239
Ue	Lagrangean Heuristic, $ZL_{(C)}$ solved to optimality	233
UeKe	Lagrangean Hybrid Heuristic, start with Ue and change to Ke	239
UFLP	Uncapacitated Facility Location Problem	32
ULP	LP Lower Bound; Upper Bound Heuristic with best upper bound from ULP+Y+S, ULP–Y+S and ULP+Y–S	231
ULP+Y+S	LP Lower Bound; Upper Bound Heuristic: With potential integers, with complementary slackness	231
ULP+Y–S	LP Lower Bound; Upper Bound Heuristic: With potential integers, no complementary slackness	231
ULP–Y+S	LP Lower Bound; Upper Bound Heuristic: No potential integers, with complementary slackness	231
ULP–Y–S	LP Lower Bound; Upper Bound Heuristic: No potential integers, no complementary slackness	231
ULPh	Heuristic ULP in the Lagrangean Heuristic	233
UNDP	Uncapacitated Network Design Problem	32
UPD	Primal-dual Lower Bound; Upper Bound Heuristic: No potential integers, with complementary slackness	231
UPDh	Heuristic UPD in the Lagrangean Heuristic	233
UPMP	Uncapacitated Planned Maintenance Problem	94
UPR	Dual Priority Rule Lower Bound; Upper Bound Heuristic: No potential integers, with complementary slackness	231
UPRh	Heuristic UPR in the Lagrangean Heuristic	233
USPP	Shortest Path Lower Bound	231
v_{it}	Dual variable and Lagrangean multiplier of i,tth constraint (P)	89
v_{it}^{init}	Initial Lagrangean multiplier of the i,tth constraint (P)	189
$(\bar{v}^{iter}, \bar{u}^{iter})$	Feasible solution to the corresponding dual problem (6.1)–(6.5) at the beginning of an iteration $iter$	176
(V)	Variable upper bound constraint	16
V	Set of nodes	72
w_j	Weight of the jth item in the KP	55

Symbol	Description	Page
W	Width of the scope in periods	213
WIDTH	Initial width of the scope	213
WUPMP	Weighted Uncapacitated Planned Maintenance Problem	4
x_{it}	$x_{it} = 1$ if the ith task is scheduled in the period t (otherwise, 0)	16
$\bar{x}_{it}$	Continuous variable x_{it} of the CPMP	173
x_j	$x_j = 1$ if the jth item is packed in the KP (otherwise, 0)	55
$x^{KP}_{\bullet t}$	Best feasible solution to the tth KP	187
(X)	Integrality constraint of x_{it}	16
X	Set of (feasible) solutions	52
X_{iq}	Lower bound on the number of open periods in $\{1,\ldots,T\}\setminus Q_q$ of maintenance activity i	106
ξ	Optimal objective function value of the Relaxed RP and of the Restricted RDP	177
(x, y)	A solution to the CPMP, UPMP or WUPMP	18
$(x, y)^{ref}$	Reference solution	165
y_t	$y_t = 1$ if a maintenance slot is established in the period t (otherwise, 0)	16
$\bar{y}_t$	Continuous variable y_t of the CPMP	173
(Y)	Integrality constraint of y_t	16
Y	Upper bound on the total capacity of an optimal solution	106
$\mathbb{Z}$	Set of all integer numbers	31
$z_{it\tau}$	$z_{it\tau} = 1$ if the arc (t, τ) is chosen from the respective arc set of the ith maintenance activity (otherwise, 0)	72
z_t	tth Knapsack subproblem of $ZL_{(P)}$	137
z_t	Objective function value of z_t	137
(Z)	Objective function of the CPMP	15
Z	The mathematical formulation of the CPMP, of an IP or of a MIP	15
Z	Optimal objective function value of the CPMP	15
$Z^{(K)}$	Problem Z with the neglected constraint (K)	40
$Z^{(K)}$	Optimal objective function value of $Z^{(K)}$	40
$Z_{(K)}$	Lagrangean dual of constraint (K)	41
$Z_{(K)}$	Optimal objective function value of $Z_{(K)}$	41
$Z_{(K)/(L)}$	Lagrangean decomposition between constraints (K) and (L)	41
$Z_{(K)/(L)}$	Optimal objective function value of $Z_{(K)/(L)}$	41
$Z^{+(K)}$	Problem Z restricted to an additional constraint (K)	40
$Z^{+(K)}$	Optimal objective function value of $Z^{+(K)}$	40
$Z^{(X)(Y)}$	LP relaxation of the CPMP	95
$Z^{(X)(Y)}$	Optimal objective function value of $Z^{(X)(Y)}$	95
$ZL_{(K)}$	Lagrangean relaxation of constraint (K)	41
$ZL_{(K)}$	Optimal objective function value of $ZL_{(K)}$	41
$ZL_{(K)/(L)}$	Lagrangean decomposition between constraints (K) and (L)	41

$ZL_{(K)/(L)}$	Optimal objective function value of $ZL_{(K)/(L)}$	41
$ZL_{(C)}$	Lagrangean relaxation of (C) of the CPMP	156
$ZL_{(C)}$	Objective function value of $ZL_{(C)}$	156
ZL^{init}	Initial Lagrangean lower bound	189
ZL^{init}	Objective function value of ZL^{init}	189
$ZL_{(P)}$	Lagrangean relaxation of (P) the CPMP	154
$ZL_{(P)}$	Objective function value of $ZL_{(P)}$	154

Chapter 1
Introduction

Maintenance comprises all technical, administrative and managerial activities as *maintenance activities*, which are carried out during the life cycle of a *system* in order to retain it in or restore it to a state in which it functions as originally intended (Comité Européen de Normalisation 2001, p. 12). Industrial production systems require maintenance as they consist of components with a limited durability. Under the conditions of production, the components reach a limiting state of use where the system functions improperly and maintenance is done during production shifts or during regular breaks (e.g., night shifts, on the weekend). Typical maintenance activities are lubrication, calibration, cleaning, repair, overhaul, replacement and inspection. However, many maintenance activities require a shut down of the production system. In order to contribute to the business goals, maintenance activities have to be carefully planned because downtimes block resources, cause production losses and lead to additional costs (e.g., an increase of the opportunity costs of the bottleneck resources). Maintenance is important in industrial asset management and positively affects the company's profit, the productivity, the quality, the overall equipment effectiveness, the service level and the efficient use of the system as well as of the personnel (Pintelon and Gelders 1992; Dekker and Scarf 1998; Brah and Chong 2004; Ashayeri et al. 1996).

Different maintenance strategies evolved during the last decades. In *Corrective Maintenance (CM)*, the system is maintained after the failure has occurred (Comité Européen de Normalisation 2001, p. 24). This strategy was very common up to 1960 where maintenance was considered as an unavoidable necessity that was hard to control (Pintelon and van Puyvelde 2006, pp. 4–8). *Preventive Maintenance (PM)* is a foresighted strategy where maintenance activities are performed prior to failure and according to prescribed criteria or at predetermined intervals (Comité Européen de Normalisation 2001, p. 23). Owing to the definition of failure, a clear distinction between CM and PM can be difficult because a component might be functioning sufficiently enough although its condition is bad (Budai et al. 2008). Although the short term cost effects are low, CM could be costly on the long run because it might cause high downtimes, high production costs due to production losses and the condition of the system could degrade over time because of a high average component wear out (Ahmad and Kamaruddin 2012). Opportunity costs arise in PM because a component could be replaced at a later point in time and it would be used longer if CM was carried out instead. The resulting trade-off requires to find an effective set of maintenance strategies for the relevant components (Joo 2009). Other maintenance strategies reduce the maintenance effort via *Design Out Maintenance (DOM)*. This can be achieved with a modular design of the system that enables to quickly replace the malfunctioning modules, which are maintained in specific depots (Joo 2009). *Maintenance scheduling* considers the production system at a shorter time scale than PM (Ma et al. 2010). It combines the scheduling of maintenance activities with machine scheduling such that either a maintenance activity or one job is scheduled at the same point in time on the considered machine. *Maintenance, Repair and Overhaul (MRO)* is mostly applied in aircraft maintenance to recover airworthiness of the respective components (Samaranayake and Kiridena 2012). Noteworthy other maintenance strategies are business centered maintenance (Kelly 1997), risk-based maintenance (Arunraj and Maiti 2007), value driven maintenance (Rosqvist et al. 2009) or a support with computerized maintenance management systems (Labib 1998). A survey of maintenance strategies is provided in Garg and Deshmukh (2006).

PM is divided into qualitative and quantitative approaches. Qualitative PM strategies, which refer to general management and engineering strategies, comprise two important approaches namely *Total Productive Maintenance (TPM)* and *Reliability Centered Maintenance (RCM)*. In TPM that was developed in Japan (McKone and Weiss 1998), the management strongly involves employees to solve maintenance problems with the quality circles method. Specifically, the machine operator's intrinsic knowledge is used to improve and to prevent maintenance activities. In a *Skill Level Upgrade (SLU)*, machine operators are trained to understand the machine better, to perform maintenance activities quicker and to detect potential problems that might cause downtimes. Cross-functional training and employee involvement is also found in total quality management as well as in just-in-time approaches. Cua et al. (2001) discuss the interdependency and propose an integrated maintenance approach. Ahuja and Khamba (2008) provide a TPM survey. RCM was established in aircraft maintenance of the United States Air Force (Rausand 1998). All components in the system with a critical reliability are identified. Afterwards, the failure effects are categorized and maintenance policies are derived from specific flow charts (Endrenyi 1999).

Quantitative PM strategies are commonly distinguished between *Condition-Based Maintenance (CBM)*[1] and other non-condition based maintenance approaches that are referred to as *Planned Maintenance (PLM)*. In CBM, the condition of the system is constantly monitored while it operates, and the collected data is evaluated with regard to specific signs and signals that allow

---

[1] A synonym for CBM is predictive maintenance.

1 Introduction

to forecast the point in time when a failure is most likely going to happen (Comité Européen de Normalisation 2001, p. 23). The most popular techniques are vibration and lubricant analysis (Carnero 2005) or oil analysis (Newell 1999). Forecast methods are surveyed in Peng et al. (2010). In PLM, maintenance activities are scheduled in accordance to established or predetermined intervals but without a previous investigation on the condition of the system (Comité Européen de Normalisation 2001, p. 23). An important planned maintenance approach is *Time-Based Maintenance (TBM)* (Ahmad and Kamaruddin 2012). In a so called bathtub curve model, the failure rate of system or a component is assumed to decrease in the early phase of the life cycle (burn-in phase), to remain almost constant afterwards (useful life phase) and to increase towards the end of the life cycle (wear-out phase). The main goals of PLM are the maximization of the system reliability and the minimization of downtimes, failures and maintenance costs (Ahmad and Kamaruddin 2012). Maintenance costs comprise variable and fixed costs. Variable costs include all costs that are caused by the execution of the respective maintenance activity such as labor costs, material costs and in some circumstances downtime costs. Fixed costs are all costs that arise if at least one maintenance activity is executed. Both cost categories can be period dependent because of different cost rates or resource availabilities.

Since a set of maintenance strategies can be applied simultaneously to the system or to components of the system, the effectiveness greatly depends on the actual composition of this set. Figure 1.1 shows maintenance strategies that are effective in the given constellation of downtime duration and the maintenance frequency (cf. Labib 1998; Scarf 2007). In general, high downtimes and high maintenance frequencies result in production losses with high cost effects. CBM and DOM are cost intensive because DOM has to be anticipated already in the design of the system and CBM requires a constant monitoring. Therefore, CBM and DOM pay off at high downtimes. SLU and DOM are long-term strategies and they are beneficial if the maintenance frequency is high. CM and SLU are advantageous at low downtimes because the short-term cost effects are low. Cost effects between CM and PM have been discussed above. PLM is a suitable strategy for moderate downtimes and

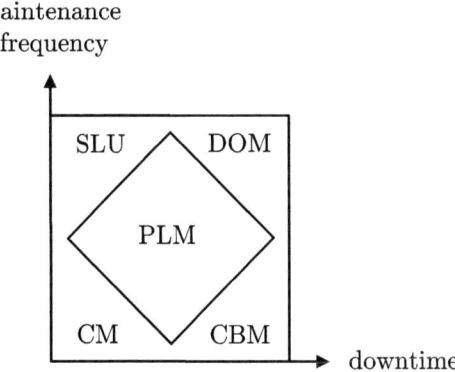

Fig. 1.1 Decision making grid for maintenance strategies (cf. Labib 1998; Scarf 2007)

maintenance frequencies. Hence, PLM plays a significant role among the presented maintenance strategies.

The presented research introduces two novel PLM approaches, namely the *Capacitated Planned Maintenance Problem (CPMP)*, which is the center of interest, and the *Weighted Uncapacitated Planned Maintenance Problem (WUPMP)*. To the author's best knowledge neither the CPMP nor the WUPMP have been discussed in literature yet. The thesis provides the following main contributions:

- The solvability of the CPMP and of the WUPMP is analyzed with respect to the computational complexity, polyhedral properties and lower bounds.
- Optimal algorithms solve the CPMP, the WUPMP and variants of both problems.
- Specific heuristics to the CPMP are developed such as three construction heuristics, three Lagrangean heuristics and a tabu search heuristic.
- Heuristics to Lagrangean relaxations of the CPMP are presented and six lower bounds and two upper bounds are provided.
- A general, problem independent hybrid approach links and alternates between two Lagrangean relaxations. A Lagrangean hybrid heuristic is developed for the CPMP.
- Computational results provide valuable insights about the performance of the proposed heuristics and lower bounds.

The thesis is organized as follows: In Chap. 2, the CPMP is defined, motivated by practical applications and discussed in the context of known PM approaches. The chapter also includes mathematical formulations, an efficient data structure and the terminology used throughout the thesis. Chapter 3 briefly introduces the applied, well-known solution techniques and theoretical concepts of operations research. Being a substructure of the CPMP, the well-known *Knapsack Problem (KP)* is introduced. In order to design specialized solution approaches, the solvability of the CPMP and the WUPMP are intensively analyzed in the following two chapters. In Chap. 4, polyhedral properties, the computational complexity status and optimal algorithms for the WUPMP and for specific problem variants are provided. Chapter 5 contains valid inequalities, polyhedral properties and an investigation on lower bounds of the CPMP. Furthermore, the computational complexity status of the CPMP and of several problem variants are resolved and optimal algorithms are provided. Chapter 6 proposes heuristics to solve the CPMP. Three construction heuristics are provided, which yield an initial solution. Afterwards, the solution is potentially improved through four metaheuristics. In particular, two Lagrangean heuristics and a tabu search heuristic are developed. A fourth Lagrangean hybrid heuristic is based on a novel, problem independent hybrid approach that links two Lagrangean heuristics with each other. Computational results are presented in Chap. 7 where the performance of the lower bounds and the heuristics is investigated. Final remarks and an outlook on further research is given in Chap. 8. Appendix A contains extended problem variants and mathematical formulations of the CPMP.

References

Ahmad, R., & Kamaruddin, S. (2012). An overview of time-based and condition-based maintenance in industrial application. *Computers & Industrial Engineering, 63*, 135–149.

Ahuja, I. P. S., & Khamba, J. S. (2008). Total productive maintenance: Literature review and directions. *International Journal of Quality & Reliability Management, 25*, 709–756.

Arunraj, N. S., & Maiti, J. (2007). Risk-based maintenance - techniques and applications. *Journal of Hazardous Materials, 142*, 653–661.

Ashayeri, J., Teelen, A., & Selen, W. (1996). A production and maintenance planning model for the process industry. *International Journal of Production Research, 34*, 3311–3326.

Brah, S. A., & Chong, W.-K. (2004). Relationship between total productive maintenance and performance. *International Journal of Production Research, 42*, 2383–2401.

Budai, G., Dekker, R., & Nicolai, R. P. (2008). Maintenance and production: A review of planning models. In K. A. H. Kobbacy & D. N. P. Murthy (Eds.), *Complex system maintenance handbook* (Chap. 13). London: Springer.

Carnero, M. C. (2005). Selection of diagnostic techniques and instrumentation in a predictive maintenance program. A case study. *Decision Support Systems, 38*, 539–555.

Comité Européen de Normalisation. (2001). EN 13306: Maintenance terminology.

Cua, K. O., McKone, K. E., & Schroeder, R. G. (2001). Relationships between implementation of TQM, JIT, and TPM and manufacturing performance. *Journal of Operations Management, 19*, 675–694.

Dekker, R., & Scarf, P. A. (1998). On the impact of optimisation models in maintenance decision making: The state of the art. *Reliability Engineering and System Safety, 60*, 111–119.

Endrenyi, J. (1999). *Impact of maintenance strategy on reliability.* New York: The Institute of Electrical and Electronics Engineers, Inc. Final report by the IEEE/PES Task Force on Impact of Maintenance Strategy on Reliability of the Reliability, Risk and Probability Applications Subcommittee.

Garg, A., & Deshmukh, S. (2006). Maintenance management: Literature review and directions. *Journal of Quality in Maintenance Engineering, 12*, 205–238.

Joo, S.-J. (2009). Scheduling preventive maintenance for modular designed components: A dynamic approach. *European Journal of Operational Research, 192*, 512–520.

Kelly, A. (1997). *Maintenance organization and systems.* Oxford: Butterworth-Heinemann.

Labib, A. W. (1998). World-class maintenance using a computerised maintenance management system. *Journal of Quality in Maintenance Engineering, 4*, 66–75.

Ma, Y., Chu, C., & Zuo, C. (2010). A survey of scheduling with deterministic machine availability constraints. *Computers & Industrial Engineering, 58*, 199–211.

McKone, K. E., & Weiss, E. N. (1998). TPM: Planned and autonomous maintenance: Bridging the gap between practice and research. *Production and Operations Management, 7*, 335–351.

Newell, G. E. (1999). Oil analysis cost-effective machine condition monitoring technique. *Industrial Lubrication and Tribology, 51*, 119–124.

Peng, Y., Dong, M., & Zuo, M. J. (2010). Current status of machine prognostics in condition-based maintenance: A review. *International Journal of Advanced Manufacturing Technology, 50*, 297–313.

Pintelon, L. M., & Gelders, L. F. (1992). Maintenance management decision making. *European Journal of Operational Research, 58*, 301–317.

Pintelon, L. M., & van Puyvelde, F. (2006). *Maintenance decision making.* Leuven: Uitgeverij Acco.

Rausand, M. (1998). Reliability centered maintenance. *Reliability Engineering and System Safety, 60*, 121–132.

Rosqvist, T., Laakso, K., & Reunanen, M. (2009). Maintenance modeling and application, value-driven maintenance planning for a production plant. *Reliability Engineering and System Safety, 94*, 97–110.

Samaranayake, P., & Kiridena, S. (2012). Aircraft maintenance planning and scheduling: An integrated framework. *Journal of Quality in Maintenance Engineering, 18*, 432–453.

Scarf, P. A. (2007). A framework for condition monitoring and condition based maintenance. *Quality Technology and Quantitative Management, 4*, 301–312.

Chapter 2
The Capacitated Planned Maintenance Problem

Section 2.1 introduces the CPMP as a novel maintenance problem. The CPMP is motivated and the assumptions are discussed. A review of related well-known maintenance approaches is given in Sect. 2.2. The mathematical formulation of the CPMP, assumptions, terminology and an efficient data structure are provided in Sect. 2.3.

2.1 Problem Definition and Motivation

Let the maintenance of a single machine require a set of maintenance activities. The planning horizon is finite and subdivided into discrete periods. A *maintenance activity* is scheduled as a *task*, which begins and ends in the same period. A period is associated with a *maintenance slot* that is *established* if it contains one or more tasks. These tasks are executed one after another by a single repairman. As the effect of maintenance is temporary, each maintenance activity has to be scheduled at least once within a deterministic *task horizon* or durability horizon. Specifically, the length of the task horizon the so called *coverage* is the maximal time that is allowed to elapse between two tasks of the same maintenance activity. All tasks of the same maintenance activity restore the system to the same state and require the same deterministic maintenance time. The maximal available maintenance time of a period is denoted as *capacity*.

The CPMP schedules all maintenance activities such that the number of periods between two tasks of the same maintenance activity is smaller than the coverage and the capacity per period is never exceeded. The obtained maintenance plan contains the minimal number of established maintenance slots.

The most important maintenance activities are component replacements and inspections. The coverage is then associated with the component durability and with the time between two inspections, respectively. As the definition of a maintenance activity is general, the CPMP provides an integrated approach to model the replacement of multiple components, multiple inspections and other maintenance activities on the production system where the capacity is limited.

In a project with an automobile supplier, the production plans showed weekly time slots that were a priori reserved for maintenance. The production was working to full capacity and required expensive extra shifts in the night and on the weekends. In this situation, a minimization of the number of maintenance slots, releases production time in the schedule. This valuable time could have been used to reduce overtime, extra shifts, to buffer production peaks or most importantly to adhere to promised delivery due dates.[1]

Another application comes from a project with a consumer goods company. The management decided to introduce PLM as a novelty. No reliable, recorded data existed from which information about the coverages could have been derived. Instead, the expert knowledge of a repairman was a valuable source to obtain the coverages. Furthermore, the opportunity costs for machine downtimes greatly exceeded the maintenance costs. This frequently occurs in process industry where the chemical or physical transformation processes are executed in batches and in succession. Maintenance was performed in between two batches and thus, a minimization of the number of maintenance slots could have provided more production time in the schedule.[2]

The limited reachability of oil rigs and offshore wind farms imposes further difficulties on maintenance. Oil rigs have maintenance teams but failures of specific components require a transport from the mainland. The maintenance of offshore wind farms require that a repairman and the equipment come from the mainland. In both cases, the opportunity costs for downtimes and the travel costs are high because traveling takes long and is usually done via helicopters or boats. A minimization of the number of maintenance slots could reduce the number of downtimes and trips, which could provide further cost savings.[3]

In what follows, basic assumptions of the CPMP are discussed.

Assumption 2.1

1. The objective function minimizes the total number of established maintenance slots.
2. Each maintenance activity has a deterministic and constant coverage.
3. The effect of a maintenance activity does not wear off during the coverage.

---

[1] Production and maintenance compete for the availability of the production system. Hence, a simultaneous optimization of both is very advantageous. Some approaches are discussed in Sect. 2.2. A combined approach of the Capacitated Lot-Sizing Problem (Bitran and Yanasse 1982) and the CPMP is presented in Appendix A.1.

[2] An introduction to process industry and case studies are found in Neumann et al. (2002), Neumann and Schwindt (2000), and Reklaitis (1991).

[3] On a short-time scale, the maintenance problem can also be related with a routing problem that optimizes the sequence in which the windmills are visited.

4. The coverage is maintenance activity dependent and the capacity is period dependent.
5. The point in time when a maintenance activity is scheduled is flexible as long as this is done before the coverage ends.

In the survey (Dekker and Wildeman 1997), it is reported that most maintenance approaches assume complete information about the maintenance costs and the life cycle of the system. However, this information is difficult to obtain in practice. The first two assumptions provide a pragmatic approach to handle incomplete information. Assumption 2.1.1 is clearly applicable if the fixed costs greatly exceed the variable costs. The quantification of maintenance costs and maintenance benefits is difficult in practice (Dekker 1996) but possible with methods surveyed in Budai et al. (2008) and Joshi and Gupta (1986). Note that many small to medium-sized enterprises do not have the infrastructure to derive this data. Dekker and Wildeman (1997) mention considerable cost savings if fixed costs of maintenance activities are reduced by simultaneously maintaining multiple components. This corresponds to the concept of establishing maintenance slots. Therefore, Assumption 2.1.1 provides a reasonable simplification that needs to be verified with respect to the cost structure of the application. Assumption 2.1.2 holds in aircraft maintenance because of a maximum flying hour limit for certain, critical components where security restrictions apply (Knotts 1999). However, many maintenance activities are liable to stochastic processes, where the coverages are dependent on distribution functions. Pintelon and Gelders (1992) and Ahmad and Kamaruddin (2012) report that the collection of reliable data is a major concern in many practical applications of maintenance optimization. Specifically, Scarf (1997) state that the collection of data is already very difficult for basic stochastic models such as one of the earliest approaches presented in Barlow and Hunter (1960). Further practical difficulties occur if a company introduces PLM as a novelty because no suitable data might be available. This can occur because of improper information systems or component manufacturers that keep distribution functions as a business secret in order to avoid a comparison between competitors in the market. Furthermore, there exist applicability problems with distribution functions provided by component manufacturers because the production system, in which the component is installed, has a significant impact on the durability but this impact is usually unknown to the component manufacturers (Ahmad and Kamaruddin 2012). Owing to the problems in collecting appropriate data, Dekker (1996) points out that expert knowledge could be used instead. However, Scarf (1997) indicates that expert knowledge is subjective because it often reflects the current practice instead of true technical understanding (cf. van Noortwijk et al. 1992). Using expert knowledge and lower bounds on coverages, Assumption 2.1.2 is a pragmatic approach that is applicable in practice if meaningful data is not available or is difficult or costly to obtain. Similar arguments could apply to reason Assumption 2.1.3. A potential violation of the Assumptions 2.1.2–2.1.4 through changing parameters in the instance data can be considered in an iterative solution approach with a rolling planning horizon. For instance, in the case of age dependency the coverages can be adapted in each iteration accordingly. Both Assumptions 2.1.2 and 2.1.3 might affect the efficiency of PLM because the maintenance costs and the probability of failure might increase. In particular, an underestimation of the coverage implies that maintenance activities are scheduled too early and hence, too often. An overestimation of the coverage or a significant influence of the age on the component performance increase the probability of failure and the number of unexpected maintenance activities. The Assumptions 2.1.4 and 2.1.5 provide more flexibility and increase the applicability of the CPMP.

Finally, the CPMP provides a problem definition and assumptions that are easy to conceive. Interestingly, the surveys (Pintelon and Gelders 1992; Dekker 1996) emphasize that most stochastic models are difficult to understand by maintenance managers. Note that this can lead to the application of inappropriate maintenance models.

2.2 Known Maintenance Problems

To the authors best knowledge the CPMP has not been discussed in literature yet. However, there exist well-known maintenance problems that cover some but not all aspects of the CPMP. In what follows, a selection of these maintenance problems is presented and the following problem classes are briefly reviewed.

- Periodic maintenance
- Machine scheduling with periodic maintenance
- Periodic maintenance inspection
- Aircraft maintenance

Unless otherwise stated, the first two problem classes have deterministic failure data and maintenance inspection has stochastic failure data. Most approaches in literature apply *periodic maintenance*, where the entire system is maintained in periodically repeating cycles. In contrast to the CPMP, these maintenance approaches do support the flexible scheduling of maintenance activities. Note that the CPMP considers a single machine with multiple maintenance activities and explicitly considers the capacity allocation per period. At the end of this section other, less related problem classes and combinations with specific production problems are presented.

2.2.1 Periodic Maintenance

The Periodic Maintenance Problem seeks a maintenance plan that consists of a repeating sequence such that all machines are maintained but at most one machine per period and the length of the sequence is minimal. Anily et al. (1998) investigate this problem for multiple machines. The operation costs for a given period linearly depend on the number of periods elapsed since the last maintenance of the machine.

However, maintaining a machine causes no costs. The long-run average operating cost per period are minimized such that at most one machine is maintained per period. They prove that an optimal cyclic maintenance plan exists, provide an exact algorithm, a heuristic with a worst case performance ratio of $2\frac{1}{2}$ and a greedy heuristic that solves the two machine case to optimality. Bar-Noy et al. (2002) extend this problem and consider a subset of machines that is maintained per period. Grigoriev et al. (2006) consider the Periodic Maintenance Problem with multiple machines and introduce machine specific fixed costs that are caused each time a machine is maintained. The objective is to minimize the total operation and maintenance costs of the maintenance plan. Joo (2009) apply periodic maintenance in aircraft maintenance. This DOM approach considers modular designed components that are replaced and send to a depot for an extensive overhaul. Maintenance is simplified because few technical skill is required and a replacement is quickly executed. The objective is to minimize the opportunity costs due to premature module replacement such that periodic due dates of planned maintenance are met and inventory constraints that reflect the availability of modules are satisfied.

2.2.2 Machine Scheduling with Periodic Maintenance

The Maintenance Scheduling Problem combines machine scheduling and maintenance such that the machine is either maintained or one job is processed at the same time. Lee (1996) consider Maintenance Scheduling Problems on a single and on multiple machines. Each machine has a single maintenance interval, which is fixed and predefined. They introduce two ways of how the job can be processed. Non-resumable job processing means that a processed job must be completed without preemption before maintenance starts. In resumable job processing, preemption is allowed and the job restarts from the point of preemption after the maintenance is done. The computational complexity and different objective functions are discussed such as minimizing the sum of completion times, the makespan, the maximal lateness and the number of tardy jobs. Liao and Chen (2003) present a single machine Maintenance Scheduling Problem with PM, non-resumable job processing and multiple maintenance intervals, which are fixed and predefined. The objective is to minimize the maximal tardiness. Chen (2009) minimize the number of tardy jobs. The approach of Chen (2008) is more flexible because the cycle length between two maintenance intervals is not rigid but can lie within certain, predetermined interval. A schedule is sought that minimizes the makespan. A simple heuristic is provided and Xu et al. (2009) prove that its worst case performance ratio is 2. Cassady and Kutanoglu (2005) solve a Maintenance Scheduling Problem where a single machine breaks down stochastically. They minimize the sum of the weighted expected completion times. Graves and Lee (1999) contribute semi-resumable job processing where the preempted job must be partially processed again. The objective is to minimize the total weighted completion time and to minimize the maximum lateness. Mauguière et al. (2005) combine resumable and non-resumable job processing and impose properties on the maintenance interval. A non-crossable maintenance interval forbids preemption of a job even if it is resumable. Whereas a crossable maintenance interval allows preemption. They study the combinations and minimize the makespan of a single machine with multiple maintenance intervals and of a job-shop scheduling problem. Maintenance Scheduling Problems with multiple machines and periodic maintenance are presented in Mati (2010), Xu et al. (2008), Aggoune and Portmann (2006), Lee and Chen (2000), and Lee (1997). A detailed survey is given in Ma et al. (2010) and Schmidt (2000).

2.2.3 Periodic Maintenance Inspection

In these inspection problems, the length of an inspection interval is sought such that the inspection is periodically executed. Barlow et al. (1963) determine the length of the inspection interval for a single machine that degrades stochastically. Maintenance comprises only inspection. The objective is to minimize the total expected costs consisting of inspection costs and the costs caused by the time elapsed between failure and detection. Baker (1990) studies a single machine that fails due to a constant probability. A failure is assumed to be hidden because it can only be identified with an inspection. Furthermore, a failure causes the production to stop. The objective is to maximize the sum of profit obtained if the machine runs subtracted the maintenance costs from inspection and from machine repair upon failure. Ben-Daya and Hariga (1998) extend this inspection model and introduce a quality shift. The system still produces after an undetected failure but the produced items have a lower quality and a smaller profit. Wang (2009) use the delay-time concept in a more general approach and introduce two types of inspection intervals. The delay time is defined as the time elapsed where a failure occurs but the machine still functions and the point in time this failure causes a breakdown. Minor inspections treat quality shift related failures whereas breakdown related failures are managed in major inspections. In contrast to many inspection models, the maintenance time is taken into account. The objective is to minimize the expected total cost per unit time. An overview of the delay-time concept is given in Wang (2012) and Christer and Waller (1984) provide an early case study. Vaurio (1995) determine lengths of inspection and of maintenance intervals for a single machine. In this context, maintenance comprises component replacement, which is done either periodically in a PM approach or because of failure as CM. The maintenance time is taken into account. The objective is to minimize the inspection and maintenance costs.

2.2.4 Aircraft Maintenance

Ahmadi and Kumar (2011) determine the length of inspection intervals for components that degrade stochastically over the time and whose failures are detected via inspection. Inspection and maintenance time is considered and the total costs are minimized. As safety systems are essential, Ahmadi and Kumar (2011), Lienhardt et al. (2008), Taghipour and Banjevic (2011) explicitly take failure finding maintenance into account because a failure is not detected under the normal use of the system. Clarke et al. (1997) investigate on the Aircraft Rotation Problem. A route for each aircraft is sought such that all daily flights are assigned and maintenance restrictions are satisfied. Different types of maintenance checks are periodically executed at predefined intervals. The approach of Sarac et al. (2006) is more detailed on the operational level. The routing decision takes available maintenance man-hours and a limited number of maintenance slots into account. Keysan et al. (2010) consider periodic maintenance in per-seat, on-demand air transportation and explicitly determine the required maintenance capacity. An aircraft is maintained if the accumulating flying hours are within a given interval. On a tactical decision level, the required maintenance capacity is determined. The operational decision level schedules aircraft for maintenance subject to capacity restrictions. In Knotts (1999), important components are inspected and maintained within regular checks that are scheduled after a maximum flying hour limit is reached. This corresponds to deterministic durability and coverages. Zhang and Jardine (1998) present a MRO approach. Components are minimally repaired upon failure, replaced after a certain durability is reached and periodic overhauls restore the component. The number of overhauls and the length of the overhaul interval is determined such that the expected unit-time cost or the total discounted cost are minimized. Gopalan and Talluri (1998) provide a survey.

2.2.5 Other Maintenance Approaches

Wagner et al. (1964) present an early approach to deterministic preventive maintenance. In the project planning problem, maintenance activities are scheduled once within a discrete planning horizon. To each maintenance activity is associated a set of periods that represent starting dates and an individual processing time of several periods. A scheduled maintenance activity consumes a single resource while it is being processed. The basic model can be extended by several objective functions. Drinkwater and Hastings (1967) propose the repair cost limit method and apply it to a fleet of vehicles. The repair cost limit is the maximal maintenance budget per vehicle. The average costs per vehicle are minimized by the decision if a vehicle is either maintained or replaced by a new one. An approximation algorithm is provided, a method to derive repair costs and a case study is presented. Lust et al. (2009) consider a maintenance problem on a single machine with multiple components where the reliability probability density function is given for all components. Maintenance is carried out in given slots. A collection of maintenance activities is selected that maximizes the system reliability such that the maximal maintenance time is not exceeded. Aghezzaf et al. (2007) combine the Capacitated Lot-Sizing Problem (Bitran and Yanasse 1982) with periodic maintenance where the failure probability density function of the production system is given. Maintenance is periodically performed in intervals of a fixed size and upon failure. Maintenance and lot-sizing are linked with each other by the available time per period. The objective is to find a production plan such that the sum of setup costs, holding costs and maintenance costs is minimal. Najid et al. (2011) generalize this lot-sizing problem and additionally consider shortage costs. Weinstein and Chung (1999) integrate maintenance in hierarchical production planning. An aggregated production plan with maintenance decisions is derived and afterwards, an uncapacitated lot-sizing model aims to meet the goals determined by the aggregated production plan. A simulation follows and determines consequences of the maintenance policy. Using this information, the aggregated production planning problem is repeatedly solved. Vassiliadis et al. (2000) consider maintenance in a multi-purpose batch processing plant and link it with aggregated production planning. This model is applied to the process industry. Each multi-purpose unit must be maintained at least once within a given durability horizon and the rate of failure increases over the time. The objective is to maximize the profit of the production plan and includes maintenance costs. Gopalakrishnan et al. (1997, 2001) consider a given maintenance plan, which covers a long planning horizon. This plan is evaluated with historical data. On the operational level, the machine failure probability is updated for all maintenance activities. In a subsequent step, a task rescheduling problem selects maintenance activities that are actually executed and maximizes the net planned maintenance savings subject to the capacity. Wang et al. (2011) investigate on a pavement maintenance problem. Contiguous road segments are clustered and each cluster is maintained by a transportation agency. A clustering with minimal cost such that each segment is maintained is advantageous. Cluster costs involve the number of assigned segments and the maximal segment maintenance cost among all segments of a cluster. The second cost factor is due to the fact that the same maintenance method is applied to all segments of a cluster. Basten et al. (2009) consider the level of repair analysis. Assume that there exists a multi-echelon repair network for a specific product design. At the lowest level there is an on-site maintenance for small repairs. If a specific component requires a complicated maintenance, it is send through the network until a repair is possible. The optimization problem decides for each component where in the network it should be repaired upon failure or if it should be discarded.

Dekker (1996) and Scarf (1997) discuss applications of stochastic maintenance optimization models and problems of the acceptance of maintenance optimization in industry. They also introduce the reader into maintenance. Elementary policies are presented in Barlow et al. (1965, pp. 84–118). In an age replacement policy, an assembly is replaced after a constant operating time has elapsed or at failure. Random age replacement policies assume that operating time is a random variable.

In contrast to this approach, a block replacement comprises a set of assemblies of the same type that is replaced at an integer multiple of the operating time and at failure. Different models and algorithms to these stochastic approaches is provided. Garg and Deshmukh (2006) survey and classify literature between 1992 and 2006. Dekker and Wildeman (1997) focus on the years 1991 up to 1997. Cho and Parlar (1991) review literature before 1991. Sherif and Smith (1981) and Valdez-Flores and Feldman (1989) survey literature after 1976. Pierskalla and Voelker (1976) provide a survey since 1965.

2.3 The Mathematical Formulation

Using the problem description from Sect. 2.1 and the definitions from Table 2.1, the mathematical formulation of the CPMP that is abbreviated as Z is stated below.[4] Note that the definition $\mathbb{N} = \{0, 1, 2, \ldots\}$ applies.

$$Z = \min \sum_{t=1}^{T} y_t \qquad \text{subject to} \tag{Z}$$

$$\sum_{\tau=t}^{t+\pi_i-1} x_{i\tau} \geq 1 \qquad \forall i = 1, \ldots, n; t = 1, \ldots, T - \pi_i + 1 \tag{P}$$

$$\sum_{i=1}^{n} r_i \cdot x_{it} \leq \bar{r}_t \cdot y_t \qquad \forall t = 1, \ldots, T \tag{C}$$

$$x_{it} \leq y_t \qquad \forall i = 1, \ldots, n; t = 1, \ldots, T \tag{V}$$

$$0 \leq x_{it}, y_t \leq 1 \qquad \forall i = 1, \ldots, n; t = 1, \ldots, T \tag{S}$$

$$x_{it} \in \{0, 1\} \qquad \forall i = 1, \ldots, n; t = 1, \ldots, T \tag{X}$$

$$y_t \in \{0, 1\} \qquad \forall t = 1, \ldots, T \tag{Y}$$

The total number of maintenance slots within the planning horizon is minimized by the objective function (Z). The sliding time window in the *period covering constraint* (P) schedules each maintenance activity i at least once in the periods $t, \ldots, t + \pi_i - 1$.[5] Otherwise, the time between two tasks is larger than the coverage, which contradicts the idea of planning maintenance prior to failure. The *capacity*

---

[4] In the appendix, extensions of the CPMP are presented in Sect. A.1. Alternative formulations of the CPMP are provided in Sect. A.2.

[5] In the discrete-time formulation of the Resource Constrained Project Scheduling Problem (Pritsker et al. 1969), the resource constraint is modeled with a sliding time window that is very similar to (P).

Table 2.1 Parameters and variables of the CPMP

Symbol	Description
$n \in \mathbb{N}$	Number of maintenance activities; index i
$T \in \mathbb{N}$	Number of periods in the planning horizon; index t
$\pi_i \in \mathbb{N}$	Coverage of the maintenance activity $i = 1, \ldots, n$
$r_i \in \mathbb{R}_+$	Maintenance time of the maintenance activity $i = 1, \ldots, n$
$\bar{r}_t \in \mathbb{R}_+$	Capacity of the period $t = 1, \ldots, T$
$y_t \in \{0, 1\}$	Represents maintenance slots. $y_t = 1$ if a maintenance slot is established in period $t = 1, \ldots, T$ (otherwise, 0)
$x_{it} \in \{0, 1\}$	Represents tasks. $x_{it} = 1$ if the maintenance activity $i = 1, \ldots, n$ is scheduled in period $t = 1, \ldots, T$ (otherwise, 0)

Table 2.2 A feasible solution to the CPMP

	t							
	1	2	3	4	5	6	7	8
y_t			1		1			1
x_{1t}			1		1			
x_{2t}			1					1
x_{3t}								1
$\sum_i r_i$			5		3			6

constraint (C) restricts the total maintenance time to the capacity of the period.[6] Summing up the maintenance time implies that tasks are executed one after another (single repairman). (C) and the *variable upper bound constraint* (V) ensure that a maintenance slot is established if at least one task is scheduled in the respective period. (V) is a valid inequality and implied by (C) and (Y). The *simple upper bound constraint* (S) is required for the LP relaxation of (X) and (Y). The decision variable domains (X) and (Y) are binary. Two additional constraints (S^x) and (S^y) are introduced.

$$0 \leq x_{it} \leq 1 \qquad \forall i = 1, \ldots, n; t = 1, \ldots, T \qquad (S^x)$$

$$0 \leq y_t \leq 1 \qquad \forall t = 1, \ldots, T \qquad (S^y)$$

Example 2.1 Table 2.2 shows a feasible solution to the CPMP where $n = 3, T = 8$, $\pi = (4, 5, 8)^T$, $r = (3, 2, 4)^T$ and $\bar{r} = (0, 0, 6, 4, 4, 2, 8, 8)^T$.

The following remark shows a very useful property that is implied by the objective function.

---

[6] The constraint (C) is identical to the capacity constraint of the Generalized Bin Packing Problem (Hung and Brown 1978).

2.3 The Mathematical Formulation

Remark 2.1 Since the objective function (Z) comprises only integer values, every lower (upper) bound is tightened by rounding it up (down) to the next integer.

2.3.1 Assumptions and Terminology

In order to avoid complicated notations and tedious special cases, some assumptions are imposed on the CPMP instance data.

Definition 2.2 $r^{min} = \min_{i=1,...,n} r_i \wedge r^{sum} = \sum_{i=1}^{n} r_i$

Definition 2.3 $\bar{r}^{max} = \max_{t=1,...,T} \bar{r}_t \wedge \bar{r}^{sum} = \sum_{t=1}^{T} \bar{r}_t$

Assumption 2.4

$$\pi_i \leq \pi_{i+1} \ \forall i = 1, \ldots, n-1 \tag{2.1}$$

$$\pi_i \geq 2 \ \forall i = 1, \ldots, n \tag{2.2}$$

$$\bar{r}^{max} \leq r^{sum} \tag{2.3}$$

$$\bar{r}_t = 0 \text{ if } \bar{r}_t < r^{min} \ \forall t = 1, \ldots, T \tag{2.4}$$

$$\pi_i \leq T \wedge r_i > 0 \ \forall i = 1, \ldots, n \tag{2.5}$$

$$x_{i0} = 1 \ \forall i = 1, \ldots, n \wedge x_{i,T+1} = 1 \ \forall i = 1, \ldots, n \wedge \bar{r}_0 = r^{sum} \wedge \bar{r}_{T+1} = r^{sum} \tag{2.6}$$

The instance data comprises only rationals and integers. In Assumption (2.1), all maintenance activities are sorted according to non-decreasing values. Maintenance activities with the smallest coverage play an important role in the presented solution approaches. Assumption (2.2) prevents the trivial solution of T open periods (i.e., $y_t = 1 \ \forall t$). If Assumption (2.3) is violated, then the capacity is set to its smallest possible value $\bar{r}_t = r^{sum}$ since it suffices to represent an uncapacitated period. Assumption (2.4) is a logical implication because no maintenance activity can be scheduled in a period t with $\bar{r}_t < r_i \ \forall i$.[7] If one maintenance activity i violates Assumption (2.5), it is excluded from consideration with $x_{it} = 0 \ \forall t$ because neither (P) nor (C) are restrictive. The assumption ensures that all maintenance activities are scheduled at least once. In order to define the initial and the final state of the machine properly, all maintenance activities are scheduled before and after the planning horizon. This yields Assumption (2.6) that is also implied by constraint (P). If Assumption (2.6) is violated, initial (target) tasks are introduced by a variable fixation in an initial (a terminal) planning horizon (see Example 2.2). This is very useful when the CPMP is embedded in a rolling planning horizon. Note that target

---
[7] Some lower bounds presented in Chap. 6 are easily tightened via the Assumptions (2.3) and (2.4).

Table 2.3 Variable fixation in an initial and a terminal planning horizon

t	1	2	3	4	5	6	7	8	9	10	11	12
y_t	1	0	1	1						1	1	1
x_{1t}	0	0	1	0								1
x_{2t}	0	0	0	1							1	0
x_{3t}	1	0	0	0						1	0	0
	Fixation				Decision					Fixation		

maintenance activities are often unknown at the specific point in time in which the decision maker optimizes planned maintenance.

Example 2.2 Given an instance with $n = 3$ and $T = 12$. Table 2.3 shows the variable fixation with initial and target tasks in an initial and a terminal planning horizon. The actual decision involves the variables that are assigned no value.

Remark 2.2 Since fractions in constraint (C) can be scaled by a multiplication with a proper factor, r_i and $\bar{r}_t$ realize integer values whenever necessary.[8] Thus, all parameters of a CPMP instance can be transformed into integers.

A period is referred to as *open*, if and only if $y_t = 1$. Otherwise, the period is *closed*. A *slot plan* is a solution of the vector $(y_1, \ldots, y_T)^T$. A *task plan* for a maintenance activity i is a solution of the vector $(x_{i1}, \ldots, x_{iT})^T$. A *maintenance plan* is a solution (x, y) that consists of one slot plan and one task plan for every maintenance activity. A *task horizon* of a maintenance activity i comprises the periods $t, \ldots, t + \pi_i - 1$. Given two task horizons $t, \ldots, t + \pi_i - 1$ and $\tau, \ldots, \tau + \pi_i - 1$ with $t < \tau$. Both task horizons are *consecutive* if and only if $t + \pi_i = \tau$. They *overlap* if and only if $t + \pi_i > \tau$. The task horizons are *complete* if and only if they do not overlap and $\{t, \ldots, t + \pi_i - 1\} \subseteq P$ as well as $\{\tau, \ldots, \tau + \pi_i - 1\} \subseteq P$ holds with $P \subseteq \{1, \ldots, T\}$. Two tasks of maintenance activity i in the periods t and τ *cover* each other, if and only if the task horizons $t, \ldots, t + \pi_i - 1$ and $\tau, \ldots, \tau + \pi_i - 1$ are consecutive or overlap. A task horizon is referred to as *covered* if and only if the respective maintenance activity i is scheduled at least once within the periods $t, \ldots, t + \pi_i - 1$. The *left neighbor (right neighbor)* of a given period t and a given maintenance activity i is the immediate previous (succeeding) period t' (t'') where a task of the maintenance activity i is scheduled.[9] Note that the leftmost (rightmost) neighbor is always associated with the period $t = 0$ ($t = T + 1$) because of Assumption (2.6). A task is *redundant*, if and only if the left and right neighbor cover each other. The *residual capacity* is the slack of constraint (C).

---

[8] This assumption is commonly imposed on the KP (Martello and Toth 1990, p. 14) and it is also applicable here.
[9] Immediate means that no third task is scheduled in between.

2.3.2 Data Structure

A slot plan is represented by an array $(A_1, \ldots, A_T)$ of pointers and each element A_t either is a null pointer or points to a *list*.[10] A list element of the referenced list A_t represents a task in period t. In order to associate a task with a maintenance activity i, the index i is an attribute of a list element. Let $left_{it}$ ($right_{it}$) be a pointer of a list element in the referenced list A_t that references to the list element associated with the left neighbor t' (right neighbor t''). A list element contains a pointer $prev_{it}$ ($next_{it}$) that references the predecessor (successor) in the referenced list A_t. The referenced lists A_0 and A_{T+1} contain tasks of all n maintenance activities (Assumption (2.6)). Let $(First_1, \ldots, First_n)$ and $(Last_1, \ldots, Last_n)$ be arrays of pointers to list elements. $First_i$ ($Last_i$) points to the list element that represents the maintenance activity i in the referenced list A_0 (A_{T+1}).[11] In order to minimize the number of memory allocation operations, two 'bins' hold elements, which are no longer used: A *stack* S^{bin} holds pointers to T empty lists and a list L^{bin} holds $n \cdot T$ list elements.[12]

The most important operations on this data structure are sketched in Remark 2.3 and Example 2.3 illustrates the data structure. Note that the residual capacity can be stored in an additional array and updated in $O(1)$, whenever a period is closed or a maintenance activity is unplanned or scheduled. If a pointer in A_t references to an empty list, an open period without any tasks is represented (i.e., $y_t = 1 \land x_{it} = 0 \ \forall i$). A null pointer in A_t stands for a closed period. A task plan is obtained by referencing the list element in $First_i$ ($Last_i$) and traversing the referenced list elements $right_{it}$ ($left_{it}$). Therefore, it can be quickly determined if a given task horizon is covered or if a task is redundant.

Remark 2.3

1. Detect if a period t is open or closed in time $O(1)$: $y_t = 1$ if A_t points to a list. Otherwise, A_t is the null pointer and $y_t = 0$ holds.
2. Detect if a maintenance activity i is scheduled in period t in time $O(1)$: Given a task i scheduled a period τ such that there is no scheduled task in between τ and t. If $right_{i\tau}$ yields a task in period t then $x_{it} = 1$ (otherwise, 0).
3. Access all tasks in period t in time $O(1)$: Reference the list A_t.

---

[10]In the remainder of the thesis, singly linked and doubly linked lists are just referred to as list. Cf. Ottmann and Widmayer (2012, pp. 29–41) for the data structure list.

[11]From the definitions follows $First_i = left_{i0}$ and $Last_i = right_{i,T+1}$.

[12]Cf. Ottmann and Widmayer (2012, pp. 41–47) for the data structure stack. Implementation detail: In this context, it is noteworthy to introduce an efficient *memory management* (Bock 2004, pp. 241–242), which minimizes the number of memory allocation operations. Assume that memory for a specific data structure is required. Let a stack consist of pointers that reference several instances of the respective data structure. The stack is initialized in the initialization phase of the algorithm. Whenever a pointer to a new instance of the data structure is required, the memory is not allocated but the top element of the stack is returned and removed from the stack. If the instance is not needed anymore, the memory is not deallocated but the pointer is inserted in the stack.

4. Access the predecessor (successor) of a task i in period t in time $O(1)$: Given a pointer that yields task i. Obtain the previous (next) task from $prev_{it}$ ($next_{it}$).
5. Access a task that belongs to the left (right) neighbor of task i in period t in time $O(1)$: Given a pointer that yields task i. Obtain the left (right) neighbor from $left_{it}$ ($right_{it}$).
6. Open a period t in time $O(1)$: Obtain a pointer p from S^{bin} and set $A_t = p$.
7. Close a period t in time $O(n)$: Empty the referenced list A_t by moving all list elements to L^{bin}, insert the pointer A_t in S^{bin} and set A_t to the null pointer. These steps require $O(1)$. Adjust the neighbors for all list elements in time $O(n)$.
8. Remove the task i from period t in time $O(1)$: Given a pointer that yields task i. Move the respective list element into L^{bin}. Adjust the neighbors of the list element.
9. Schedule a task i in period t in time $O(1)$: Given a pointer that yields the left or the right neighbor. Move one list element from L^{bin} to the end of the referenced list A_t, associate it with maintenance activity i and adjust the neighbors.
10. Schedule a task i before the task j with $i \neq j$ in period t in time $O(1)$: Given a pointer that yields task j. Execute step 9 but move the list element from L^{bin} in front of the list element associated with task j.
11. Empty the solution in time $O(n + T)$: Empty the referenced lists $(A_1, \ldots, A_T)$ by moving the list elements to L^{bin}, insert all pointers to the empty lists in S^{bin} and set $(A_1, \ldots, A_T)$ to the null pointer. These steps require $O(T)$. Adjust the neighbors in time $O(n)$.

Example 2.3 Let $n = 3$, $T = 6$ and $\pi = (4, 4, 4)^T$. Figure 2.1 depicts the aforementioned data structure for the solution $x = \begin{pmatrix} 0 & 1 & 0 & 1 & 0 & 0 \\ 0 & 0 & 0 & 1 & 0 & 0 \\ 0 & 0 & 0 & 0 & 0 & 0 \end{pmatrix}$ and $y = (0, 1, 0, 1, 1, 0)^T$. The array on the top $(A_1, \ldots, A_T)$ represents the slot plan $(y_1, \ldots, y_T)^T$. A solid arrow with an angle stands for the null pointer. A referenced list A_t represents the vector $(x_{1t}, \ldots, x_{nt})$. The arrays on the left and on the right are $(First_1, \ldots, First_n)$ and $(Last_1, \ldots, Last_n)$. A circle stands for a list element in A_t where the number is the index of a maintenance activity i. A solid arrow represents the pointers to the predecessor and successor given by $prev_{it}$ and $next_{it}$ of a list element. A dashed

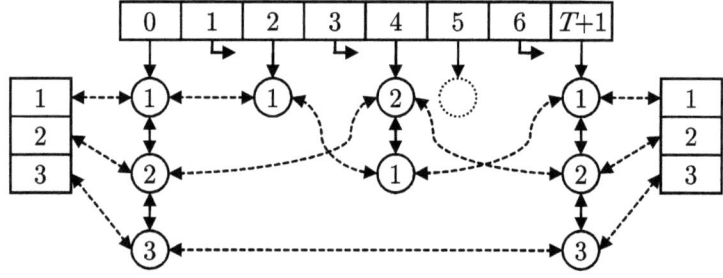

Fig. 2.1 The data structure representing a solution to the CPMP

arrow stands for the pointers *left*$_{it}$ and *right*$_{it}$ that yield the left and the right neighbor of a list element.

Observe that the tasks in a list are not ordered with respect to the maintenance activity indices such as in period $t = 4$ because lists are used. In period $t = 5$, A_5 points to an empty list, which indicates an open period 5 without scheduled tasks. Note that it takes few computational steps to verify that the task x_{21} is redundant and that the solution is infeasible because the maintenance activity $i = 3$ violates (P).

References

Aggoune, R., & Portmann, M. (2006). Flow shop scheduling problem with limited machine availability: A heuristic approach. *International Journal of Production Economics, 99*, 4–15.

Aghezzaf, E. H., Jamali, M. A., & Ait-Kadi, D. (2007). An integrated production and preventive maintenance planning model. *European Journal of Operational Research, 181*, 679–685.

Ahmad, R., & Kamaruddin, S. (2012). An overview of time-based and condition-based maintenance in industrial application. *Computers & Industrial Engineering, 63*, 135–149.

Ahmadi, A., & Kumar, U. (2011). Cost based risk analysis to identify inspection and restoration intervals of hidden failures subject to aging. *IEEE Transactions on Reliability, 60*, 197–209.

Anily, S., Glass, C. A., & Hassin, R. (1998). The scheduling of maintenance service. *Discrete Applied Mathematics, 82*, 27–42.

Baker, M. J. C. (1990). How often should a machine be inspected? *International Journal of Quality & Reliability Management, 7*, 14–18.

Bar-Noy, A., Bhataia, R., Naor, J. S., & Schieber, B. (2002). Minimizing service and operation costs of periodic scheduling. *Mathematics of Operations Research, 27*, 518–544.

Barlow, R. E., & Hunter, L. C. (1960). Optimum preventive maintenance policies. *Operations Research, 8*, 90–100.

Barlow, R. E., Hunter, L. C., & Proschan, F. (1963). Optimum checking procedures. *SIAM Journal on Applied Mathematics, 11*, 1078–1095.

Barlow, R. E., Proschan, F., & Hunter, L. C. (1965). *Mathematical theory of reliability*. New York: Wiley.

Basten, R., Schutten, J., & van der Heijden, M. (2009). An efficient model formulation for level of repair analysis. *Annals of Operations Research, 172*, 119–142.

Ben-Daya, M., & Hariga, M. (1998). A maintenance inspection model: Optimal and heuristic solutions. *International Journal of Quality & Reliability Management, 15*, 481–488.

Bitran, G. R., & Yanasse, H. H. (1982). Computational complexity of the capacitated lot size problem. *Management Science, 28*, 1174–1186.

Bock, S. (2004). *Echtzeitfähige Steuerung von Speditionsnetzwerken*. Wiesbaden: Gabler DUV.

Budai, G., Dekker, R., & Nicolai, R. P. (2008). Maintenance and production: A review of planning models. In K. A. H. Kobbacy & D. N. P. Murthy (Eds.), *Complex system maintenance handbook* (Chap. 13). London: Springer.

Cassady, C. R., & Kutanoglu, E. (2005). Integrating preventive maintenance planning and production scheduling for a single machine. *IEEE Transactions on Reliability, 54*, 304–309.

Chen, J.-S. (2008). Scheduling of nonresumable jobs and flexible maintenance activities on a single machine to minimize makespan. *European Journal of Operational Research, 190*, 90–102.

Chen, W.-J. (2009). Minimizing number of tardy jobs on a single machine subject to periodic maintenance. *Omega, 37*, 591–599.

Cho, D. I., & Parlar, M. (1991). A survey of maintenance models for multiunit systems. *European Journal of Operational Research, 51*, 1–23.

Christer, A. H., & Waller, W. M. (1984). Reducing production downtime using delay-time analysis. *Journal of the Operational Research Society, 35*, 499–512.

Clarke, L., Johnson, E., Nemhauser, G., & Zhu, Z. (1997). The aircraft rotation problem. *Annals of Operations Research, 69*, 33–46.

Dekker, R. (1996). Applications of maintenance optimization models: A review and analysis. *Reliability Engineering and System Safety, 51*, 229–240.

Dekker, R., & Wildeman, R. E. (1997). A review of multi-component maintenance models with economic dependence. *Mathematical Methods of Operations Research, 45*, 411–435.

Drinkwater, R. W., & Hastings, N. A. J. (1967). An economic replacement model. *Operational Research Quarterly, 18*, 121–138.

Garg, A., & Deshmukh, S. (2006). Maintenance management: Literature review and directions. *Journal of Quality in Maintenance Engineering, 12*, 205–238.

Gopalakrishnan, M., Ahire, S. L., & Miller, D. M. (1997). Maximizing the effectiveness of a preventive maintenance system: An adaptive modeling approach. *Management Science, 43*, 827–840.

Gopalakrishnan, M., Mohan, S., & He, Z. (2001). A tabu search heuristic for preventive maintenance scheduling. *Computers & Industrial Engineering, 40*, 149–160.

Gopalan, R., & Talluri, K. T. (1998). Mathematical models in airline schedule planning: A survey. *Annals of Operations Research, 76*, 155–185.

Graves, G. H., & Lee, C.-Y. (1999). Scheduling maintenance and semiresumable jobs on a single machine. *Naval Research Logistics, 46*, 845–863.

Grigoriev, A., van de Klundert, J., & Spieksma, F. C. R. (2006). Modeling and solving the periodic maintenance problem. *European Journal of Operational Research, 172*, 783–797.

Hung, M. S., & Brown, J. R. (1978). An algorithm for a class of loading problems. *Naval Research Logistics Quarterly, 25*, 289–297.

Joo, S.-J. (2009). Scheduling preventive maintenance for modular designed components: A dynamic approach. *European Journal of Operational Research, 192*, 512–520.

Joshi, S., & Gupta, R. (1986). Scheduling of routine maintenance using production schedules and equipment failure history. *Computers & Industrial Engineering, 10*, 11–20.

Keysan, G., Nemhauser, G. L., & Savelsbergh, M. W. P. (2010). Tactical and operational planning of scheduled maintenance for per-seat, on-demand air transportation. *Transportation Science, 44*, 291–306.

Knotts, R. M. H. (1999). Civil aircraft maintenance and support: Fault diagnosis from a business perspective. *Journal of Quality in Maintenance Engineering, 5*, 335–348.

Lee, C.-Y. (1996). Machine scheduling with an availability constraint. *Journal of Global Optimization, 9*, 395–416.

Lee, C.-Y. (1997). Minimizing the makespan in the two-machine flowshop scheduling problem with an availability constraint. *Operations Research Letters, 20*, 129–139.

Lee, C.-Y., & Chen, Z.-L. (2000). Scheduling jobs and maintenance activities on parallel machines. *Naval Research Logistics, 47*, 145–165.

Liao, C. J., & Chen, W. J. (2003). Single-machine scheduling with periodic maintenance and nonresumable jobs. *Computers & Operations Research, 30*, 1335–1347.

Lienhardt, B., Hugues, E., Bes, C., & Noll, D. (2008). Failure-finding frequency for a repairable system subject to hidden failures. *Journal of Aircraft, 45*, 1804–1809.

Lust, T., Roux, O., & Riane, F. (2009). Exact and heuristic methods for the selective maintenance problem. *European Journal of Operational Research, 197*, 1166–1177.

Ma, Y., Chu, C., & Zuo, C. (2010). A survey of scheduling with deterministic machine availability constraints. *Computers & Industrial Engineering, 58*, 199–211.

Martello, S., & Toth, P. (1990). *Knapsack problems: Algorithms and computer implementations*. New York: Wiley.

References

Mati, Y. (2010). Minimizing the makespan in the non-preemptive job-shop scheduling with limited machine availability. *Computers & Industrial Engineering, 59*, 537–543.

Mauguière, P., Billaut, J., & Bouquard, J. (2005). New single machine and job-shop scheduling problems with availability constraints. *Journal of Scheduling, 8*, 211–231.

Najid, N. M., Alaoui-Sesouli, M., & Mohafid, A. (2011). An integrated production and maintenance planning model with time windows and shortage cost. *International Journal of Production Research, 49*, 2265–2283.

Neumann, K., & Schwindt, C. (2000). Batch scheduling in process industries: An application of resource-constrained project scheduling. *OR Spectrum, 22*, 501–524.

Neumann, K., Schwindt, C., & Trautmann, N. (2002). Advanced production scheduling for batch plants in process industries. *OR Spectrum, 24*, 251–279.

Ottmann, T., & Widmayer, P. (2012). *Algorithmen und Datenstrukturen* (5th ed.). Heidelberg: Spektrum Akdemischer.

Pierskalla, W. P., & Voelker, J. A. (1976). A survey of maintenance models: The control and surveillance of deteriorating systems. *Naval Research Logistics Quarterly, 23*, 353–388.

Pintelon, L. M., & Gelders, L. F. (1992). Maintenance management decision making. *European Journal of Operational Research, 58*, 301–317.

Pritsker, A. A. B., Watters, L. J., & Wolfe, P. M. (1969). Multiproject scheduling with limited resources: A zero-one programming approach. *Management Science, 16*, 93–108.

Reklaitis, G. V. (1991). Overview of scheduling and planning of batch process operations. In *Proceedings of the NATO Advanced Study Institute on Batch Process Systems Engineering: Current Status and Future Directions* (pp. 660–705). Berlin: Springer.

Sarac, A., Batta, R., & Rump, C. (2006). A branch-and-price approach for operational aircraft maintenance routing. *European Journal of Operational Research, 175*, 1850–1869.

Scarf, P. A. (1997). On the application of mathematical models in maintenance. *European Journal of Operational Research, 99*, 493–506.

Schmidt, G. (2000). Scheduling with limited machine availability. *European Journal of Operational Research, 121*, 1–15.

Sherif, Y. S., & Smith, M. L. (1981). Optimal maintenance models for systems subject to failure - A review. *Naval Research Logistics Quarterly, 28*, 47–74.

Taghipour, S., & Banjevic, D. (2011). Periodic inspection optimization models for a repairable system subject to hidden failures. *IEEE Transactions on Reliability, 60*, 275–285.

Valdez-Flores, C., & Feldman, R. M. (1989). A survey of preventive maintenance models for stochastically detoriating single-unit systems. *Naval Research Logistics, 36*, 419–446.

van Noortwijk, J. A., Dekker, R., Cooke, R. M., & Mazucchi, T. A. (1992). Expert judgement in maintenance optimization. *IEEE Transactions on Reliability, 41*, 427–432.

Vassiliadis, C., Arvela, J., Pistikopoulos, E., & Papageorgiou, L. (2000). Planning and maintenance optimization for multipurpose plants. *Computer Aided Chemical Engineering, 8*, 1105–1110.

Vaurio, J. K. (1995). Optimization of test and maintenance intervals based on risk and cost. *Reliability Engineering and System Safety, 49*, 23–36.

Wagner, H. M., Giglio, R. J., & Glaser, R. G. (1964). Preventive maintenance scheduling by mathematical programming. *Management Science, 10*, 316–334.

Wang, I.-L., Tsai, Y.-C. J., & Li, F. (2011). A network flow model for clustering segments and minimizing total maintenance and rehabilitation cost. *Computers & Industrial Engineering, 60*, 593–601.

Wang, W. (2009). An inspection model for a process with two types of inspections and repairs. *Reliability Engineering and System Safety, 94*, 526–533.

Wang, W. (2012). An overview of the recent advances in delay-time-based maintenance modelling. *Reliability Engineering and System Safety, 106*, 165–178.

Weinstein, L., & Chung, C.-H. (1999). Integrating maintenance and production decisions in a hierarchical production planning environment. *Computers & Operations Research, 26*, 1059–1074.

Xu, D., Sun, K., & Li, H. (2008). Parallel machine scheduling with almost periodic maintenance and non-preemptive jobs to minimize makespan. *Computers & Operations Research, 35*, 1344–1349.

Xu, D., Yin, Y., & Li, H. (2009). A note on scheduling of nonresumable jobs and flexible maintenance activities on a single machine to minimize makespan. *European Journal of Operational Research, 197*, 825–827.

Zhang, F., & Jardine, A. K. S. (1998). Optimal maintenance models with minimal repair, periodic overhaul and complete renewal. *IIE Transactions, 30*, 1109–1119.

Chapter 3
Known Concepts and Solution Techniques

The reader is introduced to known mathematical concepts and solution techniques, which are used in the remainder of this thesis. This includes an introduction to computational complexity (Sect. 3.1), linear and integer programming (Sect. 3.2), Lagrangean relaxation and methods to solve the Lagrangean dual (Sect. 3.3), Benders' decomposition (Sect. 3.4), local and tabu search (Sect. 3.5) and the Knapsack Problem (Sect. 3.6).

3.1 Computational Complexity

A *decision problem* is answerable either by "yes" or "no".[1] An *optimization problem* P is written in an upright font and its optimal objective function value P is given in italics. If the optimization problem has a minimization (maximization) objective function, the objective function value of the respective decision problem is smaller (greater) or equal to K for a given non-negative integer K (Garey and Johnson 1979, pp. 18–19). The instance data comprises only integers (Martello and Toth 1990, p. 14). The *input length* or size of an instance is the number of bits required for the binary representation of the instance.[2] For a given function $g(n)$, the function $f(n)$ belongs to the set of functions $O(g(n))$ if there exist two constants $d, n_0 > 0$ such that $0 \leq f(n) \leq d \cdot g(n)$ holds for all $n \geq n_0$ (Cormen et al. 2001, p. 44). This notation is used to measure the *worst case running time* of an algorithm provided that every elementary operation in the algorithm is executed in time $O(1)$. A logarithmic factor $O(\log n)$ of some number n is given without an explicit

---

[1] Well-known decision problems are written in capital letters.
[2] The binary representation of an integer $n \in \mathbb{N}$ requires $\lfloor \log_2 n \rfloor + 1$ bits. This includes $n = 0$ that is represented by 1 bit. An integer $n \in \mathbb{Z}$ requires $\lfloor \log_2 n \rfloor + 2$ bits because 1 bit is reserved for the sign (Nemhauser and Wolsey 1988, p. 118). Fractions can be represented by pairs of integers.

base because $\log_b n = \frac{\log_d n}{\log_d b}$ to some other base d holds and $\frac{1}{\log_d b}$ is a constant (Papadimitriou and Steiglitz 1998, p. 160).

A *polynomial time algorithm* has a time complexity function that is bounded from above by a polynomial function of the input length (Garey and Johnson 1979, p. 6). A *pseudo-polynomial time algorithm* has a time complexity function that is bounded from above by a polynomial function comprising the input length and the largest integer in the instance (Garey and Johnson 1979, p. 94). The class $\mathcal{P}$ comprises all decision problems that are solvable by a deterministic Turing machine in polynomial time. A non-deterministic Turing machine solves decision problems of the class $\mathcal{NP}$ in polynomial time. Whereas $\mathcal{P} \subseteq \mathcal{NP}$ is obvious, $\mathcal{P} \supseteq \mathcal{NP}$ has not been proven yet but $\mathcal{P} \neq \mathcal{NP}$ is assumed (Garey and Johnson 1979, p. 33). This yet unsolved 'Millennium Prize Problem' in mathematics is important because if $\mathcal{P} = \mathcal{NP}$, all problems in $\mathcal{NP}$ would be polynomially solvable (Garey and Johnson 1979, pp. 37, 117; Papadimitriou and Steiglitz 1998, p. 343). A problem of $\mathcal{NP}$ is solvable by an *exponential time algorithm* with a time complexity function that is bounded from above by an exponential function of the input length (Garey and Johnson 1979, p. 32).[3]

Definition 3.1 A decision problem is in $\mathcal{NP}$ if the following condition holds: If an x is a "yes" instance to a decision problem, there exists a *certificate* for x such that its length is bounded by a polynomial in the input length of x and its validity is checked with a polynomial time algorithm (Papadimitriou and Steiglitz 1998, pp. 348–349).

A decision problem P′ *polynomially reduces* to the decision problem under investigation P that is $P' \leq_p P$ if there exists a polynomial p that transforms every instance I' of P′ into an instance $I = p(I')$ of P such that I' is a "yes" instance if and only if I is a "yes" instance (Papadimitriou and Steiglitz 1998, pp. 351–353). Hence, an algorithm that solves P also solves P′ and P is at least as hard as P′. A decision problem is $\mathcal{NP}$-*complete* if all problems belonging to $\mathcal{NP}$ polynomially reduce to it and if it is in $\mathcal{NP}$ (Garey and Johnson 1979, p. 37). A decision problem is weakly or *binary* $\mathcal{NP}$-*complete* if it is $\mathcal{NP}$-complete in the binary representation of the instance. A decision problem is unary or *strongly* $\mathcal{NP}$-*complete* if it is $\mathcal{NP}$-complete in the unary representation of the instance. Then the largest integer of the

---

[3]Given a decision problem where an instance comprises n elements and a data record $s \in \mathbb{N}^n$. Let $s^{max} = \max_{i=1,\ldots,n} s_i$. The input length is $\sum_{i=1}^{n} \lfloor \log_2 s_i \rfloor + n$. An algorithm that runs in time $O(n)$ is a polynomial time algorithm but an algorithm that runs in time $O(s^{max})$ is a pseudo-polynomial time algorithm. The binary representation of s^{max} in the input length is $\lfloor \log_2 s^{max} \rfloor + 1$. Unless s^{max} is bounded by some polynomial of the input length, the algorithm is an exponential time algorithm because $O(s^{max}) = O(2^{\lfloor \log_2 s^{max} \rfloor + 1})$.

instance is bounded from above by a polynomial function of the input length (Garey and Johnson 1979, pp. 94–95). Hence, no pseudo-polynomial time algorithm exists to solve a strongly $\mathcal{NP}$-complete problem unless $\mathcal{P} = \mathcal{NP}$ (Garey and Johnson 1979, p. 95). However, a decision problem is an *open problem* if neither polynomial solvability nor $\mathcal{NP}$-hardness has been proven. In order to prove $\mathcal{NP}$-completeness, show that the decision problem is in $\mathcal{NP}$ and provide a polynomial reduction from a $\mathcal{NP}$-complete decision problem (Garey and Johnson 1979, pp. 34–38, 45, 101–102). An optimization problem is $\mathcal{NP}$-hard if the respective decision problem is $\mathcal{NP}$-complete (Papadimitriou and Steiglitz 1998, p. 398). It can be solved with a binary search over the parameter K (Garey and Johnson 1979, pp. 116–117; Papadimitriou and Steiglitz 1998, p. 346).

Cook was the first, who proved $\mathcal{NP}$-completeness of SATISFIABILITY (Cook 1971). Karp provided 21 $\mathcal{NP}$-complete problems (Karp 1972) and Garey, Johnson investigated on strong $\mathcal{NP}$-completeness (Garey and Johnson 1978). A rich collection of $\mathcal{NP}$-complete problems and a theoretical introduction is provided in Garey and Johnson (1979), Johnson (1990).

3.2 Linear and Integer Programming

A *Linear Program (LP)* is an optimization problem that seeks an optimal solution from a set of feasible solutions. A *solution* is an encoded decision and a non-negative vector of *structure variables* $x \in \mathbb{R}_+^n$. The solution space is imposed a set of *linear constraints* of the type $A \cdot x \geq b$ with $A \in \mathbb{R}^{m \times n}$ and the *right-hand side* $b \in \mathbb{R}^m$. In what follows, assume $rank(A) = n$. Let A_i be the ith *row vector* and A^j be the jth *column vector* of A. Denote E_h as the *identity matrix* of the size h. Depending on the logical context, 0 and 1 are vectors or matrices with respective sizes. A solution is *feasible* if it satisfies all constraints and the non-negativity constraint. Otherwise, the solution is *infeasible*. The linear *objective function* evaluates a given solution and let $c \in \mathbb{R}^n$ be the *objective function coefficients*. Throughout the thesis, all matrix and vector components are rational numbers. The *solution space* S is a convex polyhedron that defines the set of feasible solutions (Nemhauser and Wolsey 1988, p. 86). Solving a LP is in $\mathcal{P}$ (Khachiyan 1979) and stated as

$$Z = \min\left\{c^T \cdot x \,\middle|\, x \in S\right\} \text{ with } S = \left\{x \in \mathbb{R}_+^n \,\middle|\, A \cdot x \geq b\right\}. \tag{3.1}$$

A *slack variable* $s_i \geq 0$ transforms $A_i \cdot x \geq b_i$ into $A_i \cdot x - s_i = b_i$. A minimization objective function is transformed into a maximization objective function by $-Z = \max(-c)^T \cdot x$. A variable $x_j \in \mathbb{R}$ is substituted by $x_j^+, x_j^- \geq 0$ and $x_j = x_j^+ - x_j^-$.[4]

---
[4] Being linearly independent in A, either x_j^+ or x_j^- can be strictly positive in the basis.

The intersection $H^= = H^\leq \cap H^\geq$ of two *halfspaces* $H^\leq = \{x \geq 0 | a^T \cdot x \leq d\}$ and $H^\geq = \{x \geq 0 | a^T \cdot x \geq d\}$ defines a *hyperplane* $H^= = \{x \geq 0 | a^T \cdot x = d\}$. An inequality $a^T \cdot x \geq d$ is a *valid inequality* if it is satisfied by each $x \in S$ (Nemhauser and Wolsey 1988, p. 88). A *face* F is a valid inequality with $F = \{x \in S | a^T \cdot x = d\}$ (Nemhauser and Wolsey 1988, p. 88). If the face F is non-empty, the valid inequality $a^T \cdot x \geq d$ *supports* S. The extreme points of an *integral face* are integer extreme points. A *facet* is a face F with the largest dimension $\dim(F) = \dim(S) - 1$ (Nemhauser and Wolsey 1988, p. 89). If a facet $F = \{x\}$ is a zero-dimensional face $\dim(F) = 0$, then x is denoted as an *extreme point* or a vertex of S (Nemhauser and Wolsey 1988, p. 93). An alternative definition states that an extreme point is a feasible solution of S that cannot be yielded by a strict convex combination of two different feasible solutions (Nemhauser and Wolsey 1988, p. 93). Another definition states that an extreme point is a basic feasible solution (Nemhauser and Wolsey 1988, p. 37 with p. 97; Chvátal 2002, p. 274).

If S is a *bounded and non-empty polyhedron* (i.e., a *polytope*), the solution space is described by a convex combination of extreme points (Nemhauser and Wolsey 1988, p. 96). Let P be the index set of all extreme points x^p. In a polytope, there exists an extreme point with a finite optimal objective function value (Nemhauser and Wolsey 1988, p. 94). Thus, the LP is called *solvable* and it follows

$$Z \leq c^T \cdot x^p \quad \forall p \in P. \tag{3.2}$$

If S is an *unbounded and non-empty polyhedron*, the convex combination of all extreme points does not fully describe the solution space. A point $r \in \mathbb{R}^n$ is denoted as a *ray* if and only if it holds that $\{y \in \mathbb{R}^n | y = x + \lambda \cdot r \wedge \lambda \geq 0\} \subseteq S$ for any $x \in S$ (Nemhauser and Wolsey 1988, p. 93). An *extreme ray* is a ray that cannot be obtained from a strict convex combination of two different rays (Nemhauser and Wolsey 1988, p. 94). Let R be the index set of all extreme rays x^r. Therefore, if Z is unbounded

$$Z \leq c^T \cdot x^p + \lambda \cdot c^T \cdot x^r \text{ with } \lambda > 0 \; \forall p \in P; r \in R \tag{3.3}$$

$$= -\infty \tag{3.4}$$

holds because there always exists an extreme ray x^r with $c^T \cdot x^r < 0$ (Nemhauser and Wolsey 1988, p. 95).[5] A polyhedron has finite sets P and R (Nemhauser and Wolsey 1988, p. 94). An *integral polyhedron* has only integer extreme points. If S is *empty*, the LP is *infeasible*.

---

[5] An extreme ray with $c^T \cdot x^r > 0$ exists if the LP is a maximization problem (Nemhauser and Wolsey 1988, p. 95).

3.2 Linear and Integer Programming

Table 3.1 Dualization rules

Primal LP		Dual LP	
Objective function	$\min c^T \cdot x$	Objective function	$\max b^T \cdot \mu$
ith constraint	$A_i \cdot x \geq b_i$	ith variable	$\mu_i \geq 0$
	$A_i \cdot x = b_i$		$\mu_i \in \mathbb{R}$
jth variable	$x_j \geq 0$	jth constraint	$(A^j)^T \cdot \mu \leq c_j$
	$x_j \in \mathbb{R}$		$(A^j)^T \cdot \mu = c_j$

Minkowski's Theorem 3.2 states that a non-empty polyhedron S is described by a convex combination of extreme points and a linear combination of extreme rays (Nemhauser and Wolsey 1988, p. 96).

Theorem 3.2 Minkowski's Theorem

$$S = \left\{ \sum_{p \in P} x^p \cdot \lambda^p + \sum_{r \in R} x^r \cdot \lambda^r \,\middle|\, \sum_{p \in P} \lambda^p = 1 \wedge \lambda^p \geq 0 \,\forall p \in P \wedge \lambda^r \geq 0 \,\forall r \in R \right\}$$

To every LP a (corresponding) *dual problem* or dual linear problem exists. In this context, (3.1) is referred to as the *primal (linear) problem*. Denote $\mu \in \mathbb{R}_+^m$ as *dual variables*[6] and x as *primal variables*. The dual problem is stated as

$$ZD = \max \left\{ b^T \cdot \mu \,\middle|\, A^T \cdot \mu \leq c \wedge \mu \geq 0 \right\}.$$

A primal constraint corresponds to a dual variable and vice versa. Table 3.1 presents dualization rules (Domschke and Drexl 2005, p. 33). It immediately follows that the dualization of a dual problem yields the primal problem (Nemhauser and Wolsey 1988, p. 28).

Table 3.2 summarizes all possibilities to combine solutions of a primal and a dual problem (Nemhauser and Wolsey 1988, p. 29; Chvátal 2002, p. 60).

Properties between the primal and the dual problem are provided in Theorem 3.3 (Nemhauser and Wolsey 1988, pp. 28–29, 97; Chvátal 2002, pp. 62–65).

---

[6]Synonyms are *shadow prices* or *opportunity costs of a constraint*.

Table 3.2 Combinations of solutions to the primal and the dual problem

	Dual LP		
Primal LP	Optimal	Infeasible	Unbounded
Optimal	Possible, $Z = ZD$	Impossible	Impossible
Infeasible	Impossible	Possible	Possible, $ZD = \infty$
Unbounded	Impossible	Possible, $Z = -\infty$	Impossible

Theorem 3.3

1. Weak duality: $Z \geq ZD$ holds for all feasible solutions x and μ.
2. Strong duality: If and only if $Z = ZD$ is finite for two feasible solutions x and μ, then x and μ are optimal.
3. Complementary slackness: If and only if x and μ are feasible solutions with $(c^T - \mu^T \cdot A) \cdot x = 0$ and $\mu^T \cdot (x \cdot A - b) = 0$, then x and μ are optimal.

Lemma 3.4 provides a straightforward way to prove if a given system of linear inequalities is solvable (Nemhauser and Wolsey 1988, p. 30).

Lemma 3.4 Farkas' Lemma
Either $\{x \in \mathbb{R}^n_+ | A \cdot x \leq b\} \neq \emptyset$ or there exists a $\mu \in \mathbb{R}^m_+$ with $b^T \cdot \mu < 0$ and $A^T \cdot \mu \geq 0$.

Proof Consider a primal problem and the corresponding dual problem. Finding any feasible solution is equivalent to $c = 0$.

$$Z = \max \{0^T \cdot x | A \cdot x \leq b \wedge x \geq 0\} \quad (3.5)$$

$$ZD = \min \{b^T \cdot \mu | A^T \cdot \mu \geq 0 \wedge \mu \geq 0\} \quad (3.6)$$

Since $\mu = 0$ is feasible, two excluding cases exist (see Table 3.2). In the first case, (3.5) has an optimal solution. Both LPs are solvable and $Z = ZD = 0$ holds. Hence, the solution space is not empty that is $\{x \geq 0 | A \cdot x \leq b\} \neq \emptyset$. In the second case, (3.5) is infeasible and the solution space of (3.6) is unbounded, which yields $ZD = -\infty$ with (3.4). Hence, there exists a feasible solution μ to (3.6) with $b^T \cdot \mu < 0$.

A *Mixed Integer Program (MIP)* has k integer variables $y \in \mathbb{N}^k$ and $B \in \mathbb{R}^{m \times k}$, $f \in \mathbb{R}^k$.

$$\min \{c^T \cdot x + f^T \cdot y | A \cdot x + B \cdot y \geq b \wedge x \geq 0 \wedge y \in \mathbb{N}^k\}$$

Solving a MIP strongly $\mathcal{NP}$-hard (Garey and Johnson 1979, p. 245) and the solution space is not necessarily convex. An *Integer Program (IP)* is a MIP with $n = 0$.

3.2 Linear and Integer Programming

Relaxing the *integrality constraint* $y \in \mathbb{N}^k$ to $y \geq 0$ yields the *LP relaxation*. Since a LP relaxation has a larger solution space, its optimal objective function value yields a *lower bound*. The objective function value of any feasible solution is referred to as *upper bound*.[7] The *convex hull* comprises the set of all points, which are obtained from convex combinations of points in the solution space of the MIP (Nemhauser and Wolsey 1988, p. 83). A MIP can be solved by a LP where the solution space is the convex hull (Nemhauser and Wolsey 1988, p. 106).

Notation 3.5 An intersection of half spaces of the constraints (1),…,(K) is denoted as $H(1)\ldots(K)$. The convex hull of the constraints (1),…,(K) is represented by $Conv(1)\ldots(K)$.

A system of linear inequalities $A \cdot x \leq b$ is called *totally dual integral*, if for all $c \in \mathbb{Z}^n$ where $\max\{c^T \cdot x \mid A \cdot x \leq b \wedge x \in \mathbb{R}^n\}$ is finite, the dual problem $\min\{b^T \cdot \mu \mid A^T \cdot \mu = c \wedge \mu \in \mathbb{R}_+^m\}$ has an optimal integer solution μ (Nemhauser and Wolsey 1988, p. 537). A polytope is integral if Lemma 3.6 holds (Nemhauser and Wolsey 1988, pp. 536–537).

Lemma 3.6 The polyhedron $\{x \in \mathbb{R}^n \mid A \cdot x \leq b\}$ is integral, if $A \cdot x \leq b$ is totally dual integral, $b \in \mathbb{Z}^m$ and the polyhedron is non-empty.

A matrix A is called *totally unimodular*, if and only if the determinant of every square submatrix is $-1, 0$ or 1 (Nemhauser and Wolsey 1988, pp. 540–542). (A, E_m), $(A, -E_m)$ as well as A^T are totally unimodular if and only if A is totally unimodular (Nemhauser and Wolsey 1988, p. 540). Lemma 3.7 provides a sufficient criterion (Nemhauser and Wolsey 1988, p. 544). A polyhedron is integral if Lemma 3.8 holds (Nemhauser and Wolsey 1988, p. 541).[8]

Lemma 3.7 Let A be a $\{-1, 0, 1\}$ matrix with no more than two non-zero elements in each column (row). Then A is totally unimodular if and only if the rows (columns) of A can be partitioned into two subsets A_1 and A_2 as follows: If a column (row) contains two non-zero elements of the same sign, both rows (columns) are in different subsets. If a column (row) contains two non-zero elements of the opposite sign, both rows (columns) are in the same subset.

Lemma 3.8 The polyhedron $\{x \in \mathbb{R}^n \mid b' \leq A \cdot x \leq b \wedge d' \leq x \leq d\}$ is integral, if A is totally unimodular, $b', b \in \mathbb{Z}^m$; $d', d \in \mathbb{Z}^n$ and the polyhedron is non-empty.

---

[7]Unless otherwise stated, denote *LB* (*UB*) as a lower (upper) bound and *LB** (*UB**) as the best known lower (upper) bound.
[8]Total unimodularity yields integral polytopes for the *Shortest Path Problem (SPP)*, the Maximum Flow Problem, the Minimum Cost Flow Problem (Ahuja et al. 1993, pp. 447–449), the Standardization Problem (Domschke and Wagner 2005) and the Shared Fixed-Cost Problem (Rhys 1970).

Let S be a polytope and let S_I with $S_I \subseteq S$ be the set of integer points. A polytope S is *quasi-integral* if every edge of the convex hull of S_I is also an edge of S (Yemelichev et al. 1984, pp. 189–190). A sufficient criterion to prove this property is stated in Yemelichev et al. (1984, p. 191) as

Lemma 3.9 Assume S_I belongs to the set of extreme points of a considered polytope S. Then, if for every pair of two integer points x_1, x_2 of the polytope S there exists an integral face containing x_1 and x_2, then S is quasi-integral (Yemelichev et al. 1984, p. 191).

Quasi-integrality holds for the polytopes of some $\mathcal{NP}$-hard problems, namely the *Uncapacitated Facility Location Problem (UFLP)* (Yemelichev et al. 1984, pp. 192–193),[9] the *Uncapacitated Network Design Problem (UNDP)* (Hellstrand et al. 1992),[10] the *Set Partitioning Problem (SetPP)* (Yemelichev et al. 1984, pp. 191–192;

---

[9] Given a set of facilities K with fixed costs $f_k \geq 0 \ \forall k \in K$ and a set of customers J with transportation costs $c_{jk} \geq 0 \ \forall j \in J; k \in K$. Let the binary variable y_k equal to 1 if the facility k is established (otherwise, 0). The binary variable x_{jk} equals one if the customer j is served from the facility k. Find the cost minimal assignment of customers to facilities such that the demand of all customers is satisfied. Cornuejols et al. (1990) show strong $\mathcal{NP}$-hardness by a polynomial reduction with the strongly $\mathcal{NP}$-complete VERTEX COVER (Garey and Johnson 1979, p. 190). The UFLP is stated as

$$\min \sum_{k \in K} f_k \cdot y_k + \sum_{j \in J} \sum_{k \in K} c_{jk} \cdot x_{jk} \quad \text{subject to}$$

$$\sum_{k \in K} x_{jk} = 1 \ \forall j \in J \wedge x_{jk} \leq y_k \ \forall j \in J; k \in K \wedge x_{jk} \in \{0,1\} \ \forall j \in J; k \in K \wedge y_k \in \{0,1\} \ \forall k \in K.$$

[10] Given a directed graph $G(V,E)$ with a node set V, an arc set E and a set of commodities K. For a commodity $k \in K$, assume that a unit demand must be send from a source node O_k to a sink node D_k. Every unit that is send on an arc $(a,b) \in E$ causes flow costs c_{kab} whereas a non-zero flow causes fixed costs f_{ab}. The flow of the commodity k on the arc (a,b) is represented by a binary variable x_{kab}. If the flow is non-zero, the binary variable y_{ab} equals one (otherwise, 0). Find a cost minimal flow for all commodities such that the flow is balanced in every node. Magnanti and Wong (1984) show strong $\mathcal{NP}$-hardness by a polynomial reduction with the strongly $\mathcal{NP}$-complete STEINER TREE IN GRAPHS (Garey and Johnson 1979, p. 208). The UNDP is formulated as

$$\min \sum_{(a,b) \in E} f_{ab} \cdot y_{ab} + \sum_{k \in K} \sum_{(a,b) \in E} c_{kab} \cdot x_{kab} \quad \text{subject to}$$

$$\sum_{(a,b) \in E} x_{kab} - \sum_{(b,a) \in E} x_{kba} = \begin{cases} 1 & \text{if } a = O_k \\ -1 & \text{if } a = D_k \\ 0 & \text{otherwise} \end{cases} \quad \forall k \in K; a \in V$$

$x_{kab} \leq y_{ab} \ \forall k \in K; (a,b) \in E \wedge x_{kab} \in \{0,1\} \ \forall k \in K; (a,b) \in E \wedge y_{ab} \in \{0,1\} \ \forall (a,b) \in E.$

3.2 Linear and Integer Programming

Trubin 1969; Balas and Padberg 1972)[11] and the One-Facility/Two-Facility One-Commodity Network Flow Problem (Sastry 2001).[12] In a quasi-integral polytope, two arbitrary integer extreme points of the polytope are connected by a path where each edge connects two integer extreme points and the edges of the path are edges of the polytope itself. Hence, a MIP where the LP relaxation has a quasi-integral polytope is solvable as the LP relaxation with a modified Simplex algorithm, which pivots only on integral basic feasible solutions (Yemelichev et al. 1984, p. 190; Trubin 1969). Based on this idea many authors developed specialized algorithms for the SetPP. The integral Simplex using decomposition algorithm is developed by Zaghrouti et al. (2014). The authors show its superiority towards CPLEX for small to large instances from aircrew scheduling as well as bus and driver scheduling. Thompson (2002) proposes another integral Simplex algorithm and states promising results for crew scheduling problems. A method for column generation with a SetPP as a master problem is presented by Rönnberg and Larsson (2014).

The remainder of the section presents methods to solve LPs. Interior point methods provide polynomial time algorithms to solve a LP. The ellipsoid method was the first algorithm and it has a worst case running time of $O(n^6 \cdot L)$ where L is the input length of m, n, A, b and c (Khachiyan 1979). Karmarkar's projection algorithm is more efficient and runs in time $O(n^{3.5} \cdot L)$ (Karmarkar 1984). Assume there are more columns than rows (i.e., $n \geq m$). Being independent of the values of b and c, Tardos developed an algorithm that requires the time $O(n^{5.5} \cdot \bar{L})$ with $\bar{L}$ as the input length of n and A (Tardos 1986) but a more efficient algorithm runs in time $O(n^{3.5} \cdot \bar{L})$ (Vavasis and Ye 1996). If the number of constraints is polynomial in m, n and if A is a $\{-1, 0, 1\}$ matrix, Tardos' algorithm solves the LP in *strongly* polynomial time. This is a significant result because many combinatorial optimization problems can be formulated this way. Orlin (1986) provides a dual variant of Tardos' algorithm. Megiddo (1989) reviews the computational complexity of LP, Bland et al. (1981) surveys the ellipsoid method and the focus of Wright (2005) lies on interior point methods.

---

[11]Given a set K and a collection C of subsets of K each associated with a weight. Find a subcollection $C' \subseteq C$ of subsets such that every element of K occurs in exactly one member of C' such that the total weight is minimal. A polynomial reduction with the strongly $\mathcal{NP}$-complete EXACT COVER BY 3-SETS (Garey and Johnson 1979, p. 221) shows that the SetPP is strongly $\mathcal{NP}$-hard. Let f_c be weights per subset $c \in C$. Let $a_{kc} = 1$ if the element k belongs to the subset c (otherwise, 0) $\forall k \in K; c \in C$. The binary variable x_c equals one if $c \in C'$ (otherwise, 0). The SetPP is

$$\min \sum_{c \in C} f_c \cdot x_c \quad \text{subject to} \quad \sum_{c \in C} a_{kc} \cdot x_c = 1 \; \forall k \in K \land x_c \in \{0, 1\} \; \forall c \in C.$$

[12]The polytopes of the Perfect Edge Matching Problem, the Maximum-Cardinality Edge Matching Problem, the Maximum-Cardinality Node Packing Problem and the Set Packing Problem are quasi-integral because the problems can be reformulated as specific SetPPs (Balas and Padberg 1976).

3.2.1 The Simplex Algorithm

The most popular algorithm to solve a LP to optimality is the (Primal) Simplex algorithm (Nemhauser and Wolsey 1988, pp. 33–34; Dantzig 1951). The slack of the dual constraints that correspond to the primal structure variables (primal slack variables) yields the *reduced costs* (corresponding dual solution). Assume that the LP (3.1) has been transformed to

$$\min \left\{ c^T \cdot x \,\middle|\, A \cdot x = b \wedge x \geq 0 \right\}. \tag{3.7}$$

Initialize the algorithm with a basic feasible solution.[13] If no basic feasible solution exists, the LP is infeasible. If all reduced costs are non-negative, terminate because the dual solution is feasible and optimality follows from Theorem 3.3 (*pricing out*). Otherwise, select a non-basic variable x_t with strictly negative reduced cost (pivot column).[14] If the tth column contains only negative constraint coefficients, then terminate because a ray is found[15] and the primal problem is unbounded. Otherwise, let the sth row provide the smallest feasible upper bound the non-basic variable x_t would receive if it was introduced in the current basis (pivot row). Perform a base change with the Gauß algorithm such that x_t enters and x_s leaves the basis. The Simplex algorithm evaluates the extreme points of a non-empty polyhedron and finds an optimal extreme point to the LP (3.7) if one exists.

Nemhauser and Wolsey (1988, pp. 37–38), Lemke (1954) present a variant that solves the dual problem. Although there exist instances showing that the worst case computational complexity grows exponentially with respect to n (Klee and Minty 1972), the Simplex algorithm is reported to have a linear average running time (Borgwardt 1982). The Simplex algorithm is surveyed in Shamir (1987).

3.2.2 The Primal-Dual Simplex Algorithm

This algorithm exploits the complementary slackness of Theorem 3.3.3 (Papadimitriou and Steiglitz 1998, pp. 104–109; Dantzig et al. 1956). Assume that the LP (3.7) has $b \geq 0$.

---

[13] Use the two-phase method (Chvátal 2002, pp. 125–132) or the big-M method (Domschke and Drexl 2005, pp. 28–30).

[14] Choosing the variable with minimal reduced cost is a straight forward rule to improve the primal solution. Bland (1977) present a rule that prevents a cycling of the Simplex algorithm caused by primal degeneracy.

[15] Let $\bar{a}_{ij}$ be an entry of the coefficient matrix in the current Simplex tableau. The pivot column t yields a ray as $x_j = -\bar{a}_{it}$ if j is a basic variable that is assigned to row i or as $x_j = 1$ if $j = t$ (otherwise, 0) for all $j = 1, \ldots, n$.

Initialize the algorithm with a feasible dual solution μ.[16] If no dual solution μ exists, the LP (3.7) is either infeasible or unbounded because the dual problem is infeasible. In each iteration, the algorithm modifies μ but maintains feasibility. Identify a subset of primal variables $J = \{j = 1, \ldots, n | \mu^T \cdot A^j = c_j\}$ that is referred to as *potential primal basic variables* and solve the *Reduced Primal Problem (RP)* where $y \in \mathbb{R}_+^m$ that is

$$\xi = \min 1^T \cdot y \quad \text{subject to}$$
$$E_m \cdot y + A \cdot x = b \wedge x_j = 0 \; \forall j \notin J \wedge x_j \geq 0 \; \forall j \in J \wedge y \geq 0.$$

If $\xi = 0$, terminate because x is feasible and thus, optimal by Theorem 3.3.3 and an extreme point. Otherwise, $\xi > 0$ holds. Denote the corresponding dual solution to the RP as $\tilde{\mu}$. If $\tilde{\mu}^T \cdot A^j \leq 0 \; \forall j \notin J$, terminate because the LP (3.7) is infeasible since the dual problem is unbounded. Otherwise, the dual solution of the next iteration is updated as $\mu = \mu + \lambda \cdot \tilde{\mu}$ with

$$\lambda = \min \left\{ \frac{c_j - \mu^T \cdot A^j}{\tilde{\mu}^T \cdot A^j} \middle| \tilde{\mu}^T \cdot A^j > 0 \; \forall j \notin J \right\}.$$

Instead of solving the RP, its dual problem the so called *Reduced Dual Problem (RDP)* can be solved to optimality. In practical applications, either the RP or the RDP is solved by a specialized algorithm.[17] A survey on primal-dual methods for approximation algorithms and network design problems is given in Goemans and Williamson (1996).

3.3 Dual Decomposition: Lagrangean Relaxation

In order to obtain a more structured subproblem that is easier to solve, a difficult constraint is relaxed in this approach (Guignard 2003; Fisher 1981). The IP (3.8) has n variables $x \in \mathbb{N}^n$ and two constraints (I) and (II). Let $c \in \mathbb{R}^n$, $A \in \mathbb{R}^{m \times n}$, $b \in \mathbb{R}^m$, $B \in \mathbb{R}^{k \times n}$ and $d \in \mathbb{R}^k$.

$$Z = \min \left\{ c^T \cdot x \middle| A \cdot x \geq b \; \text{(I)} \wedge B \cdot x \geq d \; \text{(II)} \wedge x \in \mathbb{N}^n \; \text{(III)} \right\} \quad (3.8)$$

Assume that neglecting constraint (I) completely would yield an optimization problem that is either well-studied or has a structure with advantageous properties.

---

[16] The method from Papadimitriou and Steiglitz (1998, p. 105) uses ideas of the two-phase method applied to the dual problem.
[17] Examples for efficient applications are the SPP (Papadimitriou and Steiglitz 1998, pp. 109–113; Chvátal 2002, pp. 390–400), the Maximum Flow Problem (Papadimitriou and Steiglitz 1998, pp. 114–115) and the Capacitated Transportation Problem (Ford and Fulkerson 1957).

Let the 'hard' constraint (I) extend the problem Z in a way that the specific structure of the 'easy' constraint (II) cannot be exploited directly. Weight the slack s of constraint (I) by a vector $\mu \in \mathbb{R}_+^m$ of penalty costs referred to as *Lagrangean multiplier* and subtract the product from the objective function. The *Lagrangean relaxation* $ZL_{(I)}$ for given μ becomes (Nemhauser and Wolsey 1988, pp. 323–324)

$$ZL_{(I)}(\mu) = \min\left\{c^T \cdot x + \mu^T \cdot (b - A \cdot x) \,\middle|\, B \cdot x \geq d \text{ (II)} \wedge x \in \mathbb{N}^n \text{ (III)}\right\}$$
$$= \min\left\{c^T \cdot x + \mu^T \cdot (b - A \cdot x) \,\middle|\, x \in Conv(\text{II})(\text{III})\right\}. \qquad (3.9)$$

The Lagrangean relaxation provides a lower bound $ZL_{(I)}(\mu) \leq Z$ because $\mu^T \cdot (b - A \cdot x) \leq 0$ holds for $x \in Conv(\text{I})(\text{II})(\text{III})$ (Nemhauser and Wolsey 1988, p. 324). In what follows, assume $Conv(\text{II})(\text{III}) \neq \emptyset$. Lemma 3.10 provides an optimality criterion (Nemhauser and Wolsey 1988, p. 331).

Lemma 3.10 Let x be an optimal solution to $ZL_{(I)}$. If x satisfies the Lagrangean relaxed constraint (I) and $\mu^T \cdot (b - A \cdot x) = 0$ holds, then x is optimal to Z.

Proof

$$\begin{aligned}
ZL_{(I)}(\mu) &= \min\left\{c^T \cdot x' + \mu^T \cdot (b - A \cdot x') \,\middle|\, x' \in Conv(\text{II})(\text{III})\right\} && x \text{ is optimal to } ZL_{(I)} \\
&= c^T \cdot x + \mu^T \cdot (b - A \cdot x) && \mu^T \cdot (b - A \cdot x) = 0 \\
&= c^T \cdot x && x \in Conv(\text{I})(\text{II})(\text{III}) \\
&= Z
\end{aligned}$$

A further analysis applies Minkowski's Theorem 3.2 (cf. Nemhauser and Wolsey 1988, pp. 327–328; Desrosiers and Lübbecke 2005, pp. 16–18). Let P (R) be the index set of all extreme points x^p (extreme rays x^r) of $Conv(\text{II})(\text{III})$. The solution space of $ZL_{(I)}$ is then described by

$Conv(\text{II})(\text{III})$

$$= \left\{ \sum_{p \in P} x^p \cdot \lambda^p + \sum_{r \in R} x^r \cdot \lambda^r \,\middle|\, \sum_{p \in P} \lambda^p = 1 \wedge \lambda^p \geq 0 \,\forall p \in P \wedge \lambda^r \geq 0 \,\forall r \in R \right\}.$$

Using (3.9) and (3.3), $ZL_{(I)}(\mu) \leq (c^T - \mu^T \cdot A) \cdot (x^p + \lambda \cdot x^r) + \mu^T \cdot b$ holds for all $r \in R$ and $p \in P$. Since $Conv(\text{II})(\text{III}) \neq \emptyset$ is assumed, $Conv(\text{II})(\text{III})$ is either unbounded or a finite optimal solution exists that is an extreme point. Together

3.3 Dual Decomposition: Lagrangean Relaxation

with (3.2) and (3.4), a *convexification approach* yields

$$ZL_{(I)}(\mu) \leq \begin{cases} -\infty & \text{if } \exists r \in R : (c^T - \mu^T \cdot A) \cdot x^r < 0 \\ (c^T - \mu^T \cdot A) \cdot x^p + \mu^T \cdot b \ \forall p \in P & \text{otherwise.} \end{cases}$$

Assume that *Conv*(II)(III) is a polytope. The convexified Lagrangean relaxation becomes

$$ZL_{(I)}(\mu) = \min_{p \in P} \left\{ c^T \cdot x^p + \mu^T \cdot (b - A \cdot x^p) \,\middle|\, (c^T - \mu^T \cdot A) \cdot x^r \geq 0 \ \forall r \in R \right\} \tag{3.10}$$

and the *Lagrangean dual* $Z_{(I)}$ yields the largest lower bound

$$Z_{(I)} = \max_{\mu \geq 0} ZL_{(I)}(\mu) \tag{3.11}$$

$$= \max_{\mu \geq 0} \min_{p \in P} \left\{ c^T \cdot x^p + \mu^T \cdot (b - A \cdot x^p) \,\middle|\, (c^T - \mu^T \cdot A) \cdot x^r \geq 0 \ \forall r \in R \right\}. \tag{3.12}$$

Clearly, $Z \geq Z_{(I)} \geq ZL_{(I)}(\mu)$ holds for a given μ. Linearizing the max-min objective function, (3.12) is equivalent to the *dual master problem*

$$Z_{(I)} = \max \eta \qquad \text{subject to} \tag{3.13}$$

$$\eta + \mu^T \cdot (A \cdot x^p - b) \leq c^T \cdot x^p \qquad \forall p \in P \qquad \lambda^p \tag{3.14}$$

$$\mu^T \cdot A \cdot x^r \leq c^T \cdot x^r \qquad \forall r \in R \qquad \lambda^r \tag{3.15}$$

$$\eta \in \mathbb{R} \wedge \mu \geq 0. \tag{3.16}$$

The set of constraints (3.14) are *dual cutting planes* and (3.15) are feasibility cuts. A dualization yields the *primal master problem* as the SetPP

$$Z_{(I)} = \min c^T \cdot \left(\sum_{p \in P} x^p \cdot \lambda^p + \sum_{r \in R} x^r \cdot \lambda^r \right) \qquad \text{subject to} \tag{3.17}$$

$$\sum_{p \in P} \lambda^p = 1 \qquad \eta \tag{3.18}$$

$$A \cdot \left(\sum_{p \in P} x^p \cdot \lambda^p + \sum_{r \in R} x^r \cdot \lambda^r \right) \geq b \qquad \mu \tag{3.19}$$

$$\lambda^p \geq 0 \ \forall p \in P \wedge \lambda^r \geq 0 \ \forall r \in R. \tag{3.20}$$

The constraint (3.18) ensures a convex combination of all extreme points. Observe that (3.19) represents constraint (I) that has been relaxed in a Lagrangean fashion.[18] Reformulate $Conv$(II)(III) in the SetPP with Minkowski's Theorem 3.2 to establish Geoffrion's Theorem 3.11 (Geoffrion 1974).

Theorem 3.11 Geoffrion's Theorem

$$Z_{(I)} = \min \left\{ c^T \cdot x \mid x \in H(I) \cap Conv(II)(III) \right\}$$

Proof From the arguments presented above.

Corollary 3.12 states that the Lagrangean dual where the Lagrangean LP relaxation has an integral polytope yields a lower bound that coincides with the lower bound Z^{LP} of the LP relaxation Z^{LP} of the IP of (3.8) (Geoffrion 1974). Furthermore, the corollary states that the Lagrangean dual provides a lower bound that is at least as strong as Z^{LP}.

Corollary 3.12 Integrality property If $Conv$(II)(III) $= H$(II)(IV) with (IV) as $x \geq 0$, then $Z_{(I)} = Z^{LP}$.

Proof By Theorem 3.11 with $Conv$(II)(III) $= H$(II)(IV).

The integrality property can only be evaluated relative to formulation of the IP.[19] Owing to the time consuming solution approaches to solve $Z_{(I)}$ to optimality (see Sect. 3.3.1), solving $ZL_{(I)}$ is less advantageous if the integrality property holds and the computational time to solve Z^{LP} is lower. However, fast approximations of $Z_{(I)}$ can yield satisfactory lower bounds (see Sect. 3.3.2).[20]

---

[18]This SetPP can also be obtained from a Dantzig-Wolfe decomposition of Z with $ZL_{(I)}$ as a subproblem, which shows the close relation between both decomposition techniques (Desrosiers and Lübbecke 2005, pp. 8–11, 16–18; Lübbecke and Desrosiers 2005).

[19]An improved Lagrangean dual may be obtained either by adding valid inequalities or from a reformulation of the IP (3.8). An example of the last case is presented in Desrosiers and Lübbecke (2005, pp. 23–27) that compare two different formulations of the Cutting Stock Problem that yield the same lower bound on the value of an optimal integer solution. The subproblem of the first (second) formulation is a KP (SPP) that does not have (has) the integrality property.

[20]In a branch and bound algorithm for the Traveling Salesman Problem (Held and Karp 1970), the Lagrangean relaxation is a Minimal 1-Tree that has the integrality property and the Lagrangean dual is solved via a subgradient algorithm. Since the proposed LP relaxation comprises an exponential number of constraints, the computational effort to approximate the Lagrangean relaxation is lower.

A very effective algorithm for the UFLP is a *dual-ascent heuristic* (Erlenkotter 1978). The dual problem of the LP relaxation is considered and the corresponding dual variables to the variable upper bound constraint are fixed to optimal values. Alternating between heuristics to the dual LP and the primal IP and satisfying the complementary slackness improves the obtained dual lower bound. Krarup and Pruzan (1983) showed that the best dual solution coincides with the Lagrangean dual in which the relaxation of the variable upper bound constraints leads to a Lagrangean relaxation that has the integrality property because it is a trivial selection problem.

3.3 Dual Decomposition: Lagrangean Relaxation

Introduce two decision variables $x' \in \mathbb{N}^n$ and $x'' \in \mathbb{N}^n$. Substitute x such that (I) contains x' and (II) as well as the objective function contains x''. Adding the linking constraint $x' = x''$, the IP of (3.8) is reformulated.

$$Z = \min\left\{c^T \cdot x'' \,\middle|\, A \cdot x' \geq b \text{ (I)} \wedge B \cdot x'' \geq d \text{ (II)} \wedge x'' - x' = 0 \text{ (IV)} \wedge x', x'' \in \mathbb{N}^n \text{ (III)}\right\}$$

The Lagrangean multiplier $\mu \in \mathbb{R}^n$ of (IV) are real variables because (IV) is an equality. *Dual decomposition* yields the Lagrangean relaxation (Nemhauser and Wolsey 1988, pp. 333–334)

$ZL_{(I)/(II)}(\mu)$

$$= \min\left\{c^T \cdot x'' + \mu^T \cdot (x' - x'') \,\middle|\, A \cdot x' \geq b \text{ (I)} \wedge B \cdot x'' \geq d \text{ (II)} \wedge x', x'' \in \mathbb{N}^n \text{ (III)}\right\}$$
$$= \min\left\{\mu^T \cdot x' \,\middle|\, x' \in Conv(\text{I})(\text{III})\right\} + \min\left\{(c - \mu)^T \cdot x'' \,\middle|\, x'' \in Conv(\text{II})(\text{III})\right\}. \tag{3.21}$$

The Lagrangean dual $Z_{(I)/(II)}$ is described by Theorem 3.13 (Minoux and Ribeiro 1985; Guignard and Kim 1987).

Theorem 3.13

$$Z_{(I)/(II)} = \min\left\{c^T \cdot x \,\middle|\, x \in Conv(\text{I})(\text{III}) \cap Conv(\text{II})(\text{III})\right\}$$

Proof Let P' (R') be the index set of all extreme points x'^p (extreme rays x'^r) of $Conv(\text{I})(\text{III})$. Analogously, define P'' and R'' for $Conv(\text{II})(\text{III})$.

$$ZL_{(I)/(II)}(\mu) = \min_{p \in P'}\left\{\mu^T \cdot x'^p \,\middle|\, \mu^T \cdot x'^r \geq 0 \,\forall r \in R'\right\}$$
$$+ \min_{p \in P''}\left\{(c - \mu)^T \cdot x''^p \,\middle|\, (c - \mu)^T \cdot x''^r \geq 0 \,\forall r \in R''\right\}$$

is obtained from a convexification approach. $Z_{(I)/(II)} = \max_{\mu \in \mathbb{R}^n} ZL_{(I)/(II)}(\mu)$ is linearized

$Z_{(I)/(II)} = \max \eta' + \eta''$ subject to

$\eta' - \mu^T \cdot x'^p \leq 0$	$\forall p \in P'$	λ'^p
$-\mu^T \cdot x'^r \leq 0$	$\forall r \in R'$	λ'^r
$\eta'' + \mu^T \cdot x''^p \leq c^T \cdot x''^p$	$\forall p \in P''$	λ''^p
$\mu^T \cdot x''^r \leq c^T \cdot x''^r$	$\forall r \in R''$	λ''^r
$\eta', \eta'' \in \mathbb{R} \wedge \mu \in \mathbb{R}^n$		

and dualized to yield

$$Z_{(I)/(II)} = \min c^T \cdot \left(\sum_{p \in P''} x''^p \cdot \lambda''^p + \sum_{r \in R''} x''^r \cdot \lambda''^r \right) \quad \text{subject to}$$

$$\sum_{p \in P'} \lambda'^p = 1 \qquad \eta'$$

$$\sum_{p \in P''} \lambda''^p = 1 \qquad \eta''$$

$$\sum_{p \in P''} x''^p \cdot \lambda''^p + \sum_{r \in R''} x''^r \cdot \lambda''^r - \sum_{p \in P'} x'^p \cdot \lambda'^p - \sum_{r \in R'} x'^r \cdot \lambda'^r = 0 \qquad \mu \quad (3.22)$$

$$\lambda'^p \geq 0 \; \forall p \in P' \wedge \lambda'^r \geq 0 \; \forall r \in R'$$

$$\lambda''^p \geq 0 \; \forall p \in P'' \wedge \lambda''^r \geq 0 \; \forall r \in R''.$$

Constraint (3.22) corresponds to the Lagrangean relaxed constraint (IV). The result follows from Minkowski's Theorem 3.2.

The inclusion of the solution spaces and both Theorems 3.11 and 3.13 allow to compare the relative strength of the lower bounds (Geoffrion 1974). In general, the more structure of the original IP is transferred to the Lagrangean relaxation, the better the lower bound becomes but this usually increases the computational effort to solve it. Let (IV) be the non-negativity constraint $x \geq 0$.

$$Z \geq Z_{(I)/(II)} \geq Z_{(I)} \geq Z^{LP}$$

$$Conv(I)(II)(III) \subseteq Conv(I)(III) \cap Conv(II)(III) \subseteq H(I) \cap Conv(II)(III) \subseteq H(I)(II)(IV)$$

Remark 3.1 The Lagrangean dual is polynomially solvable ($\mathcal{NP}$-hard) if and only if the Lagrangean relaxation is polynomially solvable ($\mathcal{NP}$-hard) (Chandru and Rao 2004, p. 423; Grötschel et al. 1981).

Notation 3.14 Let Z be a MIP. Let (K) and (L) be constraints. The optimization problem is written in an upright font and the optimal objective function value is given in italics. The following notation applies[21]:

- $Z^{(K)}$ where (K) is neglected yields $Z^{(K)}$.
- $Z^{+(K)}$ where Z is restricted to an additional constraint (K) yields $Z^{+(K)}$.
- The Lagrangean relaxation $ZL_{(K)}$ (Lagrangean dual $Z_{(K)}$) of (K) yields $ZL_{(K)}$ ($Z_{(K)}$).
- The Lagrangean relaxation $ZL_{(K)/(L)}$ (Lagrangean dual $Z_{(K)/(L)}$) with a Lagrangean decomposition between (K) and (L) yields $ZL_{(K)/(L)}$ ($Z_{(K)/(L)}$).

---

[21]This notation is adapted from Klose and Drexl (2005) and Cornuejols et al. (1991).

3.3 Dual Decomposition: Lagrangean Relaxation

Example 3.1

$$Z = \min \left\{ c^T \cdot x \,\middle|\, x \in H(\text{I})(\text{II})(\text{III}) \wedge x \in \mathbb{N}^n \text{ (IV)} \right\}$$

$$Z^{(\text{I})+(\text{V})} = \min \left\{ c^T \cdot x \,\middle|\, x \in H(\text{II})(\text{III})(\text{V}) \wedge x \in \mathbb{N}^n \text{ (IV)} \right\}$$

$$Z_{(\text{I})} = \min \left\{ c^T \cdot x \,\middle|\, x \in H(\text{I}) \cap Conv(\text{II})(\text{III})(\text{IV}) \right\}$$

$$Z_{(\text{I})/(\text{II})} = \min \left\{ c^T \cdot x \,\middle|\, x \in Conv(\text{I})(\text{III})(\text{IV}) \cap Conv(\text{II})(\text{III})(\text{IV}) \right\}$$

$$Z_{(\text{I})/(\text{II})(\text{III})} = \min \left\{ c^T \cdot x \,\middle|\, x \in Conv(\text{I})(\text{IV}) \cap Conv(\text{II})(\text{III})(\text{IV}) \right\}$$

3.3.1 Column Generation

Column generation solves the Lagrangean dual $Z_{(\text{I})}$ (3.11) to optimality (Lübbecke and Desrosiers 2005; Brooks and Geoffrion 1966). Because the primal master problem (3.17)–(3.20) comprises a huge number of columns, pricing out in the Simplex algorithm requires too much computational time.[22] Thus, a *Restricted Master Problem (RMP)* is formulated on a subset of extreme points $P' \subseteq P$. Introduce an artificial variable $M \in \mathbb{R}_+$. Let $f \in \mathbb{R}_+$ be a finite *big number*. The RMP is

$$Z^{RMP} = \min \sum_{p \in P'} c^T \cdot x^p \cdot \lambda^p + f \cdot M \qquad \text{subject to} \qquad (3.23)$$

$$\sum_{p \in P'} \lambda^p + M = 1 \qquad \eta \qquad (3.24)$$

$$A \cdot \sum_{p \in P'} x^p \cdot \lambda^p + b \cdot M \geq b \qquad \mu \qquad (3.25)$$

$$\lambda^p \geq 0 \; \forall p \in P' \wedge M \geq 0. \qquad (3.26)$$

If $P' = \emptyset$, then $Z^{RMP} = f$. Otherwise, an optimal solution has $M = 0$. If $P' = P$, then $Z^{RMP} = Z_{(\text{I})}$. Otherwise, $Z^{RMP} \geq Z_{(\text{I})}$ since $P' \subset P$. Because of the artificial variable M and $Conv(\text{II})(\text{III}) \neq \emptyset$, the RMP is always solvable. In contrast to this RMP, many column generation algorithms require initial columns $P' \neq \emptyset$.

---

[22] A good example is the application to the Cutting Stock Problem (Amor and Valério de Carvalho 2005).

3.1 Algorithm. A column generation algorithm
Input: Instance data and initial columns P'
Output: The Lagrangean dual $Z_{(I)}$
1 Set $P' \leftarrow \emptyset$
2 **repeat**
3   Solve RMP (3.23)–(3.26) to optimality and obtain the dual solution μ
4   Solve $ZL_{(I)}$ with μ and yield the optimal solution x
5   **if** $ZL_{(I)}(\mu) - b^T \cdot \mu - \eta < 0$ **then**
6   Add x as a new column to the RMP by $x^{|P'|+1} \leftarrow x$ and $P' \leftarrow P' \cup \{|P'|+1\}$
7 **until** $ZL_{(I)}(\mu) - b^T \cdot \mu - \eta = 0$
8 $Z_{(I)} \leftarrow Z^{RMP}$

Initialize Algorithm 3.1 with no initial columns $P' = \emptyset$.[23] In every iteration, solve the RMP and obtain the dual variables μ. Solve the Lagrangean relaxation $ZL_{(I)}$ with μ as the Lagrangean multiplier and yield the optimal solution x. Pricing out requires the reduced cost $\bar{c}^p = (c^T - \mu^T \cdot A) \cdot x^p - \eta$ that corresponds to the slack of the pth dual cutting plane (3.14). The smallest reduced cost is obtained from

$$\min_{p \in P} \bar{c}^p = \min \{ (c^T - \mu^T \cdot A) \cdot x \mid x \in Conv(\text{II})(\text{III}) \} - \eta$$

$$= ZL_{(I)}(\mu) - b^T \cdot \mu - \eta \qquad \text{with (3.9)}.$$

If $ZL_{(I)}(\mu) - b^T \cdot \mu - \eta < 0$, $Z_{(I)}$ is not solved yet because the extreme point x is not considered in P' but it improves Z^{RMP}. Hence, insert x as a new column in the RMP. The next iteration reoptimizes the RMP. Observe that Z^{RMP} decreases in the course of iterations because new column induces a dual cutting plane that restricts the solution space of the dual problem to the RMP. The algorithm terminates if $ZL_{(I)}(\mu) - b^T \cdot \mu - \eta = 0$ holds and returns $Z_{(I)} = Z^{RMP}$.

Prior to termination, Z^{RMP} is an upper bound to $Z_{(I)}$ but since $Z_{(I)} \leq Z$ holds this upper bound provides no information about Z. If the variable i with the smallest, negative reduces cost would enter the basis by $\lambda^i = 1$, the objective function value would decrease by at most $|\bar{c}^i|$. Hence, $Z^{RMP} + \bar{c}^i$ provides a lower bound other than $ZL_{(I)}(\mu)$ in every iteration. Instead of commencing with $P' = \emptyset$, trivial feasible solutions to many Lagrangean relaxations are easily determined [e.g., the Cutting Stock Problem has a KP as a Lagrangean relaxation (Gilmore and Gomory 1961). It is observed that the relative decrease of Z^{RMP} becomes smaller in the course of iterations. This *tailing off effect* slows down the convergence towards

---

[23]Other column generation algorithms are found in Desrosiers and Lübbecke (2005), Lübbecke and Desrosiers (2005), Gilmore and Gomory (1961) with Gilmore and Gomory (1963) for a particular formulation of the Cutting Stock Problem that can also be obtained via Lagrangean relaxation (Amor and Valério de Carvalho 2005).

3.3 Dual Decomposition: Lagrangean Relaxation

$Z_{(I)}$ (Desrosiers and Lübbecke 2005). Furthermore, the dual variables oscillate over the iterations and only smoothly evolve to the optimal values in the last iterations (Desrosiers and Lübbecke 2005). Two approaches prevent this dual oscillation. Boxstep methods impose bounds on the dual variables (Marsten 1975; Marsten et al. 1975) that can alternatively be violated at certain penalty cost (Du Merle et al. 1997). In a weighted Dantzig-Wolfe decomposition (Wentges 1997), a convex combination of the best known Lagrangean multiplier and the RMP Lagrangean multiplier yields novel Lagrangean multiplier to solve the subproblem with. Dynamic aggregation of the convexity constraints (3.18) decreases both the size of the primal master problem and its primal degeneracy (El Hallaoui et al. 2005). Inequalities can also be added to the dual master problem (Valério de Carvalho 2003). Kelley's cutting plane method solves the dual problem of the RMP (Kelley 1960). An effective interior point algorithm is the analytic center cutting plane method (Goffin et al. 1992, 1993). A survey is provided in Lübbecke and Desrosiers (2005) and Desaulnier et al. (2005).

3.3.2 Subgradient Optimization

Non-differentiable functions such as the Lagrangean dual $Z_{(I)}$ (3.11) can be solved by subgradient optimization (Nemhauser and Wolsey 1988, pp. 41–49; Fisher 1981; Held et al. 1974; Klose 2001, pp. 80–85). Assume the maximization of a concave function $f : \mathbb{R}_+^m \mapsto \mathbb{R}$ where $f(\mu^{iter})$ is *non-differentiable* at the position μ^{iter}.

Definition 3.15 If $f(\mu)$ is concave, $g \in \mathbb{R}^m$ is a *subgradient*[24] of $f(\mu^{iter})$ at the position μ^{iter} if $f(\mu) \leq f(\mu^{iter}) + g^T \cdot (\mu - \mu^{iter})$.

The subdifferential $\delta f(\mu^{iter})$ comprises all subgradients of $f(\mu^{iter})$ and it is a non-empty and convex set (Nemhauser and Wolsey 1988, pp. 45–46).

$$\delta f(\mu^{iter}) = \{g \in \mathbb{R}^m \, | f(\mu) \leq f(\mu^{iter}) + g^T \cdot (\mu - \mu^{iter}) \wedge \mu \geq 0\}$$

$ZL_{(I)}(\mu)$ (3.9) is piecewise linear and concave for different values of μ (Nemhauser and Wolsey 1988, p. 329). Thus,

$$ZL_{(I)}(\mu) \leq ZL_{(I)}(\mu^{iter}) + g^T \cdot (\mu - \mu^{iter}) \qquad (3.27)$$

satisfies Definition 3.15 and $ZL_{(I)}(\mu)$ is subdifferentiable. In Lemma 3.16, the negative slack of (I) of an optimal solution to $ZL_{(I)}$ yields a valid subgradient (Nemhauser and Wolsey 1988, p. 409).

---

[24] A subgradient applies to convex functions and a supergradient to concave functions but the term subgradient is widely used for concave functions in literature.

Lemma 3.16 If $x \in Conv(II)(III)$ yields $ZL_{(I)}(\mu)$, then $g = b - A \cdot x$ is a valid subgradient.

Proof Let $x'(x)$ be some optimal solution with $ZL_{(I)}(\mu')$ ($ZL_{(I)}(\mu^{iter})$) to μ' (μ^{iter}).

$$ZL_{(I)}(\mu') = c^T \cdot x' + (b - A \cdot x')^T \cdot \mu' \qquad x \text{ yields an upper bound}$$
$$\leq c^T \cdot x + (b - A \cdot x)^T \cdot \mu' \qquad \text{add a neutral } 0$$
$$= ZL_{(I)}(\mu^{iter}) + (b - A \cdot x)^T \cdot (\mu' - \mu^{iter})$$

This satisfies Definition 3.15 because $ZL_{(I)}(\mu)$ is concave for $\mu \geq 0$ (Nemhauser and Wolsey 1988, p. 329).

Example 3.2 Let $m = 2$. Solid lines in Fig. 3.1 represent $ZL_{(I)}(\mu)$ for μ_1 and $ZL_{(I)}(\mu^*)$ is the maximum. The dashed line shows the hyperplane

$$H^= = \{(z, \mu) \in (\mathbb{R} \times \mathbb{R}_+^m) \,|\, z = ZL_{(I)}(\mu^{iter}) + g^T \cdot (\mu - \mu^{iter})\}$$

with $g \in \delta ZL_{(I)}(\mu^{iter})$ for μ_1 that supports the hypograph $\{(z, \mu) \in (\mathbb{R} \times \mathbb{R}_+^m) \,|\, z \leq ZL_{(I)}(\mu)\}$ at $\mu = \mu^{iter}$ (Klose 2001, p. 81).

In what follows, g is given as defined by Lemma 3.16. Optimality of Lemma 3.10 holds if $g \leq 0$ and $\mu^T \cdot g = 0$. $\delta ZL_{(I)}(\mu^{iter})$ can be described by a convex combination of optimal solutions to the Lagrangean relaxation (Klose 2001, pp. 174–176). If $0 \in \delta ZL_{(I)}(\mu^*)$, the Lagrangean multiplier $\mu^{iter} = \mu^*$ are optimal such that $Z_{(I)} = ZL_{(I)}(\mu^*)$ because (3.27) becomes $ZL_{(I)}(\mu) \leq ZL_{(I)}(\mu^*)$. A subgradient algorithm has no such verification procedure but if $g = 0$, then

Fig. 3.1 Concavity of the Lagrangean objective function

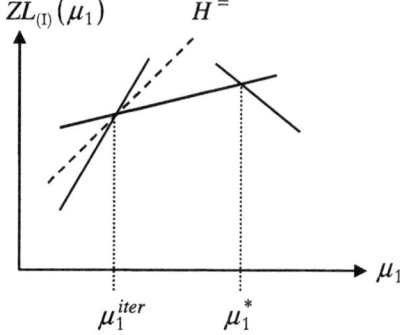

3.3 Dual Decomposition: Lagrangean Relaxation

optimality to Z immediately follows.[25] Considering (3.27), a necessary but not sufficient condition to maximize $ZL_{(I)}\left(\mu^{iter+1}\right)$ in the sense that $ZL_{(I)}\left(\mu^{iter+1}\right) > ZL_{(I)}\left(\mu^{iter}\right)$ holds is to choose the subsequent Lagrangean multiplier μ^{iter+1} such that (Nemhauser and Wolsey 1988, pp. 44–45; Held et al. 1974; Klose 2001, p. 82)

$$\mu^{iter+1} \in \left\{\mu \geq 0 \,\big|\, g^T \cdot \left(\mu - \mu^{iter}\right) > 0\right\}. \qquad (3.28)$$

By Lemma 3.17, an appropriate way is the steepest-ascent approach

$$\mu^{iter+1} = \max\left\{0, \mu^{iter} + \sigma \cdot g\right\} \quad \forall iter = 0, 1, 2, \ldots \qquad (3.29)$$

with the *step size* $\sigma \in \mathbb{R}_+$.[26]

Lemma 3.17 If $\sigma > 0$ and $g \neq 0$, then (3.29) satisfies (3.28).

Proof

$$\begin{aligned}
g^T \cdot \left(\mu^{iter+1} - \mu^{iter}\right) &\geq g^T \cdot (\sigma \cdot g) & \mu^{iter+1} - \mu^{iter} &\geq \sigma \cdot g \text{ from (3.29)} \\
&= \sigma \cdot g^T \cdot g & \sigma &> 0 \wedge g \neq 0 \\
&> 0
\end{aligned}$$

Suppose μ^{iter+1} is chosen by moving into the direction defined by the subgradient g. In the isosceles triangle from Fig. 3.2, an arrow represents the subgradient g that is normal to the hyperplane $g^T \cdot (\mu - \mu^{iter}) = 0$ and $\bar{\sigma}$ is an upper bound to σ defined by the side with the two equal angles. Because the angle between g and $\mu^* - \mu^{iter}$ is acute, there exists a μ^{iter+1} with $\mu^{iter+1} \leq \max\left\{0, \mu^{iter} + g \cdot \bar{\sigma}\right\}$ such that the Euclidian distance[27]

$$\parallel \mu^* - \mu^{iter+1} \parallel \,<\, \parallel \mu^* - \mu^{iter} \parallel \qquad (3.30)$$

to the optimal point μ^* becomes smaller. Figure 3.2 shows μ^{iter+1} for some arbitrary σ and illustrates (3.30) by red lines. Lemma 3.18 proves this intuitive result (Held and Karp 1970; Poljak 1969). The geometrical series from Lemma 3.18 converge to μ^* (Nemhauser and Wolsey 1988, p. 46).

---

[25] This result is also obtained from Lemma 3.10 with a subgradient defined by Lemma 3.16 and $g = 0$.
[26] If (I) was an equality, then $\mu^{iter+1} = \mu^{iter} + \sigma \cdot g$ $\forall iter = 0, 1, 2, \ldots$ since $\mu \in \mathbb{R}^m$.
[27] Denote $\parallel a \parallel = \sqrt{a^T \cdot a}$ as the Euclidean norm of a vector a.

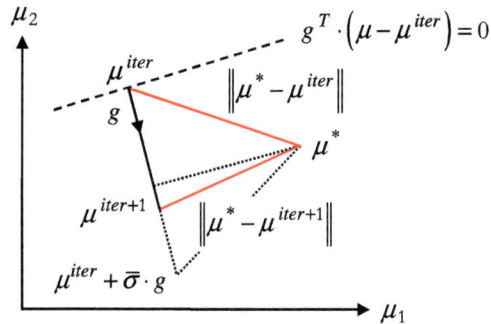

Fig. 3.2 A subgradient step

Lemma 3.18

1. If $Z_{(I)} > ZL_{(I)}(\mu^{iter})$, $g \neq 0$ and (3.27) is satisfied, $0 < \bar{\sigma}^{Poljak} \leq \bar{\sigma}$ holds with

$$\bar{\sigma}^{Poljak} = 2 \cdot \frac{Z_{(I)} - ZL_{(I)}(\mu^{iter})}{\sum_{t=1}^{m} g_t^2}. \tag{3.31}$$

2. If $0 < \sigma < \bar{\sigma}^{Poljak}$, then σ satisfies (3.30).

Proof '1.': Define $2 \cdot b = \bar{\sigma} \cdot g$ and $c = \mu^* - \mu^{iter}$. The angle $\angle g\,c$ in Fig. 3.2 is acute by definition. Trigonometry and $g^T \cdot c = \|g\| \cdot \|c\| \cdot \cos(\angle g\,c)$ yields

$$b = c \cdot \cos(\angle g\,c) = c \cdot \frac{g^T \cdot c}{\|g\| \cdot \|c\|}$$

and hence, the Euclidean norm of vector b becomes

$$\|b\| = \sqrt{b^T \cdot b} = \frac{g^T \cdot c}{\|g\| \cdot \|c\|} \cdot \sqrt{c^T \cdot c} = \frac{g^T \cdot c}{\|g\|}. \tag{3.32}$$

$$\begin{aligned}
\bar{\sigma} &= 2 \cdot \frac{\|b\|}{\|g\|} = 2 \cdot \frac{g^T \cdot c}{\|g\|^2} &&\text{because of } \angle g\,b = 0° \text{ and (3.32)} \\
&= 2 \cdot \frac{g^T \cdot (\mu^* - \mu^{iter})}{\|g\|^2} &&\text{(3.27) and } Z_{(I)} = ZL_{(I)}(\mu^*) \\
&\geq 2 \cdot \frac{Z_{(I)} - ZL_{(I)}(\mu^{iter})}{\|g\|^2} &&g \neq 0 \text{ and } Z_{(I)} > ZL_{(I)}(\mu^{iter}) \\
&> 0.
\end{aligned}$$

3.3 Dual Decomposition: Lagrangean Relaxation

'2.': For two vectors a and b it holds

$$\| a - b \|^2 = \sum_i (a_i - b_i)^2 = \sum_i a_i^2 - 2 \cdot \sum_i a_i \cdot b_i + \sum_i b_i^2$$
$$= \| a \|^2 - 2 \cdot b^T \cdot a + \| b \|^2 . \quad (3.33)$$

$\| \mu^* - \mu^{iter+1} \|^2$

$= \| \mu^* - \max\{0, \mu^{iter} + \sigma \cdot g\} \|^2$ from (3.29)

$\leq \| \mu^* - \mu^{iter} - \sigma \cdot g \|^2$ with (3.33)

$= \| \mu^* - \mu^{iter} \|^2 - 2 \cdot \sigma \cdot g^T \cdot (\mu^* - \mu^{iter}) + \| \sigma \cdot g \|^2$ $\| b \cdot a \| = |b| \cdot \| a \|$ for scalar b and vector $a \neq 0$

$= \| \mu^* - \mu^{iter} \|^2 - 2 \cdot \sigma \cdot g^T \cdot (\mu^* - \mu^{iter}) + \sigma^2 \cdot \| g \|^2$ (3.27); $Z_{(I)} = ZL_{(I)}(\mu^*)$

$\leq \| \mu^* - \mu^{iter} \|^2 - 2 \cdot \sigma \cdot (Z_{(I)} - ZL_{(I)}(\mu^{iter})) + \sigma^2 \cdot \| g \|^2$ (3.31); $0 < \sigma < \bar{\sigma}^{Poljak}$

$< \| \mu^* - \mu^{iter} \|^2$

The lemma follows from the square root of the expression.

Since $Z_{(I)}$ is not known in advance, (3.31) is modified in the *Held-Karp step size* (3.34) (Held and Karp 1970, 1971). The *step size parameter* STEPSIZE has $0 <$ STEPSIZE ≤ 2 and let UB^* be the best known upper bound to $Z_{(I)}$.

$$\sigma = \text{STEPSIZE} \cdot \frac{UB^* - ZL_{(I)}(\mu^{iter})}{\sum_{t=1}^{m} g_t^2} \quad (3.34)$$

Because of this modification, the convergence of the Lagrangean multiplier to μ^* is not guaranteed and the Lagrangean dual is approximated. Furthermore, the Euclidean distance to μ^* can be increased because of $Z_{(I)} \leq UB^*$ (*overestimation*). STEPSIZE $= 1$ is straightforward because the smallest Euclidean distance between μ^* and μ^{iter} is obtained from the altitude of the triangle of Fig. 3.2. Because of the potential overestimation, the step size parameter STEPSIZE is adjusted. The most common approach is to halve the STEPSIZE if the objective function value did not improve after a given number of iterations (*constant step size strategy*).

The *Lagrangean heuristic* from Algorithm 3.2 is a metaheuristic where a *subgradient algorithm* aims to improve $ZL_{(I)}$ and guides a subsequent *repair heuristic*, which tries to modify the Lagrangean solution such that constraint (I) is satisfied (Klose 2001, pp. 80–112; Boschetti and Maniezzo 2009).

3.2 Algorithm. A Lagrangean subgradient heuristic

Input: Instance data, an initial feasible solution if one was found, initial Lagrangean multiplier μ and an initial lower bound LB^*
Output: A lower bound LB^*, an upper bound UB^*, the best known feasible solution x^* if one was found

1 **if** *An initial feasible solution exists* **then** Initialize x^* and set $UB^* \leftarrow c^T \cdot x^*$
2 **else** Set UB^* to a reasonable upper bound
3 **while** *A termination criterion is not met* **do**
4   Solve the Lagrangean relaxation $ZL_{(I)}$ with μ and obtain the optimal solution x
5   Update the best known Lagrangean lower bound $LB^* \leftarrow \max\{ZL_{(I)}(\mu), LB^*\}$
6   Calculate the subgradient g from Lemma 3.16
7   **if** $c^T \cdot x \leq UB^* \wedge g \leq 0$ **then**
8   Update the best known feasible solution x^* and $UB^* \leftarrow c^T \cdot x$
9   **if** $g^T \cdot \mu = 0$ *or* $g = 0$ **then** Terminate because x^* is optimal
10   Update the step size with (3.34) and the Lagrangean multiplier μ with (3.29)
11   Repair the solution x such that constraint (I) is satisfied
12   Update the best known feasible solution x^* and $UB^* \leftarrow \min\{c^T \cdot x^*, UB^*\}$
13   **if** $UB^* = LB^*$ **then** Terminate because x^* is optimal

Initialize the algorithm with initial Lagrangean multiplier and a value for STEPSIZE. The best known feasible solution is denoted as x^*. Let UB^* and LB^* be the best upper and lower bound to Z. If no initial Lagrangean solution exists, set $UB^* < \infty$. In an iteration, solve $ZL_{(I)}$ to optimality, update LB^* and calculate the subgradient g with Lemma 3.16. If the Lagrangean solution improves UB^* and satisfies constraint (I), update x^* as well as UB^* and terminate if Lemma 3.10 or $g = 0$ holds because x^* is optimal to Z. Otherwise, update STEPSIZE and calculate new Lagrangean multiplier from (3.29) and (3.34). Let a repair algorithm modify the Lagrangean solution such that it satisfies constraint (I) and update x^* as well as UB^*. If $UB^* = LB^*$, terminate because x^* is optimal to Z. Commence the next iteration if some termination criterion holds (maximal number of iterations, a bound on the calculation time or the observation that LB^* is converging to some value). The algorithm returns the obtained bounds and a feasible solution to Z if one was found.

Compared to column generation, the master process in subgradient optimization requires few computational time but the lower bound does not have a guaranteed strength because the Lagrangean dual is approximated. Combinations of both approaches alternate between a column generation and a subgradient algorithm (Klose 2001, pp. 94–96; Guignard and Zhu 1994; Fu et al. 2005; Vera et al. 2003). Another combined algorithm solves a Lagrangean dual with a subgradient algorithm and a subsequent dual-ascent method solves a Lagrangean decomposition that is initialized with the best known Lagrangean multiplier of the previous subgradient algorithm (Lee and Guignard 1996). *Primal recovery* allows to obtain primal information for subgradient optimization. The arithmetic average of all optimal solutions of the Lagrangean relaxation is optimal to the Lagrangean dual if the step size strategy satisfies some criteria of which a guaranteed convergence to the Lagrangean dual is most important (Sherali and Choi 1996).

Convergence towards optimal Lagrangean multiplier is negatively influenced by several effects. For instance, the objective function value typically oscillates over the iterations (see Fig. 6.1). Guta (2003, pp. 84–86) presents an algorithm where this so called *zig-zagging* is reduced. Geometric constellations of the subgradients can yield similar Lagrangean multiplier and therefore, comparable objective function values. Deflected subgradient optimization stabilizes the Lagrangean multiplier by taking the subgradient of the previous iteration into account (Camerini et al. 1975). Bundle methods guarantee an increase of the Lagrangean lower bound and the ϵ-subdifferential is approximated with a bundle of already generated subgradients (Carraresi et al. 1997). A related algorithm is the volume algorithm (Guta 2003, pp. 91–102; Barahona and Anbil 2000).

Alternatives to Lagrangean relaxation, which is a price-directive decomposition approach, are resource-directive decomposition (Ahuja et al. 1993, pp. 674–678), surrogate relaxation, which is a linear combination of constraints and furthermore, the surrogate dual provides a tighter bound than the Lagrangean dual (Nemhauser and Wolsey 1988, p. 334), and a decomposition through Benders' reformulation that is presented in the next section.

3.4 Primal Decomposition: Benders' Reformulation

Whereas Lagrangean relaxation accesses a well-structured subproblem by a separation of constraints, Benders' reformulation separates different sets of variables (Nemhauser and Wolsey 1988, pp. 337–341; Klose 2001, pp. 66–70; Benders 1962, 2005). The MIP Z has n integer variables $x \in \mathbb{N}^n$ and k continuous variables $y \in \mathbb{R}_+^k$. Let $c \in \mathbb{R}^n, f \in \mathbb{R}^k, A \in \mathbb{R}^{m \times n}, B \in \mathbb{R}^{m \times k}, b \in \mathbb{R}^m$ and $X \subseteq \mathbb{N}^n$.

$$Z = \min \left\{ c^T \cdot x + f^T \cdot y \,\middle|\, A \cdot x + B \cdot y \geq b \wedge x \in X \wedge y \in \mathbb{R}_+^k \right\}$$

The integer variables x are referred to as 'hard' variables because the continuous variables y solely yield a LP that is relatively 'easy' to solve. Benders' decomposition commences with a fixation of the integer variables x to some value and Z becomes

$$D(x) = c^T \cdot x + \min \left\{ f^T \cdot y \,\middle|\, B \cdot y \geq b - A \cdot x \wedge y \geq 0 \right\} \tag{3.35}$$

$$= c^T \cdot x + \max \left\{ (b - A \cdot x)^T \cdot \mu \,\middle|\, \mu \in S \right\} \text{ with } S = \left\{ \mu \in \mathbb{R}_+^m \,\middle|\, B^T \cdot \mu \leq f \right\}. \tag{3.36}$$

The *Benders' subproblem* (3.36) is the corresponding dual problem of (3.35). Assume $S \neq \emptyset$. If $X = \emptyset$, (3.35) is unbounded with $D(x) = \infty$ (Table 3.2). Otherwise, (3.35) has a finite optimal solution. Let P (R) be the index set of all extreme points μ^p (extreme rays μ^r) of S. Considering (3.2) and (3.4), a convexification approach for (3.36) yields

$$D(x) \geq \begin{cases} \infty & \text{if } \exists r \in R : (b - A \cdot x)^T \cdot \mu^r > 0 \\ c^T \cdot x + (b - A \cdot x)^T \cdot \mu^p \ \forall p \in P & \text{otherwise.} \end{cases} \quad (3.37)$$

Therefore, a finite optimal solution to Z exists if all solutions $x \in X$ satisfy $(b - A \cdot x)^T \cdot \mu^r \leq 0 \ \forall r \in R$. Note that the Benders' subproblem provides an upper bound to Z. The *Benders' reformulation* of Z is obtained by a minimization of $D(x)$ over $x \in X$ as follows

$$Z = \min_{x \in X} D(x)$$

$$= \min_{x \in X} \left\{ c^T \cdot x + \max_{\mu \in S} (b - A \cdot x)^T \cdot \mu \right\} \qquad \text{from (3.36)}$$

$$= \min_{x \in X} \max_{p \in P} \left\{ c^T \cdot x + (b - A \cdot x)^T \cdot \mu^p \,\middle|\, (b - A \cdot x)^T \cdot \mu^r \leq 0 \ \forall r \in R \right\} \quad \text{from (3.37).}$$

A linearization of the min-max objective function yields the *Benders' master problem* (3.38)–(3.41) of Theorem 3.19 (Nemhauser and Wolsey 1988, p. 338). The constraints (3.39) are referred to as *Benders' cuts*. Boundedness of the polyhedron is ensured by the feasibility cuts (3.40).

Theorem 3.19

$$Z = \min \eta \qquad\qquad \text{subject to} \qquad\qquad (3.38)$$

$$\eta \geq c^T \cdot x + (b - A \cdot x)^T \cdot \mu^p \qquad\qquad \forall p \in P \qquad\qquad (3.39)$$

$$(b - A \cdot x)^T \cdot \mu^r \leq 0 \qquad\qquad \forall r \in R \qquad\qquad (3.40)$$

$$\eta \in \mathbb{R} \wedge x \in X \qquad\qquad\qquad\qquad (3.41)$$

Proof If there exists no $x \in X$ satisfying (3.40), then (3.38)–(3.41) is infeasible and $Z = \infty$. If there exists a $x \in X$ satisfying (3.40) and $S = \emptyset$, then $P = \emptyset$ and $Z = -\infty$ follows because η is unbounded. Otherwise, Z is equivalent to (3.38)–(3.41).

Owing to the large cardinality of the sets, a subset of extreme rays and extreme points can be used in (3.38)–(3.41) to iteratively solve a restricted Benders' master problem (Nemhauser and Wolsey 1988, p. 413; Klose 2001, pp. 67–68). The Benders' master problem is usually unstructured and hence, it is often difficult to solve to optimality. Instead of solving it to optimality by methods of integer programming (Côté and Laughton 1984), it suffices to find a Benders' cut that improves the Benders' master problem (Geoffrion and Graves 1974; Rei et al.

3.4 Primal Decomposition: Benders' Reformulation

2009). Relaxing the Benders' cuts in a Lagrangean fashion, simplifies the Benders' master problem significantly and provides a lower bound to Z (Côté and Laughton 1984; Aardal and Larsson 1990). Theoretical considerations about the strength of these obtained lower bounds is provided in Holmberg (1994b). Techniques to overcome convergence problems are pareto-optimal Benders' cuts (Magnanti and Wong 1981), cut richer reformulations of Z (Magnanti and Wong 1981, 1986, 1990) and initial Benders' cuts (Magnanti and Wong 1990). An introduction to Lagrangean and Benders' heuristics is found in Boschetti and Maniezzo (2009). A combined approach solves a Lagrangean dual by a dual-ascent heuristic subject to Benders' cuts (Guignard 1988).

Lagrangean relaxation and Benders' decomposition are connected through duality. Consider the primal problem on the left and its corresponding dual problem on the right.

$$\begin{aligned} \min\ & c^T \cdot x + f^T \cdot y \quad \text{subject to} & \max\ & b^T \cdot \mu \quad \text{subject to} \\ & A \cdot x + B \cdot y \geq b & & \mu \geq 0 \\ & x \geq 0 & & A^T \cdot \mu \leq c \\ & y \geq 0 & & B^T \cdot \mu \leq f \end{aligned} \quad (3.42)$$

A fixation of the variables x in the primal problem yields a Benders' subproblem

$$\max \left\{ (b - A \cdot x)^T \cdot \mu + c^T \cdot x \,\middle|\, B^T \cdot \mu \leq f \wedge \mu \geq 0 \right\}$$

that is identical to the Lagrangean relaxation of the dual constraint $A^T \cdot \mu \leq c$ (3.42) of the dual problem where the variables x are the Lagrangean multiplier. Cross-decomposition exploits this property (Roy 1983). A polyhedral analysis of this method is found in Holmberg (1994a). Table 3.3 compares the primal and dual decomposition techniques applied to a MIP. Denote SP as the subproblem and MP as the master problem.

Table 3.3 Comparison of primal and dual decomposition

	Lagrangean relaxation	Benders' reformulation
	Relax a constraint and weight its slack in the objective function.	Fix integer variables and adjust the right-hand side accordingly.
SP	Preferably an IP with a specific structure. Optimum yields a lower bound.	LP. Optimum yields an upper bound.
MP	LP. Convexification via primal solutions. Finds Lagrangean multiplier that maximize the lower bound. Lower bound available even if MP is solved heuristically.	Unstructured MIP. Convexification via dual solutions. Finds integer variables that minimize the upper bound. Converges to the optimum iff SP and MP are solved to optimality.

3.5 Local Search and Tabu Search

Let X be a set of feasible solutions of the optimization problem

$\min Z(x)$ subject to $x \in X$.

Let a given solution x be modified by an elementary operation *move*. The *neighborhood* is a set of solutions, which is admissible by the execution of move. Common moves swap or shift either single elements or entire segments of a solution. If a solution consists of several bodies (e.g., multiple machines or vehicles), single or several elements can be exchanged between the bodies. Moreover, a solution can be partially destroyed and recreated via insertion.[28]

3.5.1 Local Search

The *local search heuristic* from Algorithm 3.3 initializes the best known feasible solution x^* with an initial feasible solution x and the upper bound becomes $UB^* = Z(x)$ (Glover 1989). In an iteration, apply the move to the solution x, find the best solution $\bar{x}$ from the neighborhood and update $\bar{x}$ as well as UB^*. The next iteration commences with $x = \bar{x}$ if the move did not improve x^* or no other termination criterion is met such as a maximal number of iterations or the observation that $\bar{x}$ is infeasible. The algorithm returns the best feasible solution if one was found but it is not necessarily optimal.

3.3 Algorithm. A local search heuristic

Input: Instance data and an initial feasible solution x
Output: An upper bound UB^*, the best known feasible solution x^* if one was found
1 Set $x^* \leftarrow x$ and $UB^* \leftarrow Z(x)$
2 **while** *A termination criterion is not met* **do**
3   Obtain the best solution $\bar{x}$ from the neighborhood of x
4   **if** $\bar{x}$ *is feasible* **then**
5   Update the best known feasible solution x^* and set $UB^* \leftarrow \min\{UB^*, Z(\bar{x})\}$
6      $x \leftarrow \bar{x}$

---

[28] A swap move in the Quadratic Assignment Problem changes the assigned location between two entities (Taillard 1991) and a solution is encoded as a permutation sequence. In the Traveling Salesman Problem, the 2-Opt move changes the position of a tour segment as well as the direction of travel in this segment (Croes 1958) and the 3-Opt move shifts a segment (Lin 1965). In Vehicle Routing Problem, the cross-exchange move exchanges tour segments between two vehicles (Taillard et al. 1997) and the stringing-and-unstringing procedure removes nodes of a tour and rebuilds a solution with insertion moves (Gendreau et al. 1992).

3.5 Local Search and Tabu Search

Local search heuristics are commonly referred to as *hill-climbing heuristics* (Glover 1989). The major limitation of local search heuristics is that once a local optimum is found, the heuristic halts. Many mathematical algorithms are based on local search. In the Simplex algorithm from Sect. 3.2, the set X comprises the set of extreme points and a move is a base change that maintains feasibility. However, the definition of the move is sufficient to prove optimality if the reduced costs are non-negative.

3.5.2 Tabu Search

A *tabu search heuristic* is a metaheuristic that is able to leave a local optimum (Glover 1986; Hansen 1986). In Algorithm 3.4, initialize the best known feasible solution x^* with an initial feasible solution x and determine the upper bound $UB^* = Z(x)$ (Glover 1989). Find the best move of the neighborhood that yields a solution $\bar{x}$ with the minimal objective function value (*steepest-ascent mildest-decent strategy*).[29] Add $\bar{x}$ to a *tabu list* in order to store historical information about the search process and label the solution as *tabu*. Tabu solutions are excluded from the neighborhood. A straightforward realization in a *long-term memory tabu list* records all solutions visited so far. Cycling is not possible because a solution cannot be re-visited. The *inverse move*, which would undo the move if it was applied, is often faster to evaluate but the reduction of the neighborhood is larger because an inverse move contains too few information about the solution. To prevent that an improving solution from an inverse move that is tabu might be rejected, the tabu status is overridden and a move is admissible if an *aspiration level criterion* holds (e.g., the tabu solution has a lower objective function value than the best known feasible solution). A *short-term memory* tabu list is restricted in size and the 'oldest' moves are discarded in a first-in first-out manner. This strategic forgetting requires that the solution changed sufficiently after a couple of moves. In Algorithm 3.4, apply the best move to the solution and update x^* and UB^*. Termination criteria are the maximal number of iterations, an upper bound on the computational time or there only exist tabu moves. Upon termination the algorithm returns the best feasible solution if one was found but it is not necessarily optimal.

Because of the steepest-ascent mildest-decent strategy, tabu search algorithms intensively evaluate promising regions of the solution space. This *intensification phase* ends automatically since the tabu list forbids to re-visit solutions. Afterwards, the algorithm finds another region of the solution space in a succeeding *diversification phase*. A key factor of a successful tabu search lies in the design of the move and the complexity of the neighborhood. If the neighborhood is very complex or

---

[29]This strategy is known as steepest-ascent in terms of improving the best known feasible solution and mildest decent if only solutions with a larger objective function value exist in the neighborhood.

3.4 Algorithm. A tabu search heuristic

Input: Instance data and an initial feasible solution x
Output: An upper bound UB^*, the best known feasible solution x^* if one was found
1 Set $x^* \leftarrow x$ and $UB^* \leftarrow Z(x)$
2 Initialize the tabu list
3 **while** *A termination criterion is not met* **do**
4   Find the best move among all non-tabu moves from the neighborhood of x
5   **if** *A non-tabu move is found* **then**
6   Apply the best move to x and yield the best solution $\bar{x}$ of the neighborhood
7   Add $\bar{x}$ to the tabu list and update the tabu list
8   Update the best known feasible solution x^* and set $UB^* \leftarrow \min\{UB^*, Z(x)\}$

---

too large to be evaluated efficiently, the total number of iterations is too small and few regions of the solution space are explored. Hence, a computationally efficient, fast evaluation of the neighborhood and of the tabu status is particularly important. Moves that modify a solution only slightly, generate similar solutions. This can be beneficial in the intensification phase but not necessarily in the diversification phase if it takes too many iterations to leave the current region of the solution space. This drawback also holds if the problem contains a high degree of symmetry. Conversely, intensification is complicated by moves that alter a solution too much. Similar discussions are found in Bock (2000, pp. 211–212). Other aspiration level criteria control the search direction and override the tabu status if the move improved a solution in an earlier iteration. However, this assumes that solutions of a good quality posses similar and advantageous substructures. If there exists no allowed move because all moves are tabu, an aspiration level criterion is to accept the least worsening tabu move (Glover and Laguna 2002, p. 51). Many strategies aim to improve diversification and to accelerate the transition to another region of the solution space. Strategic oscillation directs the search beyond a boundary the tabu search would normally stop and guides the search towards a region the search would continue (Glover and Laguna 2002, pp. 102–103). In some approaches, the objective function is augmented by a term that penalizes infeasibility of a solution (e.g., slack of capacity or assignment constraints) (Glover and Laguna 2002, pp. 99–100). Different regions of the solution space are accessed via transition through infeasible regions. A second approach is a *multi-state search*. The algorithm commences from a ground state that has the highest level of intensification. The state is changed if all moves are tabu or a certain number of iterations with no improvement indicates that the current region is either fully explored or unpromising. The move applied in the next, higher state has a stronger emphasis on diversification. The algorithm proceeds with the ground state if an improving solution is found. An extensive survey on tabu search is provided in Glover (1989, 1990) and Glover and Laguna (2002).

3.6 The Knapsack Problem

This section presents well-known approaches for the *Knapsack Problem (KP)* (Martello and Toth 1990, pp. 13–15). The KP is one of the most important optimization problems and it has been intensively studied. It appears as a substructure in the CPMP because of the capacity constraint (C). Given $J \in \mathbb{N}$ items and one bin of a capacity $c \geq 0$. Each item $j = 1, \ldots, J$ is associated a profit $p_j \geq 0$ and a weight $w_j \geq 0$. The KP maximizes the total profit of all packed items such that the total weight does not exceed the bin capacity. Let the binary variable x_j equal to 1 if the jth item is packed (otherwise, 0). The mathematical formulation as an IP is

$$\max \sum_{j=1}^{J} p_j \cdot x_j \qquad \text{subject to} \qquad (3.43)$$

$$\sum_{j=1}^{J} w_j \cdot x_j \leq c \qquad (3.44)$$

$$0 \leq x_j \leq 1 \qquad \forall j = 1, \ldots, J \qquad (3.45)$$

$$x_j \in \{0, 1\} \qquad \forall j = 1, \ldots, J. \qquad (3.46)$$

The objective function (3.43) maximizes the profit of all packed items. The capacity constraint (3.44) limits the total weight and simple upper bounds on the decision variables are provided by constraint (3.45). Equation (3.46) states that the variable domains are binary. The KP is binary $\mathcal{NP}$-hard (Garey and Johnson 1979, pp. 65, 247; Karp 1972) and pseudo-polynomially solvable in time $O(J \cdot c)$ (Martello and Toth 1990, pp. 36–39). The *Minimization KP*

$$\min \sum_{j=1}^{J} p_j \cdot y_j \quad \text{subject to} \quad \sum_{j=1}^{J} w_j \cdot y_j \geq b \wedge y_j \in \{0, 1\} \; \forall j = 1, \ldots, J$$

is solvable via a KP (3.43)–(3.46) with $c = \sum_{j=1}^{J} w_j - b$ (Martello and Toth 1990, p. 15). Transform the variables by $y_j = 1 - x_j \; \forall j$. Intuitively, the Minimization KP maximizes the total profit of all non-packed items x. The *Subset Sum Problem (SSP)* is a special KP with $p_j = w_j \; \forall j$. It is binary $\mathcal{NP}$-hard and pseudo-polynomially solvable in time $O(J \cdot c)$ (Martello and Toth 1990, pp. 6, 105–107).

Assumption 3.20

$$p_j, w_j > 0 \ \forall j = 1, \ldots, J \qquad (3.47)$$

$$\frac{p_1}{w_1} \geq \ldots \geq \frac{p_J}{w_J} \qquad (3.48)$$

$$w_j \leq c \ \forall j = 1, \ldots, J \qquad (3.49)$$

$$\sum_{j=1}^{J} w_j > c \qquad (3.50)$$

Further assumptions are imposed on the instance data. If Assumption (3.47) is violated, apply a variable fixation as follows: If there exists an item j with $p_j = 0$, set $x_j = 0$. If there exists an item j with $w_j = 0$, set $x_j = 1$. All items are sorted to non-increasing efficiencies because of Assumption (3.48). The fraction $\frac{p_j}{w_j}$ is referred to as the *efficiency of item j*. If an item j violates Assumption (3.49), set $x_j = 0$. Assumption (3.50) prevents the existence of the trivial and optimal solution that all items are packed (i.e., $x_j = 1 \ \forall j$).[30]

3.6.1 Valid Inequalities

Let I be a subset of items with $\sum_{i \in I} w_i > c$. The following four inequalities are valid to the Knapsack polytope (Martello and Toth 1990, pp. 74–75; Nemhauser and Wolsey 1988, pp. 265–270; Balas 1975; Dietrich et al. 1993).

$$\sum_{i \in I} x_i \leq |I| - 1 \qquad (3.51)$$

is a *cover inequality* of the *cover I*. If $\sum_{j \in I} w_j - w_i \leq c \ \forall i \in I$, then I is a *minimal cover*.

$$\sum_{i \in I} x_i \leq 1 \qquad (3.52)$$

is a *clique inequality* of the *clique I* with $w_i + w_j > c \ \forall i, j \in I$ and $i \neq j$. If $|I| = 2$, (3.52) is referred to as a *2nd-degree clique inequality*. The *extension*

$$E(I) = I \cup \left\{ j \in \{1, \ldots, J\} \setminus I \ \middle| \ w_j \geq \max_{i \in I} w_i \right\} \qquad (3.53)$$

---

[30]In literature, $c \in \mathbb{N}$ and $w_j, p_j \in \mathbb{N} \ \forall j$ is often assumed (Martello and Toth 1990, pp. 14–15). This is not assumed here to ensure that the presented solution methods are directly applicable to CPMP subproblems where the coefficients are real numbers.

3.6 The Knapsack Problem

of a given set I tightens both inequalities and yields the *extended cover inequality*

$$\sum_{i \in E(I)} x_i \leq |I| - 1 \tag{3.54}$$

and the *extended clique inequality*

$$\sum_{i \in E(I)} x_i \leq 1. \tag{3.55}$$

A fractional feasible solution $\bar{x}$ of the *Continuous KP* (3.43)–(3.45) yields the *KP LP relaxation* and provides a source to obtain cover inequalities (Crowder et al. 1983). Using $|I| = \sum_{i \in I} 1$, (3.51) becomes $\sum_{i \in I} 1 - x_i \geq 1$. Every solution $\bar{x}$ violating a cover inequality is characterized by

$$\sum_{i \in I} 1 - \bar{x}_i < 1 \wedge \sum_{i \in I} w_i > c.$$

A cover C is obtained by solving the *separation problem* (3.56) (Crowder et al. 1983). Assume $c \in \mathbb{N}$ and $w_j \in \mathbb{N}$ $\forall j$ (Martello and Toth 1990, pp. 14–15). Let the binary variable y_j take the value 1 if the item j is contained in the cover C (otherwise, 0).

$$\bar{Z} = \min \sum_{j=1}^{J} (1 - \bar{x}_j) \cdot y_j \qquad \text{subject to} \tag{3.56}$$

$$\sum_{j=1}^{J} w_j \cdot y_j \geq c + 1 \wedge y_j \in \{0, 1\} \qquad \forall j = 1, \ldots, J$$

No violated cover inequality exists if $\bar{Z} \geq 1$. Otherwise, $C = \{j = 1, \ldots, J | y_j = 1\}$ is a cover. Adding the corresponding cover inequality to the KP LP relaxation, the solution $\bar{x}$ becomes infeasible and the solution space becomes strictly smaller. Thus, the optimal objective function value of the reoptimized KP LP relaxation either decreases or remains the same. Preferably, the extended cover $E(C)$ is used because (3.54) can define a facet of the Knapsack polytope (Nemhauser and Wolsey 1988, pp. 266–267; Balas and Jeroslow 1972). It becomes evident from the following argument that an optimal solution to (3.56) yields a minimal cover (Crowder et al. 1983). Since all objective function coefficients are strictly positive (Assumption (3.47)), the optimal solution y comprises the minimal number of non-zeros satisfying the capacity constraint (minimal cover). Moreover, the objective function maximizes the slack of the violated cover inequality $\sum_{i \in C} 1 - \bar{x}_i < 1$. Therefore, an optimal solution defines a maximally violated cover inequality (Crowder et al. 1983). If the separation problem is heuristically solved, it is not guaranteed that the generated covers are minimal and a cover might not be found although there exists one.

Algorithm 3.5 finds all extended cliques for (3.55) in time $O(J \cdot \log J)$ (Dietrich et al. 1993). The basic idea is to find and to extend 2nd-degree cliques. This approach exploits the observation that cliques with $|I| \geq 3$ are composed of 2nd-degree cliques per definition. In fact, these clique inequalities can be directly obtained as Chvátal–Gomory inequalities with the corresponding 2nd-degree cliques (Dietrich and Escudero 1994).[31] Denote $\mathcal{C}$ as a set of sets that contains all extended cliques. Set $\mathcal{C} = \emptyset$. Commence the Algorithm 3.5 with sorting all items according to a non-decreasing order of the weight $w_{j_1} \leq \ldots \leq w_{j_J}$.[32] Find the item j_h with $h < J$ and the smallest index such that $w_{j_h} + w_{j_J} > c$.[33] If no item j_h exists, there are no cliques and the algorithm terminates. Note that cliques of the cardinality 1 so called *singletons* are not obtained because they are implied by (3.45). Otherwise, find the item j_g with $g > h$ and the smallest index, which satisfies $w_{j_h} + w_{j_g} > c$. The set $I = \{j_h, j_g\}$ is a clique and the extended clique $E(I) = \{j_h, j_g, \ldots, j_J\}$ is obtained. Set $\mathcal{C} = \mathcal{C} \cup E(I)$. While $g \neq h + 1$ holds, more cliques possibly exist: Increment h by one and find the item $j_{\bar{g}}$ with $\bar{g} < g$ and the smallest index, which satisfies $w_{j_h} + w_{j_{\bar{g}}} > c$. If an item $j_{\bar{g}}$ exists, insert the clique $\{j_h, j_{\bar{g}}, \ldots, j_J\}$ in $\mathcal{C}$ and repeat the procedure. The algorithm terminates and returns the set $\mathcal{C}$.

Other classes of inequalities tighten valid inequalities by sequential lifting (Padberg 1975) where the inequality say (3.51) is lifted to yield

$$\sum_{i \in I} x_i + \sum_{j=\{1,\ldots,J\} \setminus I} a_j \cdot x_j \leq |I| - 1.$$

---

[31] Given three 2nd-degree clique inequalities $x_i + x_j \leq 1$, $x_j + x_k \leq 1$ and $x_i + x_k \leq 1$ with $i \neq j \neq k$. Multiply all inequalities by $\frac{1}{2}$ and the sum becomes $x_i + x_j + x_k \leq \frac{3}{2}$. Per definition the set $\{i, j, k\}$ is a clique and thus, rounding down the right-hand side is valid. This way, a 3rd-degree clique inequality is obtained as a *Chvátal–Gomory inequality* (Nemhauser and Wolsey 1988, pp. 209–212).

[32] Implementation detail: One of the fastest sorting algorithms is known as *introsort* or introspective sort (Musser 1997) and it combines the benefits of *quicksort* (Hoare 1962) and *heapsort* (Williams 1964; Floyd 1964). Sorting a data record of n elements, the best and the average case running time of quicksort is $O(n \cdot \log n)$ but its drawback is the worst case running time of $O(n^2)$ (Ottmann and Widmayer 2012, pp. 93–101). Heapsort has a worst case running time $O(n \cdot \log n)$ (Ottmann and Widmayer 2012, pp. 106–112). However, quicksort outperforms many other algorithms also heapsort in practice and hence, introsort commences with a median-of-three quicksort (Ottmann and Widmayer 2012, pp. 102–105). To avoid the worst case running time of quicksort, the algorithm halts the recursion and sorts a subsequence with heapsort if the depth of the quicksort recursion tree becomes too large. Bounding this depth by $O(\log n)$, introsort has a worst case running time of $O(n \cdot \log n)$.

[33] Implementation detail: This requires a *binary search* that has a worst case running time of $O(\log n)$ since it halves the set of candidates in every iteration (Ottmann and Widmayer 2012, pp. 174–176).

3.5 Algorithm. All extended cliques for the KP
Input: Instance data
Output: All extended cliques $\mathcal{C}$
Initialization: $\mathcal{C} \leftarrow \emptyset, h \leftarrow 0, g \leftarrow 0$
1 Sort all items with respect to a non-decreasing order of the weight $w_{j_1} \leq \ldots \leq w_{j_J}$
2 Find the item j_h with $h = \min\{k = 1, \ldots, J-1 | w_{j_k} > c - w_{j_J}\}$
3 **if** *An item h exists* **then**
4   Find the item j_g with $g = \min\{k = h+1, \ldots, J | w_{j_k} > c - w_{j_h}\}$
5   Add the extended clique $\mathcal{C} \leftarrow \mathcal{C} \cup \{j_h, j_g, \ldots, j_J\}$
6   **while** $g \neq h+1$ **do**
7          $h \leftarrow h+1$
8   **while** $g \neq h+1 \wedge w_{j_{g-1}} > c - w_{j_h}$ **do** $g \leftarrow g-1$
9   Add the extended clique $\mathcal{C} \leftarrow \mathcal{C} \cup \{j_h, j_g, \ldots, j_J\}$

Each additional coefficient a_j is determined from an optimal solution of a KP that maximizes the slack of the lifted inequality such that the solution stays feasible to a modified constraint (3.44) with the item j packed. It requires to sequentially solve a series of $J - |I|$ KPs to determine all coefficients. A simultaneous lifting procedure is proposed in Balas and Zemel (1978). A polynomial algorithm with a worst case running time of $O(J \cdot |I|)$ is presented providing the cover is given and the coefficients w are integer (Zemel 1989). Maximal (minimal) cardinality constraints

$$\sum_{j=1}^{J} a \cdot x_j \leq k \qquad (3.57)$$

with $a = 1, k > 0$ ($a = -1, k < 0$) impose an upper bound (a lower bound) on the number of items (Martello and Toth 1997).

3.6.2 Lower and Upper Bounds

The *critical item* s is the item with $\sum_{j=1}^{s-1} w_j \leq c \wedge \sum_{j=1}^{s} w_j > c$ and the *residual capacity* $\bar{c}$ becomes $\bar{c} = c - \sum_{j=1}^{s-1} w_j$.[34] The *Dantzig bound*

$$UB^D = \sum_{j=1}^{s-1} p_j + p_s \cdot \frac{\bar{c}}{w_s}$$

---

[34]Synonyms are *split item* or *break item*.

yields an optimal solution $\bar{x}$ to the Continuous KP as (Martello and Toth 1990, pp. 16–17; Dantzig 1957)

$$\bar{x}_j = \begin{cases} 1 & \text{if } j < s \\ \frac{\bar{c}}{w_s} & \text{if } j = s \\ 0 & \text{otherwise} \end{cases} \quad \forall j = 1, \ldots, J.$$

Example 3.3 shows that this intuitive priority rule of packing the items in the order of efficiencies does not yield the optimal integer solution of the KP.

Example 3.3 Let $n = 3$, $p = (7, 4, 4)^T$, $w = (3, 2, 2)^T$ and $c = 4$. The critical item is $s = 2$. The LP relaxation yields the optimal solution $\bar{x} = (1, \frac{1}{2}, 0)^T$ with $UB^D = 9$ but the optimal integer solution is $x = (0, 1, 1)^T$.

The *Martello-Toth bound* UB^{MT} has $UB^{MT} \leq UB^D$ (Martello and Toth 1990, pp. 20–21; Martello and Toth 1977). Either the critical item is removed but the item $s + 1$ is fractionally packed or the critical item is inserted but the item $s - 1$ takes a fractional value.

$$UB^{MT} = \max \left\{ \sum_{j=1}^{s-1} p_j + p_{s+1} \cdot \frac{\bar{c}}{w_{s+1}}, \sum_{j=1}^{s-1} p_j + p_{s-1} \cdot \frac{\bar{c} - w_s}{w_{s-1}} + p_s \right\}$$

The *Müller-Meerbach bound* UB^{MM} has $UB^{MM} \leq UB^D$ (Martello and Toth 1990, p. 23; Müller-Meerbach 1978). Modifying the continuous solution to the KP LP relaxation, an integer solution requires that either a packed item $1, \ldots, s - 1$ is removed, an excluded item $s + 1, \ldots, J$ is packed or the critical item is removed but the variable values of all other items remain unchanged.

$$UB^{MM} = \max \left\{ \max_{j=1,\ldots,s-1} UB^D - p_j + p_s \cdot \frac{w_j}{w_s}, \max_{j=s+1,\ldots,J} UB^D + p_j - p_s \cdot \frac{w_j}{w_s}, \sum_{j=1}^{s-1} p_j \right\}$$

The upper bound UB^{KP} with $UB^{KP} \leq UB^D$ is particularly interesting because there exists no dominance between UB^{MT} and UB^{MM}.[35]

$$UB^{KP} = \min \left\{ UB^{MT}, UB^{MM} \right\} \quad (3.58)$$

The *forward greedy solution* (Pisinger 1995) with LB^F inserts the items $1, \ldots, s-1$ and one item $k > s$ with the maximal profit. Let an artificial item $j = J + 1$ have $p_{J+1} = 0$, $w_{J+1} = 0$.

---

[35]This result is proven by providing an instance with $UB^{MT} < UB^{MM}$ (Martello and Toth 1990, pp. 23–24) and an instance with $UB^{MT} > UB^{MM}$ (Müller-Meerbach 1978).

3.6 The Knapsack Problem

3.6 Algorithm. The greedy solution for the KP

Input: Instance data, the position s of the critical item, the values $\sum_{j=1}^{s-1} p_j$ and $\bar{c}$
Output: A feasible solution x to KP with LB^G
Initialization: $x_j \leftarrow 0 \ \forall j, j \leftarrow 0$
1 Insert the items $1, \ldots, s-1$ by $x_j \leftarrow 1 \ \forall j = 1, \ldots, s-1$ and set $LB^G \leftarrow \sum_{j=1}^{s-1} p_j$
2 **for** $j \leftarrow s+1$ **to** J **do**
3   **if** $w_j \leq \bar{c}$ **then**
4   Insert the item j by $x_j \leftarrow 1$ and set $LB^G \leftarrow LB^G + p_j$
5        $\bar{c} \leftarrow \bar{c} - w_j$

$$LB^F = \sum_{j=1}^{s-1} p_j + p_k \text{ with } k = \underset{j=s+1,\ldots,J+1}{\arg\max} \{p_j | w_j \leq \bar{c}\}$$

The *backward greedy solution* (Pisinger 1995) with LB^B inserts the critical item s but an item $k < s$ with the minimal profit is excluded. Let an artificial item $j = 0$ have $p_0 = \infty$, $w_0 = w_s$.

$$LB^B = \sum_{j=1}^{s-1} p_j - p_k + p_s \text{ with } k = \underset{j=0,\ldots,s-1}{\arg\min} \{p_j | w_s \leq \bar{c} + w_j\}$$

One of the most popular myopic heuristics is Algorithm 3.6 that yields the *greedy solution* with LB^G (Kellerer et al. 2004, pp. 15–16). Inserting the items $1, \ldots, s-1$ in the solution and excluding the critical item, the algorithm proceeds to feasibly insert as many items $k > s$ as possible in the order of efficiencies.

The *single item solution* with LB^S comprises the item k with a maximal profit among all J items (Kellerer et al. 2004, p. 34).

$$LB^S = p_k \text{ with } k = \underset{j=1,\ldots,J}{\arg\max} \, p_j$$

The lower bound

$$LB^{KP} = \max\{LB^F, LB^B, LB^G, LB^S\} \tag{3.59}$$

has a *worst case performance ratio* of $\frac{1}{2}$ that is tight (Kellerer et al. 2004, p. 34). The single item solution is relevant to exclude a pathological case the other three lower bounds do not detect. This is demonstrated in Example 3.4, where LB^F, LB^B and LB^G become arbitrarily worse.[36]

---

[36]This example is found in a slightly modified form in Martello and Toth (1990, p. 28). An introduction to this kind of proofs is found in Martello and Toth (1990, pp. 9–11).

Example 3.4 Let $n = 3$. Given a constant K with $K > 2$. Let $p = (1, 1, K − 1)^T$, $w = (1, 1, K)^T$ and $c = K$. The instance satisfies Assumption 3.20. The critical item is $s = 3$. It follows $LB^F = 2$, $LB^B = -\infty$ and $LB^G = 2$ with the best known feasible solution $(1, 1, 0)^T$. The optimal solution $(0, 0, 1)^T$ has an objective function value of $Z = K − 1$. The worst case performance ratio is arbitrarily close to 0 because $\lim_{K \to \infty} \frac{\max\{LB^F, LB^B, LB^G\}}{Z} = \lim_{K \to \infty} \frac{2}{K-1} = 0$. $LB^S = K − 1$ yields the optimal solution.

The Martello-Toth and the Müller-Meerbach bound consider the first level of an enumeration tree and approximate the obtained upper bounds. Other approaches following that principle are surveyed in Martello and Toth (1990, pp. 22–24), Dudziński and Walukiewicz (1987). Tighter upper bounds are accessible from partial enumeration (Martello and Toth 1988) or from valid inequalities of the Knapsack polytope. For instance, adding a cardinality constraint (3.57) to the KP, an upper bound is determined by solving the Lagrangean dual (Martello and Toth 1997) or the surrogate dual (Martello et al. 1999). As previously presented in Sect. 3.3, both upper bounds dominate the Dantzig bound. An extension of the greedy algorithm yields a worst case performance of $\frac{3}{4}$ that is tight (Kellerer et al. 2004, pp. 35–36). If a pair of items does not violate the capacity constraint, determine the greedy solution for the remaining items. Apply this procedure for all possible pairs of items and return the best known feasible solution. An ϵ-approximation scheme is derived by augmenting this idea to all feasible sets consisting of k elements (Kellerer et al. 2004, pp. 37–39).

3.6.3 An Exact Algorithm

Let x^* be an optimal solution to the KP. Items with a high efficiency $1, \ldots, a − 1$ with $a = \min\{j = 1, \ldots, J | x_j^* = 0\}$ are included in x^*, whereas items with a low efficiency $b + 1, \ldots, J$ with $b = \max\{j = 1, \ldots, J | x_j^* = 1\}$ are not packed. The *core* comprises the items $a, \ldots, b$ that are meaningful to the structure of x^*. The KP is equivalently formulated by the *core problem* (Balas and Zemel 1980)

$$\max \sum_{j=1}^{a-1} p_j + \sum_{j=a}^{b} p_j \cdot y_j \quad \text{subject to}$$

$$\sum_{j=1}^{a-1} w_j + \sum_{j=a}^{b} w_j \cdot y_j \leq c \wedge y_j \in \{0, 1\} \; \forall j = a, \ldots, b.$$

An optimal solution is $x_j^* = 1$ if $j < a$ or $x_j^* = y_j$ if $a \leq j \leq b$ (otherwise, 0) $\forall j = 1, \ldots, J$. The potentially extensive variable fixation significantly speeds up exact solution approaches and much computational time is saved for large instances because only the items of the core must be sorted to satisfy Assumption (3.48). In

3.6 The Knapsack Problem

practice, the core $a, \ldots, b$ is not known in advance and chosen around the critical item s.

Algorithm 3.7 outlines one of the most successful exact algorithms. The *Combo Algorithm* is a dynamic programming algorithm with a pseudo-polynomial worst case running time of $O(J \cdot c)$ (Martello et al. 1999).[37] Assume $c \in \mathbb{N}$ and $w_j, p_j \in \mathbb{N}$ $\forall j$ (Martello and Toth 1990, pp. 14–15). Assumption (3.48) only holds for items in the core $a, \ldots, b$. All items $1, \ldots, a-1$ ($b+1, \ldots, J$) have an efficiency larger than a (smaller than b). For simplicity assume the core only consists of the critical item s. The core problem

$$z_{ab}(d) = \max \sum_{j=1}^{a-1} p_j + \sum_{j=a}^{b} p_j \cdot y_j \quad \text{subject to}$$

$$\sum_{j=1}^{a-1} w_j + \sum_{j=a}^{b} w_j \cdot y_j \leq d \wedge y_j \in \{0,1\} \; \forall j = a, \ldots, b$$

is solved with the recursive formula

$$z_{ab}(d) = \max \begin{cases} z_{a,b-1}(d) & \text{if } b \geq s \\ z_{a,b-1}(d - w_b) + p_b & \text{if } b \geq s \wedge w_b \leq d \\ z_{a+1,b}(d) & \text{if } a < s \\ z_{a+1,b}(d + w_a) - p_a & \text{if } a < s \wedge d + w_a \leq 2 \cdot c \end{cases} \qquad (3.60)$$

with $d = 0, \ldots, 2 \cdot c$, $a = 1, \ldots, s$, $b = s - 1, \ldots, J$ and $z_{s,s-1}(d) = -\infty$ if it holds that $d = 0, \ldots, \sum_{j=1}^{s-1} w_j - 1$ and $z_{s,s-1}(d) = \sum_{j=1}^{s-1} p_j$ if $d = \sum_{j=1}^{s-1} w_j, \ldots, 2 \cdot c$ (Pisinger 1997). The enumeration starts with $a = s$ and $b = s - 1$.[38] The first two cases refer to packing the unpacked item b and hence, evaluating the core

---

[37] Implementation detail: The C source code is available at Pisinger (2002).

[38] Removing a packed item j from an optimal solution, the remaining solution must be optimal to the subproblem defined by the set of items $1, \ldots, j-1, j+1, \ldots, J$ and the capacity $c - w_j$ (Cormen et al. 2001, pp. 339–344). This property suggests to use *dynamic programming* pioneered by Bellmann (Domschke and Drexl 2005, pp. 158, 162–164; Bellmann 1952). Assume a sequential decision process consists of I steps. Given the ith step with $i = 1, \ldots, I$. The ith state variable or *state* d_i is defined by a *state function* g_i that transforms the previous state d_{i-1} via a *decision variable* x_i into $d_i = g_i(d_{i-1}, x_i)$. Let D_i be the state domain at step i. Let $X_i(d_{i-1})$ be the variable domain that contains all possible decisions x_i can take subject to d_{i-1}. State costs are obtained by a *cost function* $f_i(d_{i-1}, x_i)$. An initial state d_0 is transformed into a final state d_I such that the sum of all state costs is minimal subject to domains of the state and the decision variables.

$$\max \sum_{i=1}^{I} f_i(d_{i-1}, x_i) \quad \text{subject to}$$

$$d_i = g_i(d_{i-1}, x_i) \; \forall i = 1, \ldots, I \wedge d_i \in D_i \; \forall i = 1, \ldots, I \wedge x_i \in X_i(d_{i-1}) \; \forall i = 1, \ldots, I$$

3.7 Algorithm. The Combo Algorithm for the KP
Input: Instance data
Output: An optimal solution x to KP
1 Find the critical item partially sort all items and determine the initial core
2 **while** *Optimum is not found* **do**
3   **for** *All values of d* **do**
4   Run the dynamic programming recursion (3.60), expand the core if the respective items are not in the core, reduce and prune with upper bounds
5   If $N > M'$, decrease the capacity with the greatest common divisor
6   If $N > M''$, solve the surrogate dual of the KP with a cardinality constraint
7   If $N > M'''$, for each item not in the core, combine the core problems with the respective item and improve the lower bound
8 Determine an optimal solution x

towards the last item J. The last two cases unpack an already packed item a, which corresponds to an evaluation of the core towards the first item 1. The state $z_{1J}(c)$ yields the optimal solution since the entire set of items is evaluated and the capacity value d equals the maximal available capacity c of the KP.

Since the number of core problems N is an indicator of the instance difficulty, further techniques are gradually introduced if N becomes larger than given constants $M', M'', M''' \in \mathbb{N}$ with $M' < M'' < M'''$. If $N > M'$, the capacity is feasibly decreased with the greatest common divisor of all weights $w_1, \ldots, w_J$. If $N > M''$, the surrogate dual of a cardinality constraint (3.57) is solved to optimality. If $N > M'''$, the lower bound is tightened by a pairing procedure that yields a lower bound by considering all items one after another, which are not in the core, and all core problems generated so far. Among other termination criteria, the algorithm terminates, if either all core problems have been evaluated. The optimal solution is not obtained from backtracking but reconstructed via storing incomplete information of the optimal solution.

---

is recursively solved by the *principle of optimality* for all $i = I, I-1, \ldots, 1$ by

$$z_i(d_i - 1) = \max\{f_i(d_{i-1}, x_i) + z_{i+1}(d_i) | x_i \in X_i(d_{i-1})\} \; \forall d_i \in D_i$$

with $d_i = g_i(d_{i-1}, x_i) \land z_{I+1}(d_I) = 0$ where $z_1(d_0)$ is an optimal state. The KP is solved with the Bellmann recursion formula

$$z_b(d) = \begin{cases} z_{b-1}(d) & \text{if } w_b > d \\ \max\{z_{b-1}(d), z_{b-1}(d - w_b) + p_b\} & \text{if } w_b \leq d \end{cases} \; \forall d = 0, \ldots, c \quad (3.61)$$

for all states $b = 1, \ldots, J$ and by initializing with $z_0(d) = 0 \; \forall d = 0, \ldots, c$. In the first case, an item is excluded because the residual capacity d does not suffice to pack it. The second case decides if an item is packed or not and maximizes the profit. The state $z_J(c)$ is optimal. (3.61) appeared as a special case in Bellmann (1957, pp. 45–46) for the Bounded KP in which x_j is not binary but a bounded integer. Dantzig (1957) also presented (3.61).

A recent study shows that the Combo Algorithm is one of the most successful algorithms that solves even hard instances of the KP (Pisinger 2005). The superiority of the Combo Algorithm mostly depends on the size of the core, which is dynamically expanding and derived in the enumeration process instead of guessing it beforehand. Furthermore, computationally expensive procedures such as tighter bounds and reductions are gradually introduced if the instance is detected as being difficult to solve. Three other outperformed but famous algorithms are shortly presented. The MT2 algorithm is a depth-first branch and bound and a core of a fixed size is symmetrical placed around the critical item (Martello and Toth 1988). An improvement is the MTH algorithm that combines the MT2 algorithm with dynamic programming (Martello and Toth 1990, pp. 36–45; Martello and Toth 1997). The Minknap algorithm uses the dynamic programming recursion (3.60) and expands the core such that the number of items required to find an optimal solution (core) is minimal (Pisinger 1997).

References

Aardal, K., & Larsson, T. (1990). A Benders decomposition based heuristic for the hierarchical production planning problem. *European Journal of Operational Research, 45*, 4–14.
Ahuja, R. K., Magnanti, T. L., & Orlin, J. B. (1993). *Network flows: Theory, algorithms, and applications.* Upper Saddle River, NJ: Prentice-Hall.
Amor, H. B., & Valério de Carvalho, J. M. (2005). Cutting stock problems. In G. Desaulnier, J. Desrosiers, & M. M. Solomon (Eds.), *Column generation* (pp. 132–161). New York: Springer.
Arora, S., & Barak, B. (2009). *Computational complexity: A modern approach.* Cambridge: Cambridge University Press.
Balas, E. (1975). Facets of the knapsack polytope. *Mathematical Programming, 8*, 146–164.
Balas, E., & Jeroslow, R. (1972). Canonical cuts on the unit hypercube. *SIAM Journal on Applied Mathematics, 23*, 61–69.
Balas, E., & Padberg, M. W. (1972). On the set-covering problem. *Operations Research, 20*, 1152–1161.
Balas, E., & Padberg, M. W. (1976). Set partitioning: A survey. *SIAM Review, 18*, 710–760.
Balas, E., & Zemel, E. (1978). Facets of the knapsack polytope from minimal covers. *SIAM Journal on Applied Mathematics, 34*, 119–148.
Balas, E., & Zemel, E. (1980). An algorithm for large zero-one knapsack problems. *Operations Research, 28*, 1130–1154.
Barahona, F., & Anbil, R. (2000). The volume algorithm: Producing primal solutions with a subgradient method. *Mathematical Programming Series A, 87*, 385–399.
Bellmann, R. E. (1952). On the theory of dynamic programming. *Proceedings of the National Academy of Sciences of the United States of America, 38*, 716–719.
Bellmann, R. E. (1957). *Dynamic programming.* Princeton, NJ: Princeton University Press.
Benders, J. F. (1962). Partitioning procedures for solving mixed-variables programming problems. *Numerische Mathematik, 4*, 238–252.
Benders, J. F. (2005). Partitioning procedures for solving mixed-variables programming problems. *Computational Management Science, 2*, 3–19.
Bland, R. G. (1977). New finite pivoting rules for the simplex method. *Mathematics of Operations Research, 2*, 103–107.
Bland, R. G., Goldfarb, D., & Todd, M. J. (1981). The ellipsoid method: A survey. *Operations Research, 29*, 1039–1091.

Bock, S. (2000). *Modelle und verteilte Algorithmen zur Planung getakteter Fließlinien*. Wiesbaden: Gabler DUV.
Borgwardt, K.-H. (1982). The average number of pivot steps required by the simplex-method is polynomial. *Zeitschrift für Operations Research, 26*, 157–177.
Boschetti, M., & Maniezzo, V. (2009). Benders decomposition, Lagrangean relaxation and metaheuristic design. *Journal of Heuristics, 15*, 283–312.
Brooks, R., & Geoffrion, A. (1966). Finding Everett's Lagrange multipliers by linear programming. *Operations Research, 14*, 1149–1153.
Camerini, P. M., Fratta, L., & Maffioli, F. (1975). On improving relaxation methods by modified gradient techniques. *Mathematical Programming Studies, 3*, 26–34.
Carraresi, P., Frangioni, A., & Nonato, M. (1997). *Applying bundle methods to the optimization of polyhedral functions: An applications-oriented development*. TR-96-17, Dipartimento di Informatica, Università di Pisa, Pisa.
Chandru, V., & Rao, M. R. (2004). Combinatorial optimization. In A. B. Tucker (Ed.), *Computer science handbook* (Chap. 15, 2nd ed.). Boca Raton: Chapman & Hall/CRC.
Chvátal, V. (2002). *Linear programming*. New York: W.H. Freeman and Company.
Cook, S. (1971). The complexity of theorem-proving procedures. In *Proceedings of the 3rd Annual ACM Symposium on Theory of Computing* (pp. 151–158).
Cormen, T. H., Leiserson, C. E., Rivest, R. L., & Stein, C. (2001). *Introduction to algorithms* (2nd ed.). Cambridge: The MIT Press.
Cornuejols, G., Nemhauser, G. L., & Wolsey, L. A. (1990). The uncapacitated facility location problem. In P. B. Mirchandani & R. L. Francis (Eds.), *Discrete location theory* (pp. 119–171). New York: Wiley.
Cornuejols, G., Sridharan, R., & Thizy, J. M. (1991). A comparison of heuristics and relaxations for the capacitated plant location problem. *European Journal of Operational Research, 50*, 280–297.
Côté, G., & Laughton, M. A. (1984). Large-scale mixed integer programming: Benders-type heuristics. *European Journal of Operational Research, 16*, 327–333.
Croes, G. A. (1958). A method for solving traveling-salesman problems. *Operations Research, 6*, 791–812.
Crowder, H., Johnson, E. L., & Padberg, M. (1983). Solving large-scale zero-one linear programming problems. *Operations Research, 31*, 803–834.
Dantzig, G. B. (1951). Maximization of a linear function of variables subject to linear inequalities. In T. C. Koopmans (Ed.), *Activity analysis of production and allocation* (pp. 339–347). New York: Wiley.
Dantzig, G. B. (1957). Discrete variable extremum problems. *Operations Research, 5*, 266–277.
Dantzig, G. B., Ford, L. R., & Fulkerson, D. R. (1956). A primal-dual algorithm for linear problems. In H. W. Kuhn, & A. W. Tucker (Eds.), *Linear inequalities and related systems. Annals of mathematics study* (Vol. 38, pp. 171–181). Princeton, NJ: Princeton University Press.
Desaulnier, G., Desrosiers, J., & Solomon, M. M. (2005). *Column generation*. New York: Springer.
Desrosiers, J., & Lübbecke, M. E. (2005). A primer in column generation. In Desaulnier, G., Desrosiers, J., & Solomon, M. M. (Eds.), *Column generation*, pp. 1–32. New York: Springer.
Dietrich, B. L., & Escudero, L. F. (1994). Obtaining clique, cover and coefficient reduction inequalities as Chvatal–Gomory inequalities and Gomory fractional cuts. *European Journal of Operational Research, 73*, 539–546.
Dietrich, B. L., Escudero, L. F., Garín, A., & Pérez, G. (1993). $O(n)$ procedures for identifying maximal cliques and non-dominated extensions of consecutive minimal covers and alternates. *TOP, 1*, 139–160.
Domschke, W., & Drexl, A. (2005). *Einführung in operations research* (6th ed.). Berlin: Springer.
Domschke, W., & Wagner, B. (2005). Models and methods for standardization problems. *European Journal of Operational Research, 162*, 713–726.
Du Merle, O., Villeneuve, D., Desrosiers, J., & Hansen, P. (1997). Stabilized column generation. *Discrete Mathematics, 194*, 229–237.

References

Dudziński, K., & Walukiewicz, S. (1987). Exact methods for the knapsack problem and its generalizations. *European Journal of Operational Research, 28*, 3–21.

El Hallaoui, I., Villeneuve, D., Soumis, F., & Desaulniers, G. (2005). Dynamic aggregation of set-partitioning constraints in column generation. *Operations Research, 53*, 632–645.

Erlenkotter, D. (1978). A dual-based procedure for uncapacitated facility location. *Operations Research, 26*, 992–1009.

Fisher, M. L. (1981). The Lagrangian relaxation method for solving integer programming problems. *Management Science, 27*, 1–18.

Floyd, R. W. (1964). Algorithm 245 - treesort 3. *Communications of the ACM, 7*, 701.

Ford, L. R., & Fulkerson, D. R. (1957). A primal-dual algorithm for the capacitated Hitchcock problem. *Naval Research Logistics Quarterly, 4*, 47–54.

Fu, Y., Shahidehpour, M., & Li, Z. (2005). Long-term security-constrained unit commitment: Hybrid Dantzig-Wolfe decomposition and subgradient approach. *IEEE Transactions on Power Systems, 20*, 2093–2106.

Garey, M. R., & Johnson, D. S. (1978). Strong NP - completeness results: Motivation, examples, and implications. *Journal of the Association for Computing Machinery, 25*, 499–508.

Garey, M. R., & Johnson, D. S. (1979). *Computers and intractability: A guide to the theory of NP-completeness.* New York: Freeman.

Gendreau, M., Hertz, A., & Laporte, G. (1992). New insertion and postoptimization procedures for the traveling salesman problem. *Operations Research, 40*, 1086–1094.

Geoffrion, A. M. (1974). Lagrangian relaxation for integer programming. *Mathematical Programming Study, 2*, 82–114.

Geoffrion, A. M., & Graves, G. W. (1974). Multicommodity distribution system design by Benders decomposition. *Management Science, 20*, 822–844.

Gilmore, P., & Gomory, R. (1961). A linear programming approach to the cutting stock problem. *Operations Research, 9*, 849–859.

Gilmore, P., & Gomory, R. (1963). A linear programming approach to the cutting stock problem - part II. *Operations Research, 11*, 863–888.

Glover, F. (1986). Future paths for integer programming and links to artificial intelligence. *Computers & Operations Research, 13*, 533–549.

Glover, F. (1989). Tabu search - part I. *ORSA Journal on Computing, 1*, 190–206.

Glover, F. (1990). Tabu search - part II. *ORSA Journal on Computing, 2*, 4–32.

Glover, F., & Laguna, M. (2002). *Tabu Search* (5th ed.). Norwell: Kluwer Academic Publishers.

Goemans, M. X., & Williamson, D. P. (1996). The primal-dual method for approximation algorithms and its application to network design problems. In D. S. Hochbaum (Ed.), *Approximation algorithms for NP-hard problems* (pp. 144–191). Boston: PWS Publishing Company.

Goffin, J.-L., Haurie, A., & Vial, J.-P. (1992). Decomposition and nondifferentiable optimization with the projective algorithm. *Management Science, 38*, 284–302.

Goffin, J.-L., Haurie, A., Vial, J.-P., & Zhu, D. (1993). Using central prices in the decomposition of linear programs. *European Journal of Operational Research, 64*, 393–409.

Grötschel, M., Lovász, L., & Schrijver, A. (1981). The ellipsoid method and its consequences in combinatorial optimization. *Combinatorica, 1*, 169–197.

Guignard, M. (1988). A Lagrangean dual ascent algorithm for simple plant location problems. *European Journal of Operational Research, 35*, 193–200.

Guignard, M. (2003). Lagrangean relaxation. *TOP, 11*, 151–228.

Guignard, M., & Kim, S. (1987). Lagrangean decomposition: A model yielding stronger Lagrangean bounds. *Mathematical Programming, 39*, 215–228.

Guignard, M., & Zhu, S. (1994). A hybrid algorithm for solving Lagrangean duals in mixed-integer programming. In *Proceedings of the VI CLAIO* (pp. 399–410).

Guta, B. (2003). *Subgradient Optimization Methods in Integer Programming with an Application to a Radiation Therapy Problem.* PhD thesis, Department of Mathematics, University Kaiserslautern, Kaiserslautern.

Hansen, P. (1986). The steepest ascent mildest descent heuristic for combinatorial programming. In *Proceedings of the Congress on Numerical Methods in Combinatorial Optimization*, Capri.

Held, M., & Karp, R. M. (1970). The traveling-salesman problem and minimum spanning trees. *Operations Research, 18*, 1138–1162.

Held, M., & Karp, R. M. (1971). The traveling-salesman problem and minimum spanning trees: Part II. *Mathematical Programming, 1*, 6–25.

Held, M., Wolfe, P., & Crowder, H. P. (1974). Validation of subgradient optimization. *Mathematical Programming, 6*, 62–88.

Hellstrand, J., Larsson, T., & Migdalas, A. (1992). A characterization of the uncapacitated network design polytope. *Operations Research Letters, 12*, 159–163.

Hoare, C. A. R. (1962). Quicksort. *The Computer Journal, 5*, 10–16.

Holmberg, K. (1994a). Cross decomposition applied to integer programming problems: Duality gaps and convexification in parts. *Operations Research, 42*, 657–668.

Holmberg, K. (1994b). On using approximations of the Benders master problem. *European Journal of Operational Research, 77*, 111–125.

Johnson, D. S. (1990). A catalog of complexity classes. In van Leeuwen, J. (Ed.), *Handbook of theoretical computer science: Algorithms and complexity* (Chap. 2, Vol. A). Cambridge: The MIT Press.

Karmarkar, N. (1984). A new polynomial time algorithm for linear programming. *Combinatorica, 4*, 373–395.

Karp, R. M. (1972). Reducibility among combinatorial problems. In Miller, R. E. & Thatcher, J. W. (Eds.), *Complexity of computer computations* (pp. 85–103). New York: Plenum.

Kellerer, H., Pferschy, U., & Pisinger, D. (2004). *Knapsack problems*. Berlin: Springer.

Kelley, J. E. (1960). The cutting-plane method for solving convex programs. *Journal of the SIAM, 8*, 703–712.

Khachiyan, L. G. (1979). A polynomial algorithm in linear programming. *Soviet Mathematics Doklady, 20*, 191–194.

Klee, V., & Minty, G. J. (1972). How good is the simplex algorithm? In Shisha, O. (Ed.), *Inequalities* (Vol. 3, pp. 159–175). New York: Academic.

Klose, A. (2001). *Standortplanung in distributiven Systemen*. Heidelberg: Physica.

Klose, A., & Drexl, A. (2005). Facility location models for distribution system design. *European Journal of Operational Research, 162*, 4–29.

Krarup, J., & Pruzan, P. M. (1983). The simple plant location problem: Survey and synthesis. *European Journal of Operational Research, 12*, 36–81.

Lee, H., & Guignard, M. (1996). A hybrid bounding procedure for the workload allocation problem on parallel unrelated machines with setups. *Journal of the Operational Research Society, 47*, 1247–1261.

Lemke, C. E. (1954). The dual method of solving the linear programming problem. *Naval Research Logistics Quarterly, 1*, 36–47.

Lin, S. (1965). Computer solutions of the traveling salesman problem. *Bell System Technical Journal, 44*, 2245–2269.

Lübbecke, M. E., & Desrosiers, J. (2005). Selected topics in column generation. *Operations Research, 53*, 1007–1023.

Magnanti, T. L., & Wong, R. T. (1981). Accelerating Benders decomposition: Algorithmic enhancement and model selection criteria. *Operations Research, 29*, 464–484.

Magnanti, T. L., & Wong, R. T. (1984). Network design and transportation planning: Models and algorithms. *Transportation Science, 18*, 1–55.

Magnanti, T. L., & Wong, R. T. (1986). *Decomposition Methods for Facility Location Problems*. OR 153-86, Massachusetts Institute of Technology, Cambridge.

Magnanti, T. L., & Wong, R. T. (1990). Decomposition methods for facility location problems. In P. B. Mirchandani & R. L. Francis (Eds.), *Discrete location theory* (pp. 209–262). New York: Wiley.

Marsten, R. E. (1975). The use of the boxstep method in discrete optimization. *Mathematical Programming Study, 3*, 127–144.

References

Marsten, R. E., Hogan, W. W., & Blankenship, J. W. (1975). The boxstep method for large-scale optimization. *Operations Research, 23*, 389–405.

Martello, S., Pisinger, D., & Toth, P. (1999). Dynamic programming and strong bounds for the 0-1 knapsack problem. *Management Science, 45*, 414–424.

Martello, S., & Toth, P. (1977). An upper bound for the zero-one knapsack problem and a branch and bound algorithm. *European Journal of Operational Research, 1*, 169–175.

Martello, S., & Toth, P. (1988). A new algorithm for the 0-1 knapsack problem. *Management Science, 34*, 633–644.

Martello, S., & Toth, P. (1990). *Knapsack problems: Algorithms and computer implementations*. New York: Wiley.

Martello, S., & Toth, P. (1997). Upper bounds and algorithms for hard 0-1 knapsack problems. *Operations Research, 45*, 768–778.

Megiddo, N. (1989). On the complexity of linear programming. In T. F. Bewley (Ed.), *Advances in economic theory. Fifth world congress* (Chap. 6). Cambridge: Cambridge University Press.

Minoux, M., & Ribeiro, C. C. (1985). Solving hard constrained shortest path problems by Lagrangean relaxation and branch-and-bound algorithms. *Methods of Operations Research, 53*, 303–316.

Müller-Meerbach, H. (1978). An improved upper bound for the zero-one knapsack problem. *European Journal of Operational Research, 2*, 212–213.

Musser, D. (1997). Introspective sorting and selection algorithms. *Software: Practice and Experience, 27*, 983–993.

Nemhauser, G. L., & Wolsey, L. A. (1988). *Integer and combinatorial optimization*. New York: Wiley.

Orlin, J. B. (1986). A dual version of Taros's algorithm for linear programming. *Operations Research Letters, 5*, 221–226.

Ottmann, T., & Widmayer, P. (2012). *Algorithmen und Datenstrukturen* (5th ed.). Heidelberg: Spektrum Akdemischer Verlag.

Padberg, M. W. (1975). A note on zero-one programming. *Operations Research, 23*, 833–837.

Papadimitriou, C. H., & Steiglitz, K. (1998). *Combinatorial optimization: Algorithms and complexity*. Mineola: Dover.

Pisinger, D. (1995). An expanding-core algorithm for the exact 0-1 knapsack problem. *European Journal of Operational Research, 87*, 175–187.

Pisinger, D. (1997). A minimal algorithm for the 0-1 knapsack problem. *Operations Research, 45*, 758–767.

Pisinger, D. (2002). Implementation of the combo algorithm in C. Online. Accessed August 2010.

Pisinger, D. (2005). Where are the hard knapsack problems? *Computers & Operations Research, 32*, 2271–2284.

Poljak, B. T. (1969). Minimization of unsmooth functionals. *USSR Computational Mathematics and Mathematical Physics, 9*, 14–29.

Rei, W., Cordeau, J.-F., Gendreau, M., & Soriano, P. (2009). Accelerating Benders decomposition by local branching. *INFORMS Journal on Computing, 21*, 333–345.

Rhys, J. M. W. (1970). A selection problem of shared fixed costs and network flows. *Management Science, 17*, 200–207.

Rönnberg, E., & Larsson, T. (2014). All-integer column generation for set partitioning: Basic principles and extensions. *European Journal of Operational Research, 233*, 529–538.

Roy, T. J. V. (1983). Cross decomposition for mixed integer programming. *Mathematical Programming, 25*, 46–63.

Sastry, T. (2001). One and two facility network design revisited. *Annals of Operations Research, 108*, 19–31.

Shamir, R. (1987). The efficiency of the simplex method: A survey. *Management Science, 33*, 301–334.

Sherali, H. D., & Choi, G. (1996). Recovery of primal solutions when using subgradient optimization methods to solve Lagrangian duals of linear programs. *Operations Research Letters, 19*, 105–113.

Taillard, E. (1991). Robust taboo search for the quadratic assignment problem. *Parallel Computing, 17*, 443–455.

Taillard, E., Badeau, P., Gendreau, M., & Potvin, F. G. J.-Y. (1997). A tabu search heuristic for the vehicle routing problem with soft time windows. *Transportation Science, 31*, 170–186.

Tardos, E. (1986). A strongly polynomial algorithm to solve combinatorial linear programs. *Operations Research, 34*, 250–256.

Thompson, G. L. (2002). An integral simplex algorithm for solving combinatorial optimization problems. *Computational Optimization and Applications, 22*, 351–367.

Trubin, V. (1969). On a method of solution of integer programming problems of a special kind. *Soviet Mathematics Doklady, 10*, 1544–1546.

Valério de Carvalho, J. M. (2003). Using extra dual cuts to accelerate convergence in column generation. *INFORMS Journal of Computing, 17*, 175–182.

Vavasis, S. A., & Ye, Y. (1996). A primal-dual interior point method whose running time depends only on the constraint matrix. *Mathematical Programming, 74*, 79–120.

Vera, J. R., Weintraub, A., Koenig, M., Bravo, G., Guignard, M., & Barahona, F. (2003). A Lagrangian relaxation approach for a machinery location problem in forest harvesting. *Pesquisa Operacional, 23*, 111–128.

Wentges, P. (1997). Weighted Dantzig-Wolfe decomposition for linear mixed-integer programming. *International Transactions in Operational Research, 4*, 151–162.

Williams, J. W. J. (1964). Algorithm 232 - heapsort. *Communications of the ACM, 7*, 347–348.

Wright, M. H. (2005). The interior-point revolution in optimization: History, recent developments, and lasting consequences. *Bulletin of the American Mathematical Society (New Series), 42*, 39–56.

Yemelichev, V. A., Kovalev, M. M., & Kravtsov, M. K. (1984). *Polytopes, graphs and optimisation*. Cambridge: Cambridge University Press (translated by G. H. Lawden).

Zaghrouti, A., Soumis, F., & El Hallaoui, I. (2014). Integral simplex using decomposition for the set partitioning problem. *Operations Research, 62*, 435–449.

Zemel, E. (1989). Easily computable facets of the knapsack polytope. *Mathematics of Operations Research, 14*, 760–764.

Chapter 4
The Weighted Uncapacitated Planned Maintenance Problem

Additional structural insights to the CPMP are obtained by a family of uncapacitated planned maintenance problems. Polyhedral properties, the computational complexity status and optimal algorithms of the WUPMP and of specific problem variants are presented in Sect. 4.1. In the following Sects. 4.4 and 4.5, two problem variants are discussed in greater detail and polyhedral properties and solution approaches are provided.[1]

4.1 The Mathematical Formulation

Let $f \in \mathbb{R}_+^T$ and $c \in \mathbb{R}_+^{n \times T}$ be finite objective function coefficients that represent fixed and variable maintenance costs (cf. Chap. 1). The WUPMP minimizes the total maintenance costs and is stated as

$$\min \sum_{t=1}^{T} f_t \cdot y_t + \sum_{i=1}^{n} \sum_{t=1}^{T} c_{it} \cdot x_{it} \quad \text{subject to}$$

$$\sum_{\tau=t}^{t+\pi_i-1} x_{i\tau} \geq 1 \qquad \forall i = 1, \ldots, n; t = 1, \ldots, T - \pi_i + 1 \quad \text{(P)}$$

$$x_{it} \leq y_t \qquad \forall i = 1, \ldots, n; t = 1, \ldots, T \quad \text{(V)}$$

$$x_{it} \in \{0, 1\} \qquad \forall i = 1, \ldots, n; t = 1, \ldots, T \quad \text{(X)}$$

$$y_t \in \{0, 1\} \qquad \forall t = 1, \ldots, T. \quad \text{(Y)}$$

---

[1] Some of the results of Chap. 4 are found in the presented form or in a slightly changed from in the author's publication (Kuschel and Bock 2016).

© Springer International Publishing Switzerland 2017
T. Kuschel, *Capacitated Planned Maintenance*, Lecture Notes in Economics and Mathematical Systems 686, DOI 10.1007/978-3-319-40289-5_4

Table 4.1 Coefficient matrix of (P) and its stairway structure

	x_{i1}	x_{i2}	x_{i3}	x_{i4}	x_{i5}	x_{i6}
v_{i1}	1	1	1			
v_{i2}		1	1	1		
v_{i3}			1	1	1	
v_{i4}				1	1	1

Note that the WUPMP is not a special case of the CPMP because of the objective function. The assumptions and the terminology from Sect. 2.3.1 apply. An important property of (P) is presented in the next remark.

Remark 4.1 The *stairway structure* is illustrated in Table 4.1 that shows the cascading coefficient matrix of (P) for a maintenance activity i with $\pi_i = 3$ and $T = 6$. Observe for $t = 3$ that the dual variable v_{i1} does not only affect the dual constraints $t - \pi_i + 1, \ldots, t$ but also the dual constraints $t+1, \ldots, t+\pi_i - 1$ through the variables v_{i2} and v_{i3}.

Remark 4.2 allows to transform (P) into a path finding problem. This is illustrated in Example 4.1.

Remark 4.2 The ith constraint (P) can be formulated as a path finding problem in a weighted, directed graph $G_i(V, E_i)$. The node set $V = \{0, \ldots, T+1\}$ represents the periods and contains a source node 0 and a sink node $T+1$. The arc set is defined as

$$E_i = \{(t, \tau) \in V \times V \mid \tau = t+1, \ldots, t + \pi_i\}.$$

If the path from node 0 to node $T+1$ contains the arc (t, τ), a maintenance activity i is scheduled in the periods t as well as in τ and the binary decision variable $z_{it\tau}$ equals 1, which is equivalent to $x_{it} = x_{i\tau} = 1$ (otherwise, 0). The weight of an arc $(t, \tau) \in E_i$ is successor independent and equals c_{it} if $t \neq 0$ (otherwise, 0). The ith constraint (P) becomes

$$x_{it} = \sum_{(t,\tau) \in E_i} z_{it\tau} \quad \forall t = 1, \ldots, T \tag{4.1}$$

$$\sum_{(t,\tau) \in E_i} z_{it\tau} - \sum_{(\tau,t) \in E_i} z_{i\tau t} = \begin{cases} 1 & \forall t = 0 \\ 0 & \forall t = 1, \ldots, T \\ -1 & \forall t = T+1 \end{cases} \tag{4.2}$$

$$z_{it\tau} \in \{0, 1\} \quad \forall (t, \tau) \in E_i. \tag{4.3}$$

4.1 The Mathematical Formulation

Fig. 4.1 Reformulation of (P) as a path finding problem

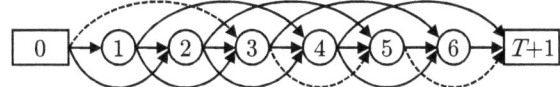

Constraint (4.1) defines the variable transformation from z to x. Equation (4.2) are the Kirchoff flow balance equations (Tomlin 1966) that represent (P). The variable domains are binary.[2]

Example 4.1 Let $\pi_i = 3$ and $T = 6$. Figure 4.1 shows $G_i(V, E_i)$ from Remark 4.2. The path that is illustrated with dashed lines has $z_{i03} = z_{i35} = z_{i57} = 1$, represents $x_{i3} = x_{i5} = 1$ and satisfies (P).

The WUPMP is closely related to two previously introduced, well-known combinatorial problems. The UFLP has been the subject of intense research (Cornuejols et al. 1990; Krarup and Pruzan 1983; Klose and Drexl 2005; ReVelle et al. 2008; Labbé et al. 1995) and the UNDP is studied in Magnanti and Wong (1984), Holmberg and Hellstrand (1998), Balakrishnan et al. (1997), and Balakrishnan et al. (1989).[3] The WUPMP is a generalization of the UFLP because the WUPMP has the stairway structure of (P) (Remark 4.1) and the WUPMP polytope contains the UFLP polytope if $\pi_i = T \; \forall i$ holds. Moreover, the WUPMP is a special case of the UNDP because the reformulation of the WUPMP with Remark 4.3 yields a specifically structured graph where the resulting polytope is included in the UNDP polytope. Therefore, any algorithm that solves the UNDP also solves the WUPMP (Holmberg and Hellstrand 1998; Balakrishnan et al. 1989).

Remark 4.3 Reformulate (P) with Remark 4.2 for each maintenance activity i. Each node $t = 1, \ldots, T$ stands for a real period and is split into two novel nodes t' and t'' such that the ingoing (outgoing) arcs of t are the ingoing (outgoing) arcs of t' (t''). The novel arc (t', t'') represents a maintenance slot and it has fixed costs f_t but zero transportation costs. The other arcs represent the covering of the periods by (P) and hence, the fixed costs are zero but the transportation costs equal c_{it}. The reformulated WUPMP is a UNDP with a specific graph (cf. network-transformation from Sect. 5.3.3.1).

---

[2] Note that the coefficient matrix of (4.2) is totally unimodular (Lemma 3.7).
[3] See Footnote 10 on page 32 for the UFLP and the UNDP.

4.2 Polyhedral Properties

Theorem 4.1 states the main polyhedral result. Together with Corollary 4.2, the polytopes of the WUPMP LP relaxation with $n = 1$ and the respective dual problem are integral. Note that the latter holds if and only if $c_{it} \in \mathbb{N}\ \forall t$.

Theorem 4.1 *Conv*(4.4)(4.5)(4.6) = H(4.4)(4.5) *holds for the ith system of linear inequalities* (4.4)–(4.6) *with* $a_{it} \in \mathbb{Z}\ \forall t$, $b_{it} \in \mathbb{N}\ \forall t$.

$$\sum_{\tau=t}^{t+\pi_i-1} x_{i\tau} \geq a_{it} \qquad \forall t = 1, \ldots, T - \pi_i + 1 \qquad (4.4)$$

$$0 \leq x_{it} \leq b_{it} \qquad \forall t = 1, \ldots, T \qquad (4.5)$$

$$x_{it} \in \mathbb{N} \qquad \forall t = 1, \ldots, T \qquad (4.6)$$

Proof '$\Rightarrow$': *Conv*(4.4)(4.5)(4.6) $\subseteq H$(4.4)(4.5) holds because H(4.4)(4.5) is convex.

'$\Leftarrow$': Assume a feasible solution $\bar{x} \in H$(4.4)(4.5) where at least one component $0 < \bar{x}_{it} < 1$ is fractional. *Conv*(4.4)(4.5)(4.6) $\supseteq H$(4.4)(4.5) is shown by the existence of a convex combination of K integer solutions that yield $\bar{x}$. Therefore, $\bar{x}$ cannot be an extreme point and it follows that $\bar{x} \in$ *Conv*(4.4)(4.5)(4.6).

Since (4.4) and (4.5) are rational linear inequalities, represent all non-zeros in $(\bar{x}_{i1}, \ldots, \bar{x}_{iT})$ as fractions.[4] Compute the *least common denominator* K among of all $(\bar{x}_{i1}, \ldots, \bar{x}_{iT})$.[5] Denote $\alpha = \frac{1}{K}$ as a *quant*.

Construct K integer solutions x^k with $x^k \in \mathbb{N}^T$ for all $k = 1, \ldots, K$ as follows (see Example 4.2): Initialize $x_t^k = 0\ \forall t = 1, \ldots, T; k = 1, \ldots, K$ and set $k = 1$, $t = 0$, $d = 0$. Repeat the following steps per iteration: While $d > 0$ holds, set $x_t^k = x_t^k + 1$, $d = d - \alpha$ and set $k = k + 1$ if $k < K$ (otherwise, $k = 1$). If $d = 0$ holds, select the next period $t = t + 1$ and set $d = \bar{x}_{it}$. Until $t > T$ holds, commence with the next iteration.

The procedure circularly assigns Q quants with $Q \in \mathbb{N}$ over the periods $t_1, \ldots, t_2$ of the K solutions. Together with K being the denominator of all non-zeros in $(\bar{x}_{i1}, \ldots, \bar{x}_{iT})$, it holds that $\lfloor Q/K \rfloor \leq \sum_{\tau=t_1}^{t_2} x_\tau^k \leq \lceil Q/K \rceil\ \forall i; k$. Two conclusions follow. From the assignment of *at most* $Q = q \cdot K$ quants in the periods $t_1, \ldots, t_2$

---

[4]This holds because of the following well-known argument (Schrijver 1998, pp. 31–32): If the system $A \cdot x = b$ of rational linear equations has a solution, it has one of a size polynomially bounded by the sizes of A and b. It follows that the solution can be represented by a rational number where the numerator and the denominator have a polynomial number of digits in the input length.

[5]The least common denominator of some fractions is the *least common multiple* of the denominators of these fractions. The least common multiple is calculated in polynomial time with the greatest common divisor, which is calculated in polynomial time by Euclid's algorithm (Cormen et al. 2001, pp. 856–862; Wolfart 2011, pp. 3–4).

4.2 Polyhedral Properties

follows that

$$\sum_{\tau=t_1}^{t_2} x_\tau^k \leq q \quad \forall k = 1, \ldots, K. \tag{4.7}$$

From the assignment of *at least* $Q = q \cdot K$ quants in the periods $t_1, \ldots, t_2$ follows that

$$\sum_{\tau=t_1}^{t_2} x_\tau^k \geq q \quad \forall k = 1, \ldots, K. \tag{4.8}$$

'(4.4)': From $\bar{x}$ being feasible and $a_{it} \in \mathbb{N}$ follows that $\lfloor \sum_{\tau=t}^{t+\pi_i-1} \bar{x}_{i\tau} \rfloor \geq a_{it}$ holds. Therefore, at least $Q = \lfloor \sum_{\tau=t}^{t+\pi_i-1} \bar{x}_{i\tau} \rfloor \cdot K$ quants are assigned in the periods $t, \ldots, t + \pi_i - 1$. Set $t_1 = t$ and $t_2 = t + \pi_i - 1$ in (4.8) to obtain $\sum_{\tau=t}^{t+\pi_i-1} x_\tau^k \geq \lfloor \sum_{\tau=t}^{t+\pi_i-1} \bar{x}_{i\tau} \rfloor \geq a_{it} \, \forall t; k$.
'(4.5)': From $\bar{x}$ being feasible and $b_{it} \in \mathbb{N}$ follows that $\lceil \bar{x}_{it} \rceil \leq b_{it}$ holds. Therefore, at most $Q = \lceil \bar{x}_{it} \rceil \cdot K$ quants are assigned in every period t. Set $t_1 = t_2$ in (4.7) to obtain $x_t^k \leq \lceil \bar{x}_{it} \rceil \leq b_{it} \, \forall t; k$.
K feasible integer solutions $x^k \in \text{Conv}(4.4)(4.5)(4.6) \, \forall k$ are obtained. Altogether,

$$\bar{x} = \sum_{k=1}^{K} \alpha \cdot x^k \tag{4.9}$$

is a *strict* convex combination since $\alpha > 0$ and $K \cdot \alpha = 1$. This contradicts that $\bar{x}$ is an extreme point (Nemhauser and Wolsey 1988, p. 93).

Example 4.2 Let $T = 10$, $\pi_i = 5$, $a_{it} = 1 \, \forall t$ and $b_{it} = 1 \, \forall t$ for the maintenance activity i. (P) and (S*) from are obtained from (4.4) and (4.5). In Table 4.2, K feasible integer solutions x^k are yielded from a given feasible fractional solution $\bar{x}$. The least common denominator $K = 4$ is calculated from the fractions $\frac{1}{4}, \frac{1}{2}, \frac{3}{4}$ and the value for the quant is $\alpha = \frac{1}{4}$. $\bar{x}$ is obtained by the convex combination (4.9).

Corollary 4.2 shows that the corresponding dual problem of a LP with the constrains from Theorem 4.1 can have an integral polytope.

Table 4.2 The convex combination (4.9) yields the fractional solution $\bar{x}$

t	1	2	3	4	5	6	7	8	9	10
$\bar{x}_{it}$	0.75		0.5	0.25	0.25			0.75		0.25
x_t^1	1		1					1		
x_t^2	1			1				1		
x_t^3	1				1					1
x_t^4					1			1		

Corollary 4.2 $Conv(4.11)(4.13) = H(4.11)(4.12)$ holds for the LP (4.10)–(4.13) where $a_{it}, c_{it} \in \mathbb{Z} \; \forall t$ and $b_{it} \in \mathbb{N} \; \forall t$ are finite.

$$\max \sum_{t=1}^{T-\pi_i+1} a_{it} \cdot v_{it} - \sum_{t=1}^{T} b_{it} \cdot \alpha_{it} \quad \text{subject to} \quad (4.10)$$

$$\sum_{\tau=\max\{1,t-\pi_i+1\}}^{\min\{t,T-\pi_i+1\}} v_{i\tau} - \alpha_{it} \leq c_{it} \quad \forall t = 1,\ldots,T \quad (4.11)$$

$$v_{it} \geq 0 \; \forall t = 1,\ldots,T-\pi_i+1 \wedge \alpha_{it} \geq 0 \; \forall t = 1,\ldots,T \quad (4.12)$$

$$v_{it} \in \mathbb{N} \; \forall t = 1,\ldots,T-\pi_i+1 \wedge \alpha_{it} \in \mathbb{N} \; \forall t = 1,\ldots,T \quad (4.13)$$

Proof Because of (S), $H(P)(V)(S)$ is bounded and clearly non-empty. Hence, the optimal objective function value of (4.10) is finite for all values of a_{it} and b_{it}. Furthermore, the corresponding dual problem

$$\min \sum_{t=1}^{T} c_{it} \cdot x_{it} \quad \text{subject to}$$

$$\sum_{\tau=t}^{t+\pi_i-1} x_{i\tau} \geq a_{it} \quad \forall t = 1,\ldots,T-\pi_i+1 \quad v_{it}$$

$$-x_{it} \geq -b_{it} \quad \forall t = 1,\ldots,T \quad \alpha_{it}$$

$$x_{it} \geq 0 \quad \forall t = 1,\ldots,T$$

has an optimal integer solution because of Theorem 4.1. Clearly, this still holds when redundant slack variables are added. Therefore, the system of inequalities (4.11) and (4.12) is totally dual integral. Using Lemma 3.6, the polytope is integral since $c_{it} \in \mathbb{Z} \; \forall t$.

The ideas presented in the proof of Theorem 4.1 are applicable to show integrality of other systems of linear inequalities such as special cases of the Minimum Cost Covering Problem (Kolen and Tamir 1990; Kolen 1983) or the system of linear inequalities of Corollary 4.2, which would give a tighter result than Corollary 4.2 does.

The WUPMP has the single-assignment property that is known from the UFLP and the UNDP (Krarup and Pruzan 1983; Balakrishnan et al. 1989). If y is binary and x is continuous, x is implicitly binary because of the following lemma.

4.2 Polyhedral Properties

Lemma 4.3 Single-assignment property: $Conv(P)(V)(X)(Y) = Conv(P)(V)(S)(Y)$

Proof '$\Rightarrow$': $Conv(P)(V)(X)(Y) \subseteq Conv(P)(V)(S)(Y)$ clearly holds.
'$\Leftarrow$': In order to show $Conv(P)(V)(X)(Y) \supseteq Conv(P)(V)(S)(Y)$, assume that $(x, y) \in Conv(P)(V)(S)(Y)$. Because of $y \in \{0, 1\}^T$ being given, the right-hand side of (V) becomes y. $Conv(P)(X) = H(P)(S^x)$ holds by Theorem 4.1 with $a_{it} = 1$ $\forall i; t$ and $b_{it} = y_t$ $\forall i; t$. Therefore, $(x, y) \in Conv(P)(V)(X)(Y)$ holds.

Observation 4.4 Preliminary computational studies revealed that the components of many optimal solutions $(\bar{x}, \bar{y})$ to the WUPMP LP relaxation are integer. Note that the single-assignment property from Lemma 4.3 allows to only consider the variables $\bar{y}$.

The observation is particularly interesting because the polytope $H(P)(V)(S)$ might contain fractional extreme points, which is shown by the following two examples. Similar observations have been made for the UFLP and the UNDP (Krarup and Pruzan 1983; Holmberg and Hellstrand 1998).

Example 4.3 Let $n = 2$, $T = 3$, $\pi = (2, 3)^T$, $f = (\frac{1}{2}, 1, \frac{1}{2})^T$, and $c = \begin{pmatrix} \frac{1}{2} & 1 & \frac{1}{2} \\ \frac{1}{2} & 1 & \frac{1}{2} \end{pmatrix}$. Extreme points of the polytope $H(P)(V)(S)$ are the optimal integer solution $x = \begin{pmatrix} 1 & 0 & 1 \\ 1 & 0 & 0 \end{pmatrix}$, $y = (1, 0, 1)^T$ and the fractional optimal solution $x = \begin{pmatrix} \frac{1}{2} & \frac{1}{2} & \frac{1}{2} \\ \frac{1}{2} & 0 & \frac{1}{2} \end{pmatrix}$, $y_t = \frac{1}{2}$ $\forall t$ that have an objective function value $2\frac{1}{2}$.

The polyhedral findings of Lemmata 4.8–4.10 mentioned below prove that Example 4.3 has the smallest possible dimensions n and T such that the extreme points are fractional. Note that π cannot be changed as well. Example 4.4 presents an instance where the fractional solution has a lower objective function value than the integer solution. Note that this example shows that the WUPMP polytope is not half-integral.

Example 4.4 Let $n = 4$, $T = 4$, $\pi_i = 4$ $\forall i$, $f_t = 1$ $\forall t$ and $c_{it} = 1$ if $i \neq t$ (otherwise, 2) $\forall i; t$. Extreme points of the polytope $H(P)(V)(S)$ are the fractional optimal solution $x_{it} = \frac{1}{3}$ if $i \neq t$ (otherwise, 0) $\forall i; t$ and $y_t = \frac{1}{3}$ $\forall t$, which has an objective function value $5\frac{1}{3}$, and the optimal integer solution $x_{it} = 1$ $\forall i$ and $y_t = 1$ if $t = 1$ (otherwise, 0) $\forall t$, which has an objective function value 6.

Since the objective function values of the fractional and the integer solution coincide, it is exemplary proven in Remark 4.4 that the fractional optimal solution from Example 4.3 is an extreme point of $H(P)(V)(S)$.

Remark 4.4 Let the basic matrix A_B to the coefficient matrix of the constraints (P) and (V) contain the columns of the basic variables $x_{11}, y_1, x_{21}, x_{12}, x_{23}, s, x_{13}, y_2$ and y_3. Denote s as the slack of (V) for $i = 2$ and $t = 2$. The vector of basic variables $x_B \in \mathbb{R}^9_+$ can be obtained by $x_B = A_B^{-1} \cdot b$ (Chvátal 2002, p. 120). This requires the inverse matrix of A_B that is

$$A_B^{-1} = \begin{pmatrix} \frac{1}{2} & -\frac{1}{2} & \frac{1}{2} & -\frac{1}{2} & 0 & -\frac{1}{2} & \frac{1}{2} & 0 & \frac{1}{2} \\ \frac{1}{2} & -\frac{1}{2} & \frac{1}{2} & \frac{1}{2} & 0 & -\frac{1}{2} & \frac{1}{2} & 0 & \frac{1}{2} \\ \frac{1}{2} & -\frac{1}{2} & \frac{1}{2} & \frac{1}{2} & 0 & -\frac{1}{2} & -\frac{1}{2} & 0 & \frac{1}{2} \\ \frac{1}{2} & \frac{1}{2} & -\frac{1}{2} & \frac{1}{2} & 0 & \frac{1}{2} & -\frac{1}{2} & 0 & -\frac{1}{2} \\ -\frac{1}{2} & \frac{1}{2} & \frac{1}{2} & -\frac{1}{2} & 0 & \frac{1}{2} & \frac{1}{2} & 0 & -\frac{1}{2} \\ \frac{1}{2} & \frac{1}{2} & -\frac{1}{2} & \frac{1}{2} & 1 & \frac{1}{2} & -\frac{1}{2} & -1 & -\frac{1}{2} \\ -\frac{1}{2} & \frac{1}{2} & \frac{1}{2} & -\frac{1}{2} & 0 & -\frac{1}{2} & \frac{1}{2} & 0 & \frac{1}{2} \\ \frac{1}{2} & \frac{1}{2} & -\frac{1}{2} & \frac{1}{2} & 1 & \frac{1}{2} & -\frac{1}{2} & 0 & -\frac{1}{2} \\ -\frac{1}{2} & \frac{1}{2} & \frac{1}{2} & -\frac{1}{2} & 0 & \frac{1}{2} & \frac{1}{2} & 0 & \frac{1}{2} \end{pmatrix}$$

as well as the vector of the right-hand side of A_B that is $b = (1, 1, 1, 0, 0, 0, 0, 0, 0)^T$. Let x_N be the non-basic variable $x_N = x_{22} = 0$. The basic feasible solution $\binom{x_B}{x_N}$ coincides with the fractional optimal solution of Example 4.3. A basic feasible solution is an extreme point and therefore, the fractional optimal solution from Example 4.3 is an extreme point of $H(P)(V)(S)$.

Remark 4.5 allows for an explanation of the Observation 4.4. A similar statement holds for the UFLP and the UNDP (Krarup and Pruzan 1983; Holmberg and Hellstrand 1998).

Remark 4.5 The extreme points of $Conv(P)(V)(X)(Y)$ are extreme points of $H(P)(V)(S)$. This holds because the $\{0, 1\}$-hypercube of (X) and (Y) contains $H(P)(V)(S)$.

Quasi-integrality is a tighter property than Remark 4.5 and it is useful for the development of integral Simplex algorithms. The UFLP and the UNDP have quasi-integral polytopes (Yemelichev et al. 1984, pp. 192–193; Hellstrand et al. 1992). In what follows, the novel result is presented that the polytope of the WUPMP $H(P)(V)(S)$ is quasi-integral. Using an approach comparable to Yemelichev et al. (1984, pp. 191–193), the proof requires the reformulation $\bar{Z}$ of the WUPMP to an extended UFLP $\bar{Z}$.[6] Constraint (P) is disaggregated such that each task horizon is explicitly modeled and the stairway structure of (P) is reflected in the transportation costs. This extended UFLP comprises $T + 1$ facilities and $J = n \cdot T$ customers. To each maintenance activity $i = 1, \ldots, n$ is assigned a set of customers $I_i = \{(i-1) \cdot T + 1, \ldots, i \cdot T\}$ that represent the T periods, which have to be covered by a real period $t = 0, \ldots, T$. The real periods are interpreted as facilities. The transportation costs between customers and facilities are $h_{jt} = 0$ if $j \in I_i$ and $t \in \{j - (i-1) \cdot T - \pi_i + 1, \ldots, j - (i-1) \cdot T\}$ (otherwise, ∞) $\forall i = 1, \ldots, n; j = 1, \ldots, J; t = 0, \ldots, T$. Specifically, $h_{jt} = 0$ represents an assignment within a task horizon and the period $(j - 1 \mod T) + 1$ of the maintenance activity $i = \lceil \frac{j}{T} \rceil$ is covered by a real period t. If the customer $j = 1, \ldots, J$ is served from facility $t = 0, \ldots, T$, the binary variable z_{jt} equals 1 (otherwise, 0) and the maintenance

---

[6]Note that the facility-transformation from Sect. 5.3.3.2 is substantially different.

4.2 Polyhedral Properties

activity $i = \lceil \frac{i}{T} \rceil$ is scheduled in period t. All other variables and parameters are taken from the WUPMP.

$$\bar{Z} = \min \sum_{t=1}^{T} f_t \cdot y_t + \sum_{i=1}^{n} \sum_{t=1}^{T} c_{it} \cdot x_{it} + \sum_{j=1}^{J} \sum_{t=0}^{T} h_{jt} \cdot z_{jt} \quad \text{subject to} \tag{4.14}$$

$$\sum_{t=0}^{T} z_{jt} = 1 \qquad \forall j = 1, \ldots, J \tag{4.15}$$

$$z_{jt} \leq x_{it} \qquad \forall i = 1, \ldots, n; j \in I_i; t = 1, \ldots, T \tag{4.16}$$

$$x_{it} \leq y_t \qquad \forall i = 1, \ldots, n; t = 1, \ldots, T \tag{4.17}$$

$$0 \leq x_{it} \leq 1 \qquad \forall i = 1, \ldots, n; t = 1, \ldots, T \tag{4.18}$$

$$0 \leq y_t \leq 1 \qquad \forall t = 1, \ldots, T \tag{4.19}$$

$$0 \leq z_{jt} \leq 1 \qquad \forall j = 1, \ldots, J; t = 0, \ldots, T \tag{4.20}$$

$$x_{it} \in \{0, 1\} \qquad \forall i = 1, \ldots, n; t = 1, \ldots, T \tag{4.21}$$

$$y_t \in \{0, 1\} \qquad \forall t = 1, \ldots, T \tag{4.22}$$

$$z_{jt} \in \{0, 1\} \qquad \forall j = 1, \ldots, J; t = 0, \ldots, T \tag{4.23}$$

The total costs are minimized by the objective function (4.14). Each customer is satisfied by exactly one facility because of (4.15). If this is done at zero cost, at least one maintenance activity appears per task horizon because of (4.16) and hence, (P) is satisfied. (4.17) represents (V). The LP relaxation requires (4.18), (4.19) and (4.20). There are integer variable domains (4.21), (4.22) and (4.23). A feasible solution to $\bar{Z}$ with a finite $\bar{Z}$ relates to a feasible solution of the WUPMP and the other way around (see Example 4.5). Note that f and c are finite.

- Every integer (fractional) solution (x, y, z) in $H(4.15)$–(4.20) corresponds to one integer (one fractional) solution (x, y) in $H(P)(V)(S)$.
- Every integer (fractional) solution (x, y) in $H(P)(V)(S)$ corresponds to one integer (one fractional) solution (x, y) in $H(4.15)$–(4.20) and at least to one integer (at least one fractional) solution (x, y, z) in $H(4.15)$–(4.20).

Example 4.5 Let $n = 2$, $T = 6$ and $\pi = (3, 5)^T$. It follows $I_1 = \{1, \ldots, 6\}$ and $I_2 = \{7, \ldots, 12\}$. A feasible solution to (4.15)–(4.23) and the transportation costs are given in Table 4.3. An entry (j, t) given in italic stands for $z_{jt} = 1$ (otherwise, 0). This yields $x_{13} = x_{14} = x_{23} = 1$ with $y_3 = y_4 = 1$.

The following property that is similar to Remark 4.5 holds for (4.15)–(4.23).

Remark 4.6 The extreme points of $Conv(4.15)$–(4.23) are extreme points of $H(4.15)$–(4.20). This holds because the $\{0, 1\}$-hypercube of (4.21)–(4.23) contains $H(4.15)$–(4.20).

Table 4.3 The transformation of the WUPMP to an extended UFLP

j	t						
	0	1	2	3	4	5	6
1	0	0	∞	∞	∞	∞	∞
2	0	0	0	∞	∞	∞	∞
3	∞	0	0	0	∞	∞	∞
4	∞	∞	0	0	0	∞	∞
5	∞	∞	∞	0	0	0	∞
6	∞	∞	∞	∞	0	0	0
7	0	0	∞	∞	∞	∞	∞
8	0	0	0	∞	∞	∞	∞
9	0	0	0	0	∞	∞	∞
10	0	0	0	0	0	∞	∞
11	∞	0	0	0	0	0	∞
12	∞	∞	0	0	0	0	0

Theorem 4.5 The polytope $H(P)(V)(S)$ is quasi-integral.

Proof The proof is conducted in two steps. First, Lemma 3.9 is applied to prove quasi-integrality for $H(4.15)$–(4.20). Since Remark 4.6 holds, it remains to show that two arbitrary integral extreme points $s^1 = (x^1, y^1, z^1)$ and $s^2 = (x^2, y^2, z^2)$ with $s^1, s^2 \in H(4.15)$–(4.20) are contained in an integral face. Define three disjoint index sets where $d = N \cdot T + T + J \cdot (T + 1)$ and

$$L_0 = \{l \in \mathbb{N}^d | s_l^1 = 0 \wedge s_l^2 = 0\}$$
$$L_1 = \{l \in \mathbb{N}^d | s_l^1 = 1 \wedge s_l^2 = 1\}$$
$$L_F = \{l \in \mathbb{N}^d | s_l^1 \neq s_l^2\}.$$

The hyperplanes $\{s \in H(4.15)$–$(4.20) | s_l = 0 \; \forall l \in L_0\}$ and $\{s \in H(4.15)$–$(4.20) | s_l = 1 \; \forall l \in L_1\}$ support the polytope $H(4.15)$–(4.20).

$$F = \{s \in H(4.15)\text{–}(4.20) | s_l = 0 \; \forall l \in L_0 \wedge s_l = 1 \; \forall l \in L_1\}$$

is a face of the polytope.

Integrality of F is proven with total unimodularity. Let A be the coefficient matrix of (4.15)–(4.20) that contains no slack variables and just the columns from L_F. In other words, all columns are removed where $s_l^1 = s_l^2$ holds. A is a $\{-1, 0, 1\}$ matrix.

In what follows it is shown that every row of A contains at most two non-zero elements. In constraint (4.15), there exist two indices $\tau, \theta = 0, \ldots, T$ such that $z_{j\tau}^1 = z_{j\theta}^2 = 1$ because s^1 and s^2 are feasible integer solutions. If $\tau = \theta$, then $z_{j\tau}^1 = z_{j\tau}^2 \; \forall t = 0, \ldots, T$ holds. The respective row contains only zero elements because A does not contain columns where the variables are equal. Otherwise, $\tau \neq \theta$

4.2 Polyhedral Properties

holds and $z^1_{j\tau} = 1 - z^2_{j\tau} = 1$ and $z^2_{j\theta} = 1 - z^1_{j\theta} = 1$ follows. Both variables are unequal and the row has two "+1"-elements.

Constraint (4.16) is stated as $z_{jt} - x_{it} \leq 0$ in A. The following four cases apply to an index $t = 0, \ldots, T$. If $z^1_{jt} = 1 - z^2_{jt}$ and $x^1_{it} = 1 - x^2_{it}$ holds, the row has one "+1"-element and one "−1"-element. If $z^1_{jt} = z^2_{jt}$ and $x^1_{it} = 1 - x^2_{it}$ holds, the row has one "−1"-element. If $z^1_{jt} = 1 - z^2_{jt}$ and $x^1_{it} = x^2_{it}$ holds, the row has one "+1"-element. Otherwise, $z^1_{jt} = z^2_{jt}$ and $x^1_{it} = x^2_{it}$ holds and the row contains only zero elements because A does not have columns with equal variables.

The same arguments apply to (4.17). The constraints (4.18), (4.19), and (4.20) clearly have either rows with one "+1"-element or with only zero elements.

The columns are partitioned as follows. Let A_1 comprise all columns $l \in L_F$ with $s^1_l = 1$ and $s^2_l = 0$. Let A_2 contain all columns with $s^1_l = 0$ and $s^2_l = 1$. Note that the remaining columns are not in A because the variables are equal. In constraint (4.15), the two "+1"-elements per row are in different sets A_1 and A_2 because each "+1"-element comes from a different integer solution. Consider the constraints (4.16) and (4.17). The "+1"-element and the "−1"-element are assigned to the same set that is either A_1 or A_2 because both elements come from the same integer solution. By Lemma 3.7, A is totally unimodular that still holds if slack variables are added. Therefore, Lemma 3.8 holds and the face F is integral. Because of Lemma 3.9, H(4.15)–(4.20) is quasi-integral.

Consider the WUPMP. Quasi-integrality for $H(P)(V)(S)$ is proven with the result from above and Lemma 3.9. Since Remark 4.5 holds, let s'^1 and s'^2 be two integral extreme points of $H(P)(V)(S)$. Let the face F' be constructed as stated above for H(4.15)–(4.20). Assuming that F' is not an integral face, let $(\bar{x}, \bar{y}) \in F'$ be an extreme point where at least one component $0 < \bar{x}_{it} < 1$ or $0 < \bar{y}_t < 1$ is fractional. Since $(\bar{x}, \bar{y})$ is feasible, the corresponding solutions (x, y, z) of $\bar{Z}$ have a finite objective function value. Therefore, there exists an extreme point (x, y, z) in F with fractional components $0 < x_{it} < 1$ or $0 < y_t < 1$. This contradicts the previously proven fact that the face F is integral. Thus, $(\bar{x}, \bar{y})$ does not exist and F' is integral. Using Lemma 3.9, $H(P)(V)(S)$ is quasi-integral.

The symmetry in (4.14)–(4.23) is broken by the following two constraints.

$$x_{it} = z_{t+(i-1) \cdot T, t} \quad \forall i = 1, \ldots, n; t = 1, \ldots, T \quad (4.24)$$

$$z_{j-1,t} \geq z_{jt} \quad \begin{array}{l} \forall i = 1, \ldots, n; t = 0, \ldots, T; \\ j = \max\{2, t+1\} + (i-1) \cdot T, \ldots, i \cdot T \end{array} \quad (4.25)$$

Constraint (4.24) is implied because a scheduled maintenance activity i opens a facility t that serves some customers in I_i at zero cost. If one or more tasks are scheduled such that a customer can be served by more than one facility at zero costs, the customer is served by the rightmost facility because of constraint (4.25). Therefore, to every integer point (x, y) corresponds one integer point (x, y, z). In Example 4.5, set $z_{53} = 0$ and $z_{54} = 1$ to satisfy (4.24) and (4.25).

The WUPMP can be transformed to a SetPP by applying (4.14)–(4.23). Hence, the WUPMP is solvable by any algorithm that solves the SetPP. Define complementary variables $x_{it}^c = 1 - x_{it}$ $\forall i; t$, $y_t^c = 1 - y_t$ $\forall t$, two slack variables s'_{ijt}, s''_{it} (Krarup and Pruzan 1983) and obtain

$$\min \sum_{t=1}^{T} f_t \cdot y_t + \sum_{i=1}^{n} \sum_{t=1}^{T} c_{it} \cdot x_{it} + \sum_{j=1}^{J} \sum_{t=0}^{T} h_{jt} \cdot z_{jt} \quad \text{subject to} \qquad (4.26)$$

$$\sum_{t=0}^{T} z_{jt} = 1 \qquad \forall j = 1, \ldots, J \qquad (4.27)$$

$$z_{jt} + x_{it}^c + s'_{ijt} = 1 \qquad \forall i = 1, \ldots, n; j \in I_i; t = 1, \ldots, T \qquad (4.28)$$

$$x_{it} + y_t^c + s''_{it} = 1 \qquad \forall i = 1, \ldots, n; t = 1, \ldots, T \qquad (4.29)$$

$$x_{it} + x_{it}^c = 1 \qquad \forall i = 1, \ldots, n; t = 1, \ldots, T \qquad (4.30)$$

$$y_t + y_t^c = 1 \qquad \forall t = 1, \ldots, T \qquad (4.31)$$

$$s'_{ijt} \in \{0, 1\} \qquad \forall i = 1, \ldots, n; j \in I_i; t = 1, \ldots, T \qquad (4.32)$$

$$s''_{it} \in \{0, 1\} \qquad \forall i = 1, \ldots, n; t = 1, \ldots, T \qquad (4.33)$$

$$x_{it}, x_{it}^c \in \{0, 1\} \qquad \forall i = 1, \ldots, n; t = 1, \ldots, T \qquad (4.34)$$

$$y_t, y_t^c \in \{0, 1\} \qquad \forall t = 1, \ldots, T \qquad (4.35)$$

$$z_{jt} \in \{0, 1\} \qquad \forall j = 1, \ldots, J; t = 0, \ldots, T. \qquad (4.36)$$

4.3 Computational Complexity

Table 4.4 presents findings about the complexity of the WUPMP.[7] The WUPMP is strongly $\mathcal{NP}$-hard even if $\pi_i = T$ $\forall i \wedge f_t = f_1$ $\forall t$ holds (Lemma 4.6). Let L be the input length of n, T and the $\{-1, 0, 1\}$ coefficient matrix. The case $T = 2$ is solvable by Lemma 4.10.

Lemma 4.6 The WUPMP is strongly $\mathcal{NP}$-hard even if $\pi_i = T$ $\forall i = 1, \ldots, n \wedge f_t = f_1$ $\forall t = 1, \ldots, T$.

Proof A polynomial reduction from the strongly $\mathcal{NP}$-complete SET COVER shows strong $\mathcal{NP}$-hardness of the WUPMP that is SET COVER $\leq_p$ WUPMP$_{\pi_i=T \wedge f_i=f_1}$.

---

[7]Improved results such as $O(n)$ for $T \in O(1)$ and $O(T \cdot \log T)$ for $n = 1$ are found in Kuschel and Bock (2016).

4.3 Computational Complexity

Table 4.4 Computational complexity of uncapacitated planned maintenance problems

Problem variant	Computational complexity	Proof
$\pi_i = T \ \forall i \wedge f_t = f_1 \ \forall t$	Strongly $\mathcal{NP}$-hard	Lemma 4.6
General case	$O(n \cdot T^{n+1} \cdot 2^n)$	Lemma 4.7
$n = k \in O(1)$	$O(T^{k+1})$	Lemma 4.7
$n = 1$	$O(T^2)$	Lemma 4.8
$\pi_i = 2 \ \forall i$	$O(n^{3.5} \cdot T^{3.5} \cdot L)$	Lemma 4.10
$f_t = 0 \ \forall t$	$O(n \cdot T^2)$	Lemma 4.11
$c_{it} = 0 \ \forall i; t$	$O(n \cdot T + T^2)$	Lemma 4.12
$f_t = f_1 \ \forall t \wedge c_{it} = 0 \ \forall i; t$	$O(n \cdot T)$	Lemma 4.13

SET COVER (Garey and Johnson 1979, p. 222): Given S elements, P sets $C_1, \ldots, C_P$ with $C_t \subseteq \{1, \ldots, S\} \ \forall t$ and a positive integer K with $K \leq P$. Exists an index set A with $A \subseteq \{1, \ldots, P\}$ and $|A| \leq K$ such that $\bigcup_{t \in A} C_t = \{1, \ldots, S\}$?

An instance of the WUPMP is obtained in polynomial time with $n = S$, $T = P$, $\pi_i = T \ \forall i, f_t = 1 \ \forall t$ and $c_{it} = 0$ if $i \in C_t$ (otherwise, $K+1$) $\forall i; t$. The WUPMP instance is called solvable if it is feasible and the objective function value is smaller or equal to K. Equivalence of both instances is proven by showing that the instance to SET COVER is solvable if and only if the instance to the WUPMP is solvable.

'$\Rightarrow$': Assume that the instance to SET COVER is solvable. Hence, there exists a set A that satisfies SET COVER with $|A| \leq K$. Let $x_{it} = 1$ if $i \in C_t$ (otherwise, 0) $\forall i; t$ and $y_t = 1$ if $t \in A$ (otherwise, 0) $\forall t$ be a solution to the WUPMP. The solution has an objective function value of $|A|$. In SET COVER, each element $1, \ldots, S$ is included in at least one set C_t where $t \in A$. Since $\pi_i = T$, (P) is satisfied. (V) holds because the ith element is either included in the tth set C_t or not. Therefore, (x, y) is feasible to the WUPMP and has an objective function value smaller of equal to K.

'$\Leftarrow$': Assume that the instance to the WUPMP is solvable. Hence, a feasible solution (x, y) to the WUPMP exists that has an objective function value smaller of equal to K. Let $A = \{t = 1, \ldots, T \mid y_t = 1\}$ be a solution to SET COVER. This solution clearly has $|A| \leq K$. Assume the existence of an element i with $\{i\} \notin \bigcup_{t \in A} C_t$. Feasibility to the WUPMP with respect to (P) requires that there exists a period t where $x_{it} = 1$ and $c_{it} > K$. This contradicts that (x, y) has an objective function value smaller or equal to K and thus, A satisfies SET COVER.

Lemma 4.7 The WUPMP is solvable in time $O(n \cdot T^{n+1} \cdot 2^n)$. If additionally $n = k \in O(1)$, the WUPMP is strongly polynomially solvable in time $O(T^{k+1})$. Moreover, there exists an optimal integral extreme point in $H(P)(V)(S)$.

Proof '$O(n \cdot T^{n+1} \cdot 2^n)$': The WUPMP is transformed into a SPP.[8] Let $G(V, E)$ be a weighted, directed graph with V as the node set and E as the arc set. A node is a partial solution $(t, \alpha_1, \ldots, \alpha_n)$ with $1 \leq \alpha_i \leq \pi_i$ $\forall i = 1, \ldots, n$ and $t = 1, \ldots, T$. An entry α_i stands for the remaining coverage of a maintenance activity i. The remaining coverage in period t is the number of periods that the preceding maintenance activity in period τ covers (i.e., $\alpha_i = \pi_i + \tau - t$). Each node is evaluated by the function $F(t, \alpha_1, \ldots, \alpha_n)$ that yields the minimal total maintenance cost to attain the constellation $\alpha_1, \ldots, \alpha_n$. An initial source node $(0, \alpha_1^{init} = \pi_1, \ldots, \alpha_n^{init} = \pi_n)$ that has zero costs and a final sink node $(T + 1, \alpha_1^{final}, \ldots, \alpha_n^{final})$ are introduced. Because of $\pi_i \leq T$ $\forall i$, each layer $t = 1, \ldots, T$ contains at most T^n nodes and the number of nodes $|V|$ in the graph is asymptotically upper bounded by $O(T^{n+1})$.

Each node $(t, \alpha_1, \ldots, \alpha_n)$ per layer $t = 0, \ldots, T - 1$ has a set of successors $S^{succ}(t, \alpha_1, \ldots, \alpha_n)$ that contains reachable feasible solutions of the next layer $t + 1$. In particular, define $S^{succ}(t, \alpha_1, \ldots, \alpha_n) = \{(t + 1, \beta_1, \ldots, \beta_n) \mid \beta_i \in \{\pi_i, \alpha_i - 1\}$ $\forall i \wedge \beta_i \geq 1$ $\forall i\}$. Note that $\beta_i \geq 1$ $\forall i$ ensures that (P) is satisfied and subsequently, every period $1, \ldots, t+1$ is covered. If $\beta_i = \pi_i$ holds, the maintenance activity i is scheduled in period $t + 1$. The maintenance vector $M(t + 1, \beta) = (m_1(\beta), \ldots, m_n(\beta))$ contains the binary components $m_i(\beta) = 1$ if $\beta_i = \pi_i$ (otherwise, 0) $\forall i$. Let $C(t + 1, \beta)$ be the arc weight of the arc that connects the node $(t, \alpha_1, \ldots, \alpha_n)$ with the succeeding node $(t + 1, \beta_1, \ldots, \beta_n)$. The total maintenance costs in period $t+1$ yield $C(t+1, \beta) = \max_{i=1,\ldots,n} m_i(\beta) \cdot f_{t+1} + \sum_{i=1}^n m_i(\beta) \cdot c_{i,t+1}$. Furthermore, each node $(T, \alpha_1, \ldots, \alpha_n)$ in the layer T yields a feasible solution and is connected with the final sink node $(T + 1, \alpha_1^{final}, \ldots, \alpha_n^{final})$ by an arc with an arc weight that is zero.

Consider the number of arcs $|E|$ in the graph. The total number to execute k maintenance activities with $k \leq n$ is calculated by $\binom{n}{k}$.[9] Since there is no restriction on the maximal number of maintenance activities per period, $\sum_{k=0}^n \binom{n}{k} = 2^n$ is an upper bound on the number of successors $|S^{succ}(t, \alpha_1, \ldots, \alpha_n)|$ of a given node in the layer $t = 0, \ldots, T - 1$. Thus, the number of arcs $|E|$ is asymptotically upper bounded by $O(|V| \cdot 2^n) = O(T^{n+1} \cdot 2^n)$.

A node is generated in time $O(n)$ and a single successor is obtained from a given node in time $O(n)$. Therefore, the entire graph $G(V, E)$ is constructed in time $O(n \cdot |V| + n \cdot |E|)$. Calculate a shortest path from the source node $(0, \alpha_1^{init}, \ldots, \alpha_n^{init})$ to the sink node $(T+1, \alpha_1^{final}, \ldots, \alpha_n^{final})$ in time $O(|V| + |E|)$ because $G(V, E)$ is cycle-free

---

[8] Given a directed graph $G(V, E)$ with a node set V, a set E of directed arcs, arc costs f_{ij} $\forall (i, j) \in E$ and two nodes s and t with $s, t \in V$. Find a path of minimal costs in $G(V, E)$ that leads from the node s to t. The *Shortest Path Problem (SPP)* in graphs without negative cycles is solvable in polynomial time (Ahuja et al. 1993, pp. 121–123, pp. 154–157) but it is $\mathcal{NP}$-complete for general cost structures (Garey and Johnson 1979, p. 213).

[9] The *binomial coefficient* $\binom{n}{k}$ yields the number of distinct subsets with k elements that can be formed from a set with n elements. It holds that $\binom{n}{k} = \frac{n!}{k! \cdot (n-k)!}$ if $k \leq n$ (otherwise, 0) (Mood et al. 1974, p. 529). A special case of the binomial theorem is $\sum_{k=0}^n \binom{n}{k} = 2^n$ (Mood et al. 1974, p. 530).

4.3 Computational Complexity

(Ahuja et al. 1993, pp. 107–108). Using the shortest path, an optimal solution (x, y) to the WUPMP is generated in time $O(n \cdot T)$. Thus, the WUPMP is solvable in time $O(n \cdot |V| + n \cdot |E|) + O(|V| + |E|) + O(n \cdot T) = O(n \cdot |V| + n \cdot |E| + n \cdot T) = O(n \cdot T^{n+1} \cdot 2^n)$.

'$O(T^{k+1})$': If $n = k \in O(1)$, the running time $O(T^{k+1})$ is strongly polynomial.

Because of Remark 4.5, the optimal solution (x, y) is an extreme point of $Conv(P)(V)(X)(Y)$ and therefore, an extreme point of $H(P)(V)(S)$. □

Lemma 4.8 If $n = 1$, the WUPMP is strongly polynomially solvable in time $O(T^2)$ and $Conv(P)(V)(X)(Y) = H(P)(V)(S)$ holds.

Proof The WUPMP is transformed with Remark 4.2 into a SPP that is solvable in time $O(|V| + |E_1|)$ because $G_1(V, E_1)$ is cycle-free (Ahuja et al. 1993, pp. 107–108). Let $f_t + c_{1t}$ $\forall t$ be the weight of an arc $(t, \tau) \in E_1$ if $t \neq 0$ (otherwise, 0). An optimal integer solution x obtained from the shortest path and calculated in time $O(T + |E_1|)$ because the number of nodes $|V|$ is asymptotically upper bounded by $O(T)$. Consider the number of arcs $|E_1|$. Let the nodes $0, \ldots, T+1$ have the maximal out-degree π_1 and subtract the number of arcs that do not exist. It follows that π_1 outgoing arcs apply to the nodes $0, \ldots, T - \pi_1 + 1$ and $T + 1 - t$ outgoing arcs to a node $t = T - \pi_1 + 2, \ldots, T + 1$.

$$|E_1| = (T + 2) \cdot \pi_1 - \sum_{t=1}^{\pi_1} t = T \cdot \pi_1 + \frac{3}{2} \cdot \pi_1 - \frac{1}{2} \cdot \pi_1^2$$

holds because of $\sum_{t=1}^{\pi_1} t = \frac{\pi_1^2 + \pi_1}{2}$. The Assumptions (2.2) and (2.5) state $2 \leq \pi_1 \leq T$. The largest number of arcs $|E_1|$ is obtained for $\pi_1 = T$. Thus, the WUPMP is solvable in time $O(T + |E_1|) = O(\frac{1}{2} \cdot T^2 + \frac{5}{2} \cdot T) = O(T^2)$.

'$\Rightarrow$': Since $H(P)(V)(S)$ is convex, $Conv(P)(V)(X)(Y) \subseteq H(P)(V)(S)$ holds.

'$\Leftarrow$': Assume a feasible extreme point $(\bar{x}, \bar{y}) \in H(P)(V)(S)$ where at least one component $0 < \bar{y}_t < 1$ or $0 < \bar{x}_{1t} < 1$ is fractional. (S^y) becomes redundant because of $n = 1$. Hence, $\bar{y}_t \in \{0, 1\}$ holds and Theorem 4.1 with $a_{1t} = 1$ $\forall t$ and $b_{1t} = 1$ $\forall t$ shows that $H(P)(S^x)$ has integer extreme points. Hence, a fractional component $0 < \bar{x}_{1t} < 1$ yields a contradiction. Consequently, $(\bar{x}, \bar{y})$ is an integer extreme point and $(\bar{x}, \bar{y}) \in Conv(P)(V)(X)(Y)$ holds. Thus, each feasible solution in $H(P)(V)(S)$ is obtained from a convex combination of integer extreme points of $Conv(P)(V)(X)(Y)$ and $Conv(P)(V)(X)(Y) \supseteq H(P)(V)(S)$ holds. □

Lemma 4.9 If $n = 2 \wedge \pi_i = T$ $\forall i = 1, \ldots, n$, $Conv(P)(V)(X)(Y) = H(P)(V)(S)$ holds.

Proof Denote A as the coefficient matrix of (P) and (V) that is presented in Table 4.5. The matrix A is a $\{-1, 0, 1\}$ matrix and it has at most two non-zero elements per column. An asterisk in Table 4.5 represents the assignment of the rows of A to the sets A_1 and A_2. By Lemma 3.7, A is totally unimodular. The coefficient matrix of (P), (V), (S) including all slack variables satisfies Lemma 3.8 and thus, $Conv(P)(V)(X)(Y) = H(P)(V)(S)$. □

Table 4.5 Coefficient matrix of (P) and (V) for $n = 2$ and $\pi_i = T \ \forall i$

	x_{11}	...	x_{1T}	x_{21}	...	x_{2T}	y_1	...	y_T	A_1	A_2
(P)	1	...	1							*	
				1	...	1					*
(V)	−1						1			*	
		⋱						⋱			⋮
			−1						1	*	
				−1			1				*
					⋱			⋱			⋮
						−1			1		*

The result of Lemma 4.9 allows to obtain an optimal integer solution for $n = 2 \wedge \pi_i = T \ \forall i$ by solving the LP relaxation with a variant of Tardos' algorithm (Vavasis and Ye 1996) in time $O(n^{3.5} \cdot T^{3.5} \cdot L) = O(T^{3.5} \cdot L)$ (see Lemma 4.10). However, Lemma 4.7 yields a significantly lower running time of $O(T^3)$ for $n = 2$.

Lemma 4.10 If $\pi_i = 2 \ \forall i = 1, \ldots, n$, the WUPMP is strongly polynomially solvable in time $O(n^{3.5} \cdot T^{3.5} \cdot L)$ where L is the input length of n, T and of the $\{-1, 0, 1\}$ coefficient matrix. Moreover, $Conv$(P)(V)(X)(Y) = H(P)(V)(S) holds.

Proof Denote A as the coefficient matrix of (P) and (V) that is shown in Table 4.6 for the ith maintenance activity. The matrix A is a $\{-1, 0, 1\}$ matrix and it has at most two non-zero elements per row. An asterisk in Table 4.6 represents the assignment of the rows of A to the sets A_1 and A_2. By Lemma 3.7, A is totally unimodular. The coefficient matrix of (P), (V), (S) including all slack variables satisfies Lemma 3.8 and thus, $Conv$(P)(V)(X)(Y) = H(P)(V)(S). The LP relaxation consists of $n \cdot T + T$ structure variables and $3 \cdot n \cdot T + T - n$ slack variables as well as rows. Therefore, the coefficient matrix has more columns than rows and the LP relaxation of the WUPMP is solvable by a variant of Tardos' algorithm (Vavasis and Ye 1996). Considering, the worst case running time with the total number of variables, it holds that

$$(4 \cdot n \cdot T + 2 \cdot T - n)^{3.5} \cdot L \leq d \cdot n^{3.5} \cdot T^{3.5} \cdot L$$

with the constant $d \geq \left(4 + \frac{2}{n} - \frac{1}{T}\right)^{3.5} > 0$ for all $n, T \geq 1$. Thus, the WUPMP is solvable in time $O((4 \cdot n \cdot T + 2 \cdot T - n)^{3.5} \cdot L) = O(n^{3.5} \cdot T^{3.5} \cdot L)$.

Lemmas 4.9 and 4.10 show that Example 4.3 provides the smallest, possible instance with fractional extreme points.

Lemma 4.11 If $f_t = 0 \ \forall t = 1, \ldots, T$, the WUPMP is strongly polynomially solvable in time $O(n \cdot T^2)$ and there exists an optimal integer extreme point in H(P)(V)(S).

4.3 Computational Complexity

Table 4.6 Coefficient matrix of (P) and (V) for $\pi_i = 2\ \forall i$

		x_{i1}	x_{i2}	x_{i3}	x_{i4}	...	y_1	y_2	y_3	y_4	...
(P)		1	1								
			1	1							
				1	1						
					1	⋱					
						⋱					
(V)		-1					1				
			-1					1			
				-1					1		
					-1					1	
						⋱					⋱
	A_1	*		*		...	*		*		...
	A_2		*		*	...		*		*	...

Proof Because of $f_t = 0\ \forall t$, the variables y are redundant and hence, there exists an optimal solution to the WUPMP where (V) is relaxed completely. Omitting (V), yields n subproblems and each subproblem is solved to optimality in time $O(T^2)$ (Lemma 4.7). Therefore, values for x are calculated in time $n \cdot O(T^2) = O(n \cdot T^2)$. Set $y_t = \max_{i=1,\ldots,n} x_{it}\ \forall t$ in time $O(n \cdot T)$. The solution (x, y) is optimal and integer. Thus, the WUPMP is solvable in time $O(n \cdot T^2) + O(n \cdot T) = O(n \cdot T^2)$.

(x, y) is an extreme point of $Conv(P)(V)(X)(Y)$ and thus, also of $H(P)(V)(S)$ (Remark 4.5).

Lemma 4.12 If $c_{it} = 0\ \forall i = 1,\ldots,n; t = 1,\ldots,T$, the WUPMP is strongly polynomially solvable in time $O(n \cdot T + T^2)$ and there exists an optimal integer extreme point in $H(P)(V)(S)$. For the constraint

$$x_{it} = y_t \quad \forall i = 1,\ldots,n; t = 1,\ldots,T \tag{4.37}$$

it holds that $Conv(P)(4.37)(X)(Y) = H(P)(4.37)(S)$ and all extreme points of $Conv(P)(4.37)(X)(Y)$ are extreme points of $H(P)(V)(S)$.

Proof Because of $c_{it} = 0\ \forall i; t$, the variables x are redundant and hence, there exists an optimal solution to the WUPMP where (4.37) is satisfied. Construct an instance to the WUPMP with $n = 1$ with the objective function coefficients $f_t\ \forall t$. Solve this WUPMP in time $O(T^2)$ and yield values for y (Lemma 4.7). Set $x_{it} = y_t\ \forall i; t$ in accordance to (4.37) in time $O(n \cdot T)$. The solution (x, y) is optimal and integer. Thus, the WUPMP is solvable in time $O(T^2) + O(n \cdot T) = O(T^2 + n \cdot T)$.

(x, y) is an extreme point of $Conv(P)(V)(X)(Y)$ and thus, also of $H(P)(V)(S)$ (Remark 4.5).

Using constraint (4.37), substitute the variables x by the variables y. $Conv(P)(4.37)(X)(Y) = H(P)(4.37)(S)$ follows from Theorem 4.1 with $a_{it} = 1\ \forall t$ and $b_{it} = 1\ \forall t$. Since $Conv(P)(4.37)(X)(Y) \subseteq Conv(P)(V)(X)(Y)$ holds, an extreme point of Conv(P)(4.37)(X)(Y) is an extreme point of Conv(P)(V)(X)(Y) and thus, also an extreme point of H(P)(V)(S) from Remark 4.5.

Strong polynomial solvability as presented in Lemmas 4.11 and 4.12 can be obtained from polyhedral arguments that use Theorem 4.1. Furthermore, similar results hold for the UFLP and the UNDP. The UNDP yields Shortest Path Problems if the fixed costs are zero. Unlike the WUPMP, the UNDP with zero flow costs remains strongly $\mathcal{NP}$-hard because of the strongly $\mathcal{NP}$-complete STEINER TREE IN GRAPHS (Magnanti and Wong 1984). However, the UNDP with zero flow costs is polynomially solvable for the two- and three-commodity case (Ng et al. 2004; Sastry 2000) and for a complete demand pattern (Magnanti and Wong 1984). Trivial selection problems are obtained for the UFLP with zero fixed costs or zero transportation costs. The transformation (4.14)–(4.23) shows the close relation between both problems. If $f_t = 0 \ \forall t$, the WUPMP decomposes into n UFLPs

$$\sum_{i=1}^{n} \min \sum_{t=1}^{T} c_{it} \cdot x_{it} + \sum_{j \in I_i} \sum_{t=0}^{T} h_{jt} \cdot z_{jt} \quad \text{subject to } (4.15) \wedge (4.16) \wedge (4.21) \wedge (4.23).$$

If $c_{it} = 0 \ \forall i; t$, (4.37) holds and the WUPMP becomes the UFLP

$$\min \sum_{t=1}^{T} f_t \cdot y_t + \sum_{j=1}^{J} \sum_{t=0}^{T} h_{jt} \cdot z_{jt} \quad \text{subject to}$$

$z_{jt} \leq y_t \ \forall j = 1, \ldots, J; t = 1, \ldots, T \wedge (4.15) \wedge (4.22) \wedge (4.23).$

These UFLPs are polynomially solvable because of the specific cost structure (Jones et al. 1995).

Lemma 4.13 *If $f_t = f_1 \ \forall t = 1, \ldots, T \wedge c_{it} = 0 \ \forall i = 1, \ldots, n; t = 1, \ldots, T$, the WUPMP is strongly polynomially solvable in time $O(n \cdot T)$ and there exists an optimal integer extreme point in H(P)(V)(S). The optimal objective function value is $f_1 \cdot \lfloor \frac{T}{\pi_1} \rfloor$.*

Proof For a given maintenance activity i constraint (P) implies that there are $\bar{Z}_i = \lfloor \frac{T}{\pi_i} \rfloor$ complete task horizons in the planning horizon $1, \ldots, T$ (cf. (A) from Sect. 5.1). Using (V), it follows $\bar{Z}_i \leq \sum_{t=1}^{T} x_{it} \leq \sum_{t=1}^{T} y_t$ and a lower bound $f_1 \cdot \bar{Z}_1$ is obtained. The largest lower bound is $f_1 \cdot \bar{Z}_1$ because $i = 1$ has the smallest coverage (Assumption (2.1)).

Schedule a task for every maintenance activity $i = 1, \ldots, n$ exactly every $\bar{\pi}_i$ periods to yield x. Set $y_t = \max_{i=1,\ldots,n} x_{it} \ \forall t$ and obtain a feasible integer solution (x, y) with an objective function value $f_1 \cdot \bar{Z}_1$.

Optimality of (x, y) follows because the upper and the lower bounds coincide. Moreover, (x, y) is obtained in time $O(n \cdot T)$ and an extreme point of Conv(P)(V)(X)(Y) and thus, also of H(P)(V)(S) (Remark 4.5).

4.4 The Single Weighted Uncapacitated Planned Maintenance Problem

The *Single Weighted Uncapacitated Planned Maintenance Problem (SWUPMP)* is a special case of the WUPMP with $n = 1$. Note that the fix costs can be defined as a part of the variable costs because $n = 1$ holds. In order to avoid complicated notations, the index i is kept in the SWUPMP formulation that is

$$\min \sum_{t=1}^{T} c_{it} \cdot x_{it} \qquad \text{subject to}$$

$$\sum_{\tau=t}^{t+\pi_i-1} x_{i\tau} \geq 1 \qquad \forall t = 1, \ldots, T - \pi_i + 1 \qquad v_{it} \qquad \text{(P)}$$

$$-x_{it} \geq -1 \qquad \forall t = 1, \ldots, T \qquad \alpha_{it} \qquad (S^x)$$

$$x_{it} \in \{0, 1\} \qquad \forall t = 1, \ldots, T. \qquad \text{(X)}$$

The corresponding dual problem of the SWUPMP LP relaxation is stated as

$$\max \sum_{t=1}^{T-\pi_i+1} v_{it} - \sum_{t=1}^{T} \alpha_{it} \qquad \text{subject to} \qquad (4.38)$$

$$\sum_{\tau=\max\{1, t-\pi_i+1\}}^{\min\{t, T-\pi_i+1\}} v_{i\tau} - \alpha_{it} \leq c_{it} \qquad \forall t = 1, \ldots, T \qquad x_{it} \qquad (4.39)$$

$$v_{it} \geq 0 \ \forall t = 1, \ldots, T - \pi_i + 1 \wedge \alpha_{it} \geq 0 \ \forall t = 1, \ldots, T. \qquad (4.40)$$

Using the results from Theorem 4.1 and Corollary 4.2, the polytopes of the SWUPMP LP relaxation and of the dual problem (4.38)–(4.40) are integral. Note that the latter holds if and only if $c_{it} \in \mathbb{N} \ \forall t$.

4.4.1 An Optimal Solution to the Primal Problem

A reformulation of the SWUPMP with Remark 4.2 yields the SPP (4.41). Denote SP_{it} as the length of a shortest path from a node 0 to a node t for a given maintenance activity i. The variable transformation (4.1) yields an optimal integer solution.

$$SP_{i,T+1} = \min \sum_{(t,\tau) \in E_i} c_{it} \cdot z_{it\tau} \quad \text{subject to } (4.2) \wedge (4.3) \qquad (4.41)$$

The directed graph $G_i(V, E_i)$ has a specific structure with two main aspects.

- $G_i(V, E_i)$ is a cycle-free and topologically sorted graph.
- The arc weights c are non-negative and independent of the successor.

Lemma 4.8 shows that the SWUPMP is solvable as the SPP (4.41) in time $O(T^2)$ with the reaching algorithm.[10] The structure of $G_i(V, E_i)$ allows two pruning criteria from Lemma 4.14 that make the reaching algorithm even more efficient.

Lemma 4.14 Given three nodes $a, a^*, b \in V$ with a^*, b being direct successors of a in $G_i(V, E_i)$ that is $a < a^*$ and $b \leq a + \pi_i$. It holds that $SP_{ia} + c_{ia} \geq SP_{ib}$ if

1. $SP_{ia} + c_{ia} \geq SP_{ia^*}$ with $b \in \{a+1, \ldots, a^*\}$ or
2. $SP_{ia} + c_{ia} \geq SP_{ia^*} + c_{ia^*}$ with $b \in \{a^*, \ldots, a + \pi_i\}$.

Proof Recall the topological order of V and that $G_i(V, E_i)$ is cycle-free. Since $c_{it} \geq 0 \; \forall t$ and the arc weights are successor independent, $SP_{it} \leq SP_{i\tau}$ and $SP_{it} + c_{it} \geq SP_{i\tau}$ hold for all direct predecessors t of τ or for $t = \tau$.
'1.': Assuming $SP_{ia} + c_{ia} \geq SP_{ia^*}$ with a, b as direct predecessors of a^*, yields $SP_{ia} + c_{ia} \geq SP_{ib}$ because $SP_{ia^*} \geq SP_{ib}$ holds.
'2.': Assuming $SP_{ia} + c_{ia} \geq SP_{ia^*} + c_{ia^*}$ with a, a^* as direct predecessors of b, yields $SP_{ia} + c_{ia} \geq SP_{ib}$ because $SP_{ia^*} + c_{ia^*} \geq SP_{ib}$ holds.

Algorithm 4.1 is a modified reaching algorithm that solves the SWUPMP. Initialize SP_{ia} for all $a = 1, \ldots, T + 3$ with $SP_{ia} = 0$ if $a = 1, \ldots, \pi_i$ (otherwise, ∞) and $c_{i,T+1} = -\infty$, $c_{i,T+2} = -\infty$. Set $s = 1$ and $a^* = \arg\min_{t=1,\ldots,\pi_i} c_{it}$ for the upper bound SP_{ia^*}. In an iteration, consider all nodes $a = a^*, a^* - 1, \ldots, s$ and if $SP_{ia} + c_{ia} < R = \min_{h=a+1,\ldots,a^*}(SP_{ih} + c_{ih})$, then update the shortest path SP_{ib} of all successors $b = a^* + 1, \ldots, a + \pi_i$ of the node a by testing against $SP_{ia} + c_{ia} < SP_{ib}$. Note that the nodes $s, \ldots, a^*$ satisfying Lemma 4.14.2 are excluded from the update and that the order $a^*, a^* - 1, \ldots, s$, tightens the applied upper bound R. Afterwards, set $s = a^*$ and find a new node a^* with $\min_{a=s,\ldots,a^*-1}(SP_{ia} + c_{ia}) \geq SP_{ia^*}$. Note that the nodes $s, \ldots, a^* - 1$ that satisfy Lemma 4.14.1 are excluded in the next iteration. Commence with the next iteration. The algorithm terminates if $s > T$. Remove the nodes $0, T + 1$ and redundant periods from the shortest path. The algorithm returns the shortest path and its length $SP_{i,T+1}$.

---

[10]The *reaching algorithm* uses a breadth-first search and is outlined as follows. Given a graph $G(V, E)$ that is cycle-free and let c_{ab} be the arc weights of $(a, b) \in E$. Denote the shortest path from the source node to the node b as SP_b. Set all nodes as unlabeled. In an iteration, first select a node a that has the smallest SP_a among all non-labeled nodes. This is ensured by topologically ordering the nodes and selecting them in an increasing order of subscripts. Second, examine whether the shortest path to all its successors can be improved. If this is the case because $SP_a + c_{ab} < SP_b$ holds for a successor b, then update SP_b and label the node a. The algorithm terminates when all nodes are labeled and returns shortest paths from the source node to all nodes. The reaching algorithm runs in time $O(|V| + |E|)$.

4.4 The SWUPMP

4.1 Algorithm. An optimal solution to the SWUPMP
Input: $c_{it} \geq 0 \; \forall t$ and an instance to the SWUPMP
Output: The ith shortest path with the nodes $Path_i$ and the length $SP_{i,T+1}$
Initialization: $c_{i,T+1} \leftarrow -\infty, c_{i,T+2} \leftarrow -\infty, R \leftarrow \infty, a^* \leftarrow \arg\min_{t=1,\ldots,\pi_i} c_{it}$,
$SP_{ia} \leftarrow \infty \; \forall a = 1,\ldots,T+3, Pre_a \leftarrow n+1 \; \forall a = 1,\ldots,T+1$,
$Path_i \leftarrow \emptyset, s \leftarrow 1, t \leftarrow 0, \tau \leftarrow 0, a \leftarrow 0, b \leftarrow 0$

1. $SP_{ia} \leftarrow 0 \; \forall a = 1,\ldots,\pi_i \wedge Pre_a \leftarrow 0 \; \forall a = 1,\ldots,\pi_i$
2. **while** $s \leq T$ **do**
3. $\quad R \leftarrow \infty$
4. $\quad$ **for** $a = a^*$ **down to** s **do**
5. $\quad\quad$ **if** $SP_{ia} + c_{ia} < R$ **then**
6. $\quad\quad\quad R \leftarrow SP_{ia} + c_{ia}$
7. $\quad\quad\quad$ **for** $b = a^* + 1$ **to** $\min\{a + \pi_i, T+1\}$ **do** [11]
8. $\quad\quad\quad\quad$ **if** $SP_{ia} + c_{ia} < SP_{ib}$ **then** $SP_{ib} \leftarrow SP_{ia} + c_{ia} \wedge Pre_b \leftarrow a$
9. $\quad R \leftarrow \infty \wedge s \leftarrow a^*$ [12]
10. $\quad$ **do** $a^* \leftarrow a^* + 1 \wedge R \leftarrow \min\{R, SP_{ia^*} + c_{ia^*}\}$
11. $\quad$ **while** $R \geq SP_{i,a^*+1}$
12. Recursively construct the shortest path $Path_i$ from Pre such that $Path_i$ does neither include the nodes $0, T+1$ nor redundant periods

The initialization and line 1 are important for correctness and for termination. Having executed line 11, it holds that $s < a^*$, $a^* \leq s + \pi_i$ since $SP_{i,s+\pi_i+1} = \infty$ is set and $a^* \leq T+1$ because of $c_{it} = -\infty \; \forall t > T$ and $SP_{it} = \infty \; \forall t > T$. If no pruning criterion from Lemma 4.14 applies, $a^* = s + 1$ holds in every iteration and the Algorithm 4.1 obviously reduces to the reaching algorithm.

4.4.2 An Optimal Solution to the Corresponding Dual Problem of the LP Relaxation

In the SWUPMP LP relaxation, a simple upper bound $x_{it} \leq 1 \; \forall t$ is redundant because the right-hand side in (P) is 1 and the objective function minimizes with $c_{it} \geq 0 \; \forall t$.[13] Hence, the dual problem (4.38)–(4.40) simplifies to

---

[11] Addendum: Another pruning criterion reduces the computational effort of the loop to **for** $b = \max\{a^*, Pre_a + \pi_i\} + 1$ **to** $\min\{a + \pi_i, T+1\}$ **do**
[12] Addendum: Set $s \leftarrow a^* + 1$ to reduce the computational effort.
[13] Another argument for the redundancy of $x_{it} \leq 1 \; \forall t$ is presented in Remark 6.1.

$$\max \sum_{t=1}^{T-\pi_i+1} v_{it} \qquad \text{subject to} \qquad (4.42)$$

$$\sum_{\tau=\max\{1,t-\pi_i+1\}}^{\min\{t,T-\pi_i+1\}} v_{i\tau} \leq c_{it} \qquad \forall t = 1,\ldots,T \qquad x_{it} \qquad (4.43)$$

$$v_{it} \geq 0 \qquad \forall t = 1,\ldots,T-\pi_i+1 \qquad (4.44)$$

that is solved with Lemma 4.15. This yields the optimal objective function value of the SWUPMP because the optimal objective function value of the SWUPMP LP relaxation and thus, of the SWUPMP are obtained (Theorem 4.1). If and only if $c_{it} \in \mathbb{N} \;\forall t$, the dual solution is integral by Corollary 4.2.[14]

Lemma 4.15 An optimal solution to (4.42)–(4.44) is

$$v_{it} = \min_{\tau=t,\ldots,t+\pi_i-1} c_{i\tau} - \sum_{\theta=\max\{1,\tau-\pi_i+1\}}^{t-1} v_{i\theta} \quad \forall t = 1,\ldots,T-\pi_i+1. \qquad (4.45)$$

Proof Assume an optimal solution exists with a higher objective function value than defined by (4.45). Hence, there exists one v_{it} with

$$v_{it} > c_{i\tau} - \sum_{\theta=\tau-\pi_i+1}^{t-1} v_{i\theta} \quad \forall \tau = t,\ldots,t+\pi_i-1. \qquad (4.46)$$

Let Δ_τ be the slack of the τth constraint. Feasibility requires that (4.43) is satisfied as

$$v_{it} \leq c_{i\tau} - \sum_{\theta=\tau-\pi_i+1 \wedge \theta \neq t}^{\tau} v_{i\theta} \quad \forall \tau = t,\ldots,t+\pi_i-1$$

because of Remark 4.1. From both inequalities follows for all $\tau = t,\ldots,t+\pi_i-1$

$$v_{it} = c_{i\tau} - \sum_{\theta=\tau-\pi_i+1}^{t-1} v_{i\theta} + \Delta_\tau \leq c_{i\tau} - \sum_{\theta=\tau-\pi_i+1 \wedge \theta \neq t}^{\tau} v_{i\theta}$$

$$\Leftrightarrow \Delta_\tau \leq -\sum_{\theta=t+1}^{\tau} v_{i\theta}.$$

---

[14] It is straightforward to show via induction that Lemma 4.15 yields an integer solution if $c_{it} \in \mathbb{N} \;\forall t$.

4.4 The SWUPMP

Note that (4.46) implies $\Delta_\tau > 0$. If $\tau = t$, then $\Delta_\tau \leq 0$ gives a contradiction. Otherwise, there exists a $\theta = t+1, \ldots, \tau$ with $v_{i\theta} < 0$, which contradicts feasibility because (4.44) is violated.

A direct implementation of Lemma 4.15 is not efficient because of several recalculations in the recursion.

Corollary 4.16 An optimal solution to (4.42)–(4.44) is obtained from the recursive equations

$$v_{it} = \min_{\tau = t, \ldots, t + \pi_i - 1} s^t_{i\tau} \quad \forall t = 1, \ldots, T - \pi_i + 1 \quad (4.47)$$

with the initial states $s^0_{i\tau} = c_{i\tau} \;\forall \tau = 1, \ldots, \pi_i - 1 \wedge v_{i0} = 0$ and the states

$$s^t_{i\tau} = \begin{cases} c_{i\tau} & \text{if } \tau = t + \pi_i - 1 \\ s^{t-1}_{i\tau} - v_{i,t-1} & \text{otherwise} \end{cases} \quad \forall \tau = t, \ldots, t + \pi_i - 1.$$

Proof The recursion state $s^t_{i\tau}$ becomes

$$s^t_{i\tau} = s^{t-1}_{i\tau} - v_{i,t-1}$$
$$= s^{t-2}_{i\tau} - v_{i,t-1} - v_{i,t-2}$$
$$\vdots$$
$$= c_{i\tau} - v_{i,t-1} - v_{i,t-2} - \ldots - v_{i,(\tau-\pi_i+2)-1}$$
$$= c_{i\tau} - \sum_{\theta = \tau - \pi_i + 1}^{t-1} v_{i\theta}.$$

Apply (4.47) and obtain (4.45). Optimality follows from Lemma 4.15.

Remark 4.7 There exists an optimal slack of constraint (4.43) $d_{i\tau}$ with $d_{i\tau} = 0$ and $\tau = t, \ldots, t + \pi_i - 1$ for a given period t. Otherwise, there would exist a $v_{it} > 0$ in (4.45) and optimality would be contradicted.

Algorithm 4.2 applies Corollary 4.16. Set $d_{i\tau} = c_{i\tau} \;\forall \tau = 1, \ldots, \pi_i$. Start with an initial task horizon $\{a, \ldots, b\} = \{1, \ldots, \pi_i\}$. An iteration commences with the largest period θ that satisfies $d_{i\theta} = \min_{\tau = a, \ldots, b} d_{i\tau}$. From Corollary 4.16 follows that $v_{ia} = d_{i\theta}$ is optimal. Update the slack with $d_{i\tau} = d_{i\tau} - v_{ia} \;\forall \tau = a, \ldots, b$. Clearly, $d_{i\theta} = 0$ holds (Remark 4.7) and $v_{i\tau} = 0 \;\forall \tau = a + 1, \ldots, \theta$ follows in (4.43) because of the stairway structure of (P) (Remark 4.1). If and only if the optimal slack of constraint (4.43) is an output, calculate line 6.[15] Set $s_{i\tau} = c_{i\tau} \;\forall \tau = b + 1, \ldots, \theta + \pi_i$. The next iteration commences with $\{a, \ldots, b\} = \{\theta + 1, \ldots, \theta + \pi_i\}$.

---

[15] Algorithm 6.5 requires the optimal slack of (4.43).

4.2 Algorithm. An optimal solution to the corresponding dual problem of the SWUPMP LP relaxation

   Input: $c_{it} \geq 0$ $\forall t$ and an instance to the SWUPMP
   Output: An optimal solution v_{it} $\forall t$ to (4.42)–(4.44) and if necessary, the optimal slack
      d_{it} $\forall t$ of (4.43)
   Initialization: $v_{it} \leftarrow 0$ $\forall t, d_{it} \leftarrow 0$ $\forall t, a \leftarrow 1, b \leftarrow \pi_i, \tau \leftarrow 0, \theta \leftarrow 0$
1 $d_{i\tau} \leftarrow c_{i\tau}$ $\forall \tau = 1, \ldots, \pi_i$
2 do
3 $\quad$ Determine the largest period θ that satisfies $d_{i\theta} = \min_{\tau = a, \ldots, b} d_{i\tau}$
4 $\quad v_{ia} \leftarrow d_{i\theta}$
5 $\quad v_{i\tau} \leftarrow 0$ $\forall \tau = a+1, \ldots, \min\{\theta, T - \pi_i + 1\}$
6 $\quad d_{i\tau} \leftarrow d_{i\tau} - v_{ia}$ $\forall \tau = a, \ldots, \theta$ $\quad$ // Calculate iff d_{it} $\forall t$ is an output
7 $\quad d_{i\tau} \leftarrow d_{i\tau} - v_{ia}$ $\forall \tau = \theta + 1, \ldots, b$
8 $\quad d_{i\tau} \leftarrow c_{i\tau}$ $\forall \tau = b + 1, \ldots, \min\{\theta + \pi_i, T\}$
9 $\quad a \leftarrow \theta + 1 \land b \leftarrow \theta + \pi_i$
10 while $a \leq T - \pi_i + 1$

The algorithm terminates if $a > T - \pi_i + 1$ and returns an optimal solution to (4.42)–(4.44) and the optimal slack of (4.43) if desired.

4.5 The Uncapacitated Planned Maintenance Problem

A special case of the WUPMP with $f_t = 1$ $\forall t$ and $c_{it} = 0$ $\forall i; t$ is the *Uncapacitated Planned Maintenance Problem (UPMP)* that is the uncapacitated variant of the CPMP.

$$Z^{(C)} = \min \sum_{t=1}^{T} y_t \qquad \text{subject to} \tag{Z}$$

$$\sum_{\tau=t}^{t+\pi_i-1} x_{i\tau} \geq 1 \qquad \forall i = 1, \ldots, n; t = 1, \ldots, T - \pi_i + 1 \tag{P}$$

$$x_{it} \leq y_t \qquad \forall i = 1, \ldots, n; t = 1, \ldots, T \tag{V}$$

$$x_{it} \in \{0, 1\} \qquad \forall i = 1, \ldots, n; t = 1, \ldots, T \tag{X}$$

$$y_t \in \{0, 1\} \qquad \forall t = 1, \ldots, T \tag{Y}$$

The UPMP is strongly polynomially solvable by Lemma 4.13 in time $O(n \cdot T)$. Lemma 4.17 shows that the UPMP provides a lower bound to the CPMP LP relaxation $Z^{(X)(Y)}$. Let $\bar{\pi}_i$ be an integer multiple of π_1 with $\bar{\pi}_i \leq \pi_i$.[16]

---

[16] $\bar{\pi}_i = a \cdot \pi_1$ with $a \in \{1, \ldots, \lfloor \frac{\pi_i}{\pi_1} \rfloor\}$ satisfies $\bar{\pi}_i \leq \pi_i$. Note that $\bar{\pi}_1 = \pi_1$.

4.5 The UPMP

Lemma 4.17 It holds that $Z^{(C)} = Z^{(C)(X)(Y)} = \lfloor \frac{T}{\pi_1} \rfloor \leq Z^{(X)(Y)}$. ∎

Proof Lemma 4.13 introduces the lower bound $\bar{Z}_i = \lfloor \frac{T}{\pi_i} \rfloor$ to the UPMP as the number of complete task horizons in $1, \ldots, T$. Using (V), it follows $\bar{Z}_i \leq \sum_{t=1}^{T} x_{it} \leq \sum_{t=1}^{T} y_t \leq Z^{(C)(X)(Y)}, Z^{(C)}$. Together with $Conv(P)(V)(X)(Y) \subseteq H(P)(V)(S)$ since $H(P)(V)(S)$ is convex, it follows $\bar{Z}_1 \leq Z^{(C)(X)(Y)} \leq Z^{(C)}$. Thus, $\bar{Z}_1 = Z^{(C)(X)(Y)} = Z^{(C)}$ (Lemma 4.13) and $Z^{(C)(X)(Y)} \leq Z^{(X)(Y)}$ because of $H(P)(V)(S) \subseteq H(P)(C)(V)(S)$.

The UPMP provides a lower bound to the CPMP that is simple to compute. However, the lower bound becomes arbitrarily bad, which is demonstrated by a series of CPMP instances.

Example 4.6 Given a constant $K \in \mathbb{N}$. Let $n = K$, $T = K$, $\pi_i = K$ $\forall i$, $r_i = 1$ $\forall i$ and $\bar{r}_t = 1$ $\forall t$. An instance satisfies Assumption 2.4. It follows $Z^{(C)} = 1$ and $Z = K$. The worst case performance ratio is arbitrarily close to 0 because $\lim_{K \to \infty} \frac{Z^{(C)}}{Z} = \lim_{K \to \infty} \frac{1}{K} = 0$.

The following example demonstrates that the UPMP allows maintenance plans with an infinite planning horizon.

Example 4.7 Let $n = 3$ and $\pi = (3, 7, 11)^T$. Set $\bar{\pi}$ is as large as possible to obtain $\bar{\pi} = (3, 6, 9)^T$. Table 4.7 depicts a partial solution that covers 18 periods which equals the least common multiple of $(\bar{\pi}_1, \ldots, \bar{\pi}_n)$. In order to yield an optimal solution, the partial solution recurs every 18 periods.

Example 4.8 shows that a feasible solution, which is constructed according to Lemma 4.17 does not minimize the total number of tasks.

Example 4.8 Let $n = 2$, $T = 9$ and $\pi = (3, 5)^T$ and it follows $\bar{\pi} = (3, 3)^T$. In Table 4.8, the optimal solution on the left is constructed with Lemma 4.17 and comprises six tasks but the optimal solution on the right has four tasks.

Observation 4.18 In a computational study with 1000 randomly generated instances, all optimal solutions to the UPMP LP relaxation were integer extreme points.[17]

Table 4.7 Example of an optimal solution to the UPMP

	t																	
	1	2	3	4	5	6	7	8	9	10	11	12	13	14	15	16	17	18
y_t			1			1			1			1			1			1
x_{1t}			1			1			1			1			1			1
x_{2t}						1						1						1
x_{3t}									1									1

---

[17]The instances were generated as presented in Sect. 7.1 and solved with the Simplex algorithm.

Table 4.8 Example that Lemma 4.17 does not minimize the total number of tasks

t	1	2	3	4	5	6	7	8	9
y_t		1			1			1	
x_{1t}		1			1			1	
x_{2t}		1			1			1	

t	1	2	3	4	5	6	7	8	9
y_t		1		1			1		
x_{1t}		1		1			1		
x_{2t}				1					

The UPMP possesses all polyhedral properties of the WUPMP, where the less strong Observation 4.4 holds. Although the optimal objective function values of a solution to the LP relaxation and to the integer solution coincide that is $Z^{(C)(X)(Y)} = Z^{(C)}$ (Lemma 4.17), the Observation 4.18 is interesting because the polytope $H(P)(V)(S)$ might contain fractional extreme points (cf. Examples 4.3 and 4.4). Lemma 4.19 allows for a partial explanation of the observation.

Lemma 4.19 *If $\pi_i = T \; \forall i = 2, \ldots, T$, the smallest convex set of all optimal solutions to $H(P)(V)(S)$ has integer extreme points.*

Proof An optimal solution (x, y) to the UPMP has a specific structure. Assume there exists a period τ with $x_{1\tau} < y_\tau$. Constraint (P) implies that there exist $\lfloor \frac{T}{\pi_1} \rfloor$ complete task horizons in the planning horizon $1, \ldots, T$ (cf. (A) from Sect. 5.1).

$$\left\lfloor \frac{T}{\pi_1} \right\rfloor \leq \sum_{t=1}^{T} x_{1t} < \sum_{t=1 \land t \neq \tau}^{T} x_{1t} + y_\tau \leq \sum_{t=1}^{T} y_t$$

follows that contradicts optimality since $\sum_{t=1}^{T} y_t = \lfloor \frac{T}{\pi_1} \rfloor$ holds by Lemma 4.17. Hence, the period τ does not exist and it holds that

$$x_{1t} = y_t \quad \forall t = 1, \ldots, T. \tag{4.48}$$

Consequently, from (4.48) and (V) follows

$$x_{it} \leq x_{1t} \quad \forall i = 2, \ldots, n; t = 1, \ldots, T. \tag{4.49}$$

Assume an optimal solution $(\bar{x}, \bar{y}) \in H(P)(V)(S)$ where at least one component $0 < \bar{x}_{it} < 1$ or $0 < \bar{y}_t < 1$ is fractional. In what follows, a convex combination of K optimal integer solutions yield $(\bar{x}, \bar{y})$. Therefore, each optimal solution in $H(P)(V)(S)$ is obtained from a convex combination of optimal integer solutions and $(\bar{x}, \bar{y})$ cannot be an extreme point.

4.5 The UPMP

Since (P), (V) and (S) are rational linear inequalities, represent all non-zeros in $(\bar{x}, \bar{y})$ as fractions.[18] Compute the least common denominator K among of all values of $(\bar{x}, \bar{y})$ and denote $\alpha = \frac{1}{K}$ as a quant.

Construct K integer solutions (x^k, y^k) with $x^k \in \{0, 1\}^{n \times T}$, $y^k \in \{0, 1\}^T$ for all $k = 1, \ldots, K$ as follows (see Example 4.9): Initialize $x_{it}^k = 0$ $\forall i = 1, \ldots, n; t = 1, \ldots, T; k = 1, \ldots, K$ and $y_t^k = 0$ $\forall t = 1, \ldots, T; k = 1, \ldots, K$. Obtain the vector $(x_{11}^k, \ldots, x_{1T}^k)$ for all $k = 1, \ldots, K$ as presented in the proof of Theorem 4.1 for $a_{1t} = 1$ $\forall t$, $b_{1t} = 1$ $\forall t$. Consequently, (P) holds for $i = 1$. For every $i = 2, \ldots, n$ execute two steps: First, assign one quant per vector $(x_{i1}^k, \ldots, x_{iT}^k)$ to an arbitrary period where $x_{1t}^k = 1$ for all $k = 1, \ldots, K$. This satisfies (P) for the ith maintenance activity. Note that such an assignment exists because (4.49) holds for $(\bar{x}, \bar{y})$ and it can be found in polynomial time.[19] Second, for each period $t = 1, \ldots, T$ with $\bar{x}_{it} > \alpha$, arbitrarily assign the remaining $(\bar{x}_{it} - \alpha) \cdot K$ quants over the K vectors $(x_{i1}^k, \ldots, x_{iT}^k)$ such that $x_{it}^k \leq 1$ holds. Consequently, (4.49) is satisfied. Finally set $y_t^k = x_{1t}^k$ $\forall t = 1, \ldots, T; k = 1, \ldots, K$, which satisfies (V) because of (4.48).

K feasible integer solutions $(x^k, y^k) \in Conv(P)(V)(X)(Y)$ $\forall k$ are obtained. Since $\sum_{t=1}^{T} \bar{y}_t = \lfloor \frac{T}{\pi_1} \rfloor$ and $K \cdot \alpha = 1$ holds, exactly $K \cdot \lfloor \frac{T}{\pi_1} \rfloor$ quants are assigned over the K vectors $(y_1^k, \ldots, y_T^k)$. Therefore, the kth integer solution (x^k, y^k) has $\sum_{t=1}^{T} y_t^k = \lfloor \frac{T}{\pi_1} \rfloor$ and therefore, is optimal. Thus,

$$(\bar{x}, \bar{y}) = \sum_{k=1}^{K} \alpha \cdot (x^k, y^k) \qquad (4.50)$$

is a *strict* convex combination since $\alpha > 0$ and $K \cdot \alpha = 1$. This contradicts that $(\bar{x}, \bar{y})$ is an extreme point (Nemhauser and Wolsey 1988, p. 93).

Example 4.9 Let $n = 2$, $T = 10$ and $\pi = (5, 10)^T$. In Table 4.9, K feasible integer solutions (x^k, y^k) are yielded from a given feasible fractional solution $(\bar{x}, \bar{y})$. Considering Example 4.2, the least common denominator is $K = 4$ and the quant is $\alpha = 0.25$. $(\bar{x}, \bar{y})$ is obtained by the convex combination (4.50).

Lemma 4.20 proposes a set of constraints such that any extreme point is integer and optimal to the UPMP. Note that Lemma 4.12 presented for the WUPMP proposes another set of such constraints.[20]

---

[18] See Footnote 4 on page 74.

[19] The respective Linear Assignment Problem is polynomially solvable (Papadimitriou and Steiglitz 1998, pp. 247–248).

[20] Similar results are known for other combinatorial optimization problems. For the Uncapacitated Lot-Sizing Problem valid inequalities exist that yield a *tight formulation* such that the polyhedron of the LP relaxation coincides with the convex hull of the MIP (Pochet and Wolsey 1994; Vyve and Wolsey 2006).

Table 4.9 The convex combination (4.50) yields the fractional solution $(\bar{x}, \bar{y})$

	t									
	1	2	3	4	5	6	7	8	9	10
$\bar{y}_t, \bar{x}_{1t}$	0.75		0.5	0.25	0.25			0.75		0.25
$\bar{x}_{2t}$	0.5			0.25				0.5		
y^1_t, x^1_{1t}	1		1					1		
x^1_{2t}	1							1		
y^2_t, x^2_{1t}	1			1				1		
x^2_{2t}				1	1					
y^3_t, x^3_{1t}	1				1					1
x^3_{2t}	1									
y^4_t, x^4_{1t}			1					1		
x^4_{2t}								1		

Lemma 4.20 $Conv(4.51)(4.52)(V)(X)(Y) = H(4.51)(4.52)(V)(S)$ and all extreme points of $Conv(4.51)(4.52)(V)(X)(Y)$ are extreme points of $H(P)(V)(S)$. Let $p_i \in \mathbb{N}$ with $p_i \leq \bar{\pi}_i$ hold for all $i = 1, \ldots, n$.

$$\sum_{\tau=t}^{t+\bar{\pi}_i-1} x_{i\tau} = 1 \quad \forall i = 1, \ldots, n; t = 1, \ldots, T - \bar{\pi}_i + 1 \tag{4.51}$$

$$x_{it} = y_t \quad \forall i = 1, \ldots, n; t = p_i, \ldots, p_i + \pi_1 - 1 \tag{4.52}$$

Proof Let (x, y) be a continuous or integer solution that is feasible to (4.51), (4.52), (V), (S). Substitute the variables x with the variables y in (4.51) for $i = 1$ and obtain

$$\sum_{\tau=t}^{t+\pi_1-1} y_\tau = 1 \quad \forall t = 1, \ldots, T - \pi_1 + 1.$$

This equality has exactly $\lfloor \frac{T}{\pi_1} \rfloor$ complete task horizons in the planning horizon $1, \ldots, T$ (cf. (A) from Sect. 5.1). Thus, (x, y) is optimal because of Lemma 4.17.

It holds that $Conv(4.51)(4.52)(V)(X)(Y) \subseteq Conv(P)(V)(X)(Y)$. Consequently, an extreme point of $Conv(4.51)(4.52)(V)(X)(Y)$ is an extreme point of $Conv(P)(V)(X)(Y)$ and thus, also an extreme point of $H(P)(V)(S)$ (Remark 4.5).

Due to the specific structure of a feasible solution (see Example 4.10), a feasible vector $(x_{i1}, \ldots, x_{iT})$ is generated from a partial solution $(x_{i1}, \ldots, x_{i\bar{\pi}_i})$. This is discussed in what follows. Constraint (4.52) implies $x_{it} = x_{1t}$ $\forall i; t = p_i, \ldots, p_i + \pi_1 - 1$ and subsequently, it follows that $x_{it} = 0$ $\forall i; t = 1, \ldots, p_i - 1, p_i + \pi_1, \ldots, \bar{\pi}_i$

4.5 The UPMP

from (4.51). Consequently, it holds that

$$(x_{i1},\ldots,x_{i\bar{\pi}_i}) = (0,\ldots,0,x_{1p_i},\ldots,x_{1,p_i+\pi_1-1},0,\ldots,0) \quad \forall i = 1,\ldots,n. \tag{4.53}$$

A feasible vector $(x_{i1},\ldots,x_{iT})$ is constructed from the partial solution $(x_{i1},\ldots,x_{i\bar{\pi}_i})$ (4.53) by applying

$$x_{it} = x_{i,t-\bar{\pi}_i} \quad \forall t = \bar{\pi}_i + 1,\ldots,T. \tag{4.54}$$

This holds because a feasible solution (x,y) satisfies (4.51) and it follows for $i = 1,\ldots,n$ and $t = 1,\ldots,T-\bar{\pi}_i$

$$1 = \sum_{\tau=t+1}^{t+\bar{\pi}_i} x_{i\tau} + x_{it} - x_{it} \qquad \text{add a neutral 0}$$

$$= \sum_{\tau=t}^{t+\bar{\pi}_i-1} x_{i\tau} + x_{i,t+\bar{\pi}_i} - x_{it} \qquad \text{because of (4.51)}$$

$$= 1 + x_{i,t+\bar{\pi}_i} - x_{it}$$

$$x_{it} = x_{i,t+\bar{\pi}_i}.$$

'$\Rightarrow$': $H(4.51)(4.52)(V)(S)$ is convex and hence, $Conv(4.51)(4.52)(V)(X)(Y) \subseteq H(4.51)(4.52)(V)(S)$ holds.

'$\Leftarrow$': Assume a feasible solution $(\bar{x},\bar{y}) \in H(4.51)(4.52)(V)(S)$ where at least one component $0 < \bar{x}_{it} < 1$ or $0 < \bar{y}_t < 1$ is fractional. $Conv(4.51)(4.52)(V)(X)(Y) \supseteq H(4.51)(4.52)(V)(S)$ is shown by the existence of a convex combination of K integer solutions that yield $(\bar{x},\bar{y})$. Therefore, $(\bar{x},\bar{y})$ cannot be an extreme point and it follows that $(\bar{x},\bar{y}) \in Conv(4.51)(4.52)(V)(X)(Y)$.

Let K be the number of non-zeros in $(\bar{x}_{i1},\ldots,\bar{x}_{i\bar{\pi}_i})$ for a given $i = 1,\ldots,n$. Let the function $\tau(k,i)$ yield the period t where the non-zero $\bar{x}_{it} > 0$ is the kth non-zero in $(\bar{x}_{i1},\ldots,\bar{x}_{i\bar{\pi}_i})$ for a given $i = 1,\ldots,n, k = 1,\ldots,K$.

Construct K integer solutions (x^k, y^k) with $x^k \in \{0,1\}^{n \times T}$, $y^k \in \{0,1\}^T$ and a scalar α_k for all $k = 1,\ldots,K$ as follows (see Example 4.10): Let $k = 1$ and $i = 1$. In every iteration, set $(x_{11}^k,\ldots,x_{1\pi_1}^k) = (0,\ldots,0,x_{1,\tau(k,1)}^k,0,\ldots,0)$ with $x_{1,\tau(k,1)}^k = 1$ for the first task horizon and repeat this pattern with $x_{1t}^k = x_{1,t-\pi_1}^k \; \forall t = \pi_1+1,\ldots,T$ to yield a vector $(x_{11}^k,\ldots,x_{1T}^k)$. Set $x_{it}^k = 1$ if $x_{1t}^k = 1 \wedge \bar{x}_{it} > 0$ (otherwise, 0) $\forall i = 2,\ldots,n; t = 1,\ldots,T$. Set $y_t^k = x_{1t}^k \; \forall t$ and $\alpha_k = \bar{x}_{1,\tau(k,1)}$. Commence the next iteration with $k = k+1$ until $k > K$.

K feasible integer solutions $(x^k, y^k) \in Conv(4.51)(4.52)(V)(X)(Y) \; \forall k$ are obtained. If and only if $\bar{x}_{it}$ is a non-zero there exists exactly one k with $x_{it}^k = 1$ and $t = \tau(k,i)$ for all $i;t = 1,\ldots,\bar{\pi}_i$. (4.51) and (4.52) are satisfied because x^k is constructed with a partial solution (4.53) and with (4.54). (V) is satisfied because of $y_t^k = x_{1t}^k \ \forall t$. Furthermore, it follows for the partial solution with $i = 1, \ldots, n; t = 1, \ldots, \bar{\pi}_i$

$$\sum_{k=1}^{K} \alpha_k \cdot x_{it}^k = \sum_{k=1}^{K} \bar{x}_{1,\tau(k,1)} \cdot x_{it}^k \qquad \text{because of } \alpha_k = \bar{x}_{1,\tau(k,1)}$$

$$= \sum_{k=1}^{K} \bar{x}_{i,\tau(k,i)} \cdot x_{it}^k \qquad \text{from (4.53)}$$

$$= \begin{cases} \bar{x}_{it} & \text{if } \bar{x}_{it} > 0 \\ 0 & \text{otherwise} \end{cases} \qquad \text{for } \bar{x}_{it} > 0 \text{ exists only one } k \text{ with } x_{it}^k = 1$$

that also holds for all $t = 1, \ldots, T$ because of (4.54). Hence,

$$(\bar{x}, \bar{y}) = \sum_{k=1}^{K} \alpha_k \cdot (x^k, y^k) \qquad (4.55)$$

is a *strict* convex combination because of $\alpha_k > 0 \ \forall k$ and $\sum_{k=1}^{K} \alpha_k = 1$ since $i = 1$ satisfies (4.51). This contradicts that $(\bar{x}, \bar{y})$ is an extreme point (Nemhauser and Wolsey 1988, p. 93).

Example 4.10 Let $n = 3, T = 16, \pi = (4, 8, 8)^T, \bar{\pi} = \pi$ and $p = (2, 3, 5)^T$. In Table 4.10, K feasible integer solutions (x^k, y^k) are yielded from a given feasible fractional solution $(\bar{x}, \bar{y})$. It follows $K = 3, \alpha = (0.2, 0.5, 0.3)^T$ and $(\bar{x}, \bar{y})$ is obtained by the convex combination (4.55).

Example 4.11 shows that the feasible solutions defined by both lemmata are complementary in the sense that Lemma 4.20 restricts (P) but only slightly (V), whereas Lemma 4.12 restricts (V) but not (P).

Example 4.11 Let $n = 3, T = 8$ and $\pi = (4, 8, 8)^T$. The solution on the left of Table 4.11 is feasible to Lemma 4.20 with $\bar{\pi} = \pi$ and $p = (1, 1, 5)^T$ but infeasible to Lemma 4.12 because of $x_{34} = 0, y_4 = 1$. The solution on the right is feasible to Lemma 4.12. The task horizon $4, \ldots, 7$ has two tasks for $i = 1$, which is infeasible to Lemma 4.20.

Table 4.10 The convex combination (4.55) yields the fractional solution $(\bar{x}, \bar{y})$

t	1	2	3	4	5	6	7	8	9	10	11	12	13	14	15	16
$\bar{y}_t$	0.2	0.5	0.3			0.2	0.5	0.3		0.2	0.5	0.3		0.2	0.5	0.3
$\bar{x}_{1t}$	0.2	0.5	0.3			0.2	0.5	0.3		0.2	0.5	0.3		0.2	0.5	0.3
$\bar{x}_{2t}$		0.5	0.3			0.2					0.5	0.3		0.2		
$\bar{x}_{3t}$						0.2	0.5	0.3						0.2	0.5	0.3
y_t^1	1					1				1				1		
x_{1t}^1	1					1				1				1		
x_{2t}^1						1								1		
x_{3t}^1						1								1		
y_t^2		1					1				1				1	
x_{1t}^2		1					1				1				1	
x_{2t}^2		1									1					
x_{3t}^2							1								1	
y_t^3			1					1				1				1
x_{1t}^3			1					1				1				1
x_{2t}^3			1									1				
x_{3t}^3								1								1

Table 4.11 Optimal solutions to Lemmata 4.20 and 4.12

t	1	2	3	4	5	6	7	8
y_t		1					1	
x_{1t}		1					1	
x_{2t}		1					1	
x_{3t}								1

t	1	2	3	4	5	6	7	8
y_t				1	1			
x_{1t}				1	1			
x_{2t}				1	1			
x_{3t}				1	1			

References

Ahuja, R. K., Magnanti, T. L., & Orlin, J. B. (1993). *Network flows: Theory, algorithms, and applications*. Englewood Cliffs: Prentice-Hall.

Balakrishnan, A., Magnanti, T. L., & Mirchandani, P. (1997). Network design. In M. Dell'Amico, F. Maffioli, & S. Martello (Eds.), *Annotated bibliographies in combinatorial optimization* (Chap. 18). Chichester: Wiley.

Balakrishnan, A., Magnanti, T. L., & Wong, R. T. (1989). A dual-ascent procedure for large-scale uncapacitated network design. *Operations Research, 37*, 716–740.

Chvátal, V. (2002). *Linear programming*. New York: W.H. Freeman and Company.

Cormen, T. H., Leiserson, C. E., Rivest, R. L., & Stein, C. (2001). *Introduction to algorithms* (2nd ed.). Cambridge: MIT.

Cornuejols, G., Nemhauser, G. L., & Wolsey, L. A. (1990). The uncapacitated facility location problem. In P. B. Mirchandani & R. L. Francis (Eds.), *Discrete location theory* (pp. 119–171). New York: Wiley.

Garey, M. R., & Johnson, D. S. (1979). *Computers and intractability: A guide to the theory of NP-completeness*. New York: Freeman.

Hellstrand, J., Larsson, T., & Migdalas, A. (1992). A characterization of the uncapacitated network design polytope. *Operations Research Letters, 12*, 159–163.

Holmberg, K., & Hellstrand, J. (1998). Solving the uncapacitated network design problem by a Lagrangean heuristic and branch-and-bound. *Operations Research, 46*, 247–259.

Jones, P. C., Lowe, T. J., Müller, G., Xu, N., Ye, Y., & Zydiak, J. L. (1995). Specially structured uncapacitated facility location problems. *Operations Research, 43*, 661–669.

Klose, A., & Drexl, A. (2005). Facility location models for distribution system design. *European Journal of Operational Research, 162*, 4–29.

Kolen, A. (1983). Solving covering problems and the uncapacitated plant location problem on trees. *European Journal of Operational Research, 12*, 266–278.

Kolen, A., & Tamir, A. (1990). Covering problems. In P. B. Mirchandani & R. L. Francis (Eds.), *Discrete location theory* (pp. 263–304). New York: Wiley.

Krarup, J., & Pruzan, P. M. (1983). The simple plant location problem: Survey and synthesis. *European Journal of Operational Research, 12*, 36–81.

Kuschel, T., & Bock, S. (2016). The weighted uncapacitated planned maintenance problem: Complexity and polyhedral properties. *European Journal of Operational Research, 250*, 773–781.

Labbé, M., Peeters, D., & Thisse, J.-F. (1995). Location on networks. In M. O. Ball, T. L. Magnanti, C. L. Monma, & G. L. Nemhauser (Eds.), *Network Routing*. Handbooks in operations research and management science (Vol. 8, Chap. 7). Amsterdam: Elsevier Science B.V.

Magnanti, T. L., & Wong, R. T. (1984). Network design and transportation planning: Models and algorithms. *Transportation Science, 18*, 1–55.

Mood, A. F., Graybill, F. A., & Boes, D. C. (1974). *Introduction to the theory of statistics* (3rd ed.). New York: McGraw-Hill.

Nemhauser, G. L., & Wolsey, L. A. (1988). *Integer and combinatorial optimization*. New York: Wiley.

Ng, A. S., Sastry, T., Leung, J. M. Y., & Cai, X. Q. (2004). On the uncapacitated k-commodity network design problem with zero flow-costs. *Naval Research Logistics, 51*, 1149–1172.

Papadimitriou, C. H., & Steiglitz, K. (1998). *Combinatorial optimization: Algorithms and complexity*. Mineola: Dover.

Pochet, Y., & Wolsey, L. A. (1994). Polyhedra for lot-sizing with Wagner-Whitin costs. *Mathematical Programming, 67*, 297–323.

ReVelle, C. S., Eiselt, H. A., & Daskin, M. S. (2008). A bibliography for some fundamental problems categories in discrete location science. *European Journal of Operational Research, 184*, 817–848.

Sastry, T. (2000). A characterization of the two-commodity network design problem. *Networks, 36*, 9–16.

Schrijver, A. (1998). *Theory of linear and integer programming*. New York: Wiley.

Tomlin, J. A. (1966). Minimum-cost multicommodity network flows. *Operations Research, 14*, 45–51.

Vavasis, S. A., & Ye, Y. (1996). A primal-dual interior point method whose running time depends only on the constraint matrix. *Mathematical Programming, 74*, 79–120.

Vyve, M. V., & Wolsey, L. A. (2006). Approximate extended formulations. *Mathematical Programming Series B, 105*, 501–522.

Wolfart, J. (2011). *Einführung in die zahlentheorie und algebra* (2nd ed.). Wiesbaden: Vieweg + Teubner.

Yemelichev, V. A., Kovalev, M. M., & Kravtsov, M. K. (1984). *Polytopes, graphs and optimisation*. Cambridge: Cambridge University Press; Translated by G.H. Lawden.

Chapter 5
Analyzing the Solvability of the Capacitated Planned Maintenance Problem

Valid inequalities and polyhedral properties of the CPMP that are used in the next sections are presented in Sect. 5.1. Section 5.2 provides the complexity status and optimal algorithms for the CPMP and for specific problem variants. An exhaustive analysis of lower bounds, the relative strength and the computational complexity is provided in Sect. 5.3.

5.1 Valid Inequalities and Polyhedral Properties

The extended cover inequality (3.54) and the extended clique inequality (3.55) of the KP are valid inequalities to the CPMP and tightened by taking the variable coupling between x and y into account.[1] In the *extended cover inequality* (O), denote C'_t as the index set of extended cover inequalities that are valid for a period t. The cth inequality (O) of a period t is defined by its cover I'_{tc} and by its extension E'_{tc}.[2]

$$\sum_{i \in E'_{tc}} x_{it} \leq \left(|I'_{tc}| - 1\right) \cdot y_t \quad \forall t = 1, \ldots, T; c \in C'_t \qquad (O)$$

The *extended clique inequality* (L) is defined in a similar way. The set C''_t denotes the index set of extended clique inequalities, which are valid for a period t. Given

---

[1] Divide (3.54) and (3.55) by the right-hand side, apply (V) and obtain (O) and (L).
[2] For I'_{tc} holds $\sum_{i \in I'_{tc}} r_i > \bar{r}_t$. Note that the cover I'_{tc} is minimal if $\sum_{j \in I'_{tc}} r_j - r_i \leq \bar{r}_t \; \forall i \in I'_{tc}$ holds. According to (3.53), the extension of a set I is $I \cup \{i \in \{1, \ldots, n\} \setminus I \mid r_i \geq \max_{k \in I} r_k\}$.

the cth clique I''_{tc} of a period t, let the set E''_{tc} be the extension of this clique.[3]

$$\sum_{i \in E''_{tc}} x_{it} \leq y_t \quad \forall t = 1, \ldots, T; c \in C''_t \qquad (L)$$

The *task redundancy constraint* (T) prevents redundancies in a task plan. Given a feasible solution where the i, tth constraint (T) is violated. Removing the redundant task yields a solution with the same or a lower objective function value. Therefore, (T) is a valid inequality. Let $x_{i0} = x_{i,T+1} = 1 \ \forall i$ (Assumption (2.6)).

$$\sum_{\tau=t}^{t+\pi_i} x_{i\tau} \leq 2 \quad \forall i = 1, \ldots, n; t = 0, \ldots, T - \pi_i + 1 \qquad (T)$$

The *slot covering constraint* (P^y) is implied by (P) and (V) for $i = 1$. The corresponding inequalities for $i = 2, \ldots, n$ are valid but less restrictive.

$$\sum_{\tau=t}^{t+\pi_1-1} y_\tau \geq 1 \quad \forall t = 1, \ldots, T - \pi_1 + 1 \qquad (P^y)$$

The following two valid inequalities (Q) and (R) partition (P). These constraints are relevant for Sect. 5.3 and are not restrictive if the formulation contains (P). The planning horizon $1, \ldots, T$ is partitioned into $m = \lceil \frac{T}{\pi_n} \rceil$ sections. Each *section q* comprises a set of periods Q_q with at most π_n elements. Let $Q_q^{min} = \min_{t \in Q_q} t$ be the smallest period and let $Q_q^{max} = \max_{t \in Q_q} t$ be the largest period of a given set Q_q. The sections are

$$Q_q = \{Q_{q-1}^{max} + 1, \ldots, Q_{q-1}^{max} + \pi_n\} \quad \forall q = 1, \ldots, m-1,$$

$$Q_m = \{Q_{m-1}^{max} + 1, \ldots, T\}$$

with $Q_0 = \{0\}$ as a dummy section. Let the set S_i comprise the first periods of the successive task horizons in all sections $Q_1, \ldots, Q_m$ for the maintenance activity i. The *section covering constraint* (Q) comprises all complete task horizons defined by S_i in (P). Constraint (R) consists of the remaining task horizons of (P) without those defined by (Q). Thus, $H(P) = H(Q) \cap H(R)$ holds and (P) can be equivalently replaced by (Q) and (R). An Example 5.1 of these inequalities is given below.

$$\sum_{\tau=t}^{t+\pi_i-1} x_{i\tau} \geq 1 \quad \forall i = 1, \ldots, n; t \in S_i \qquad (Q)$$

$$\sum_{\tau=t}^{t+\pi_i-1} x_{i\tau} \geq 1 \quad \forall i = 1, \ldots, n; t = 1, \ldots, T - \pi_i + 1; t \notin S_i \qquad (R)$$

---

[3]For I''_{tc} holds $\sum_{i \in I''_{tc}} r_i > \bar{r}_t$ and $r_i + r_j > \bar{r}_t \ \forall i, j \in I''_{tc}$ with $i \neq j$. Note that all extended clique inequalities are obtained with Algorithm 3.5. See (O) and (3.53) for the definition of the extension.

5.1 Valid Inequalities and Polyhedral Properties

Table 5.1 Coefficient matrix of (Q) and (R)

		x_{11}	x_{12}	x_{13}	x_{14}	x_{15}	x_{16}	x_{21}	x_{22}	x_{23}	x_{24}	x_{25}	x_{26}
(Q)	v_{11}	1	1										
	v_{13}			1	1								
	v_{15}					1	1						
	v_{21}							1	1	1	1		
(R)	v_{12}		1	1									
	v_{14}				1	1							
	v_{22}								1	1	1	1	
	v_{23}									1	1	1	1

Example 5.1 Let $n = 2$, $T = 6$, $\pi = (2,4)^T$. It follows $m = 2$, $Q_1 = \{1,\ldots,4\}$, $Q_2 = \{5,6\}$, $S_1 = \{1,3,5\}$ and $S_2 = \{1\}$. Table 5.1 shows the coefficient matrix of (Q) and (R), which yield (P). The respective dual variables $v \in \mathbb{R}_+^{n \times T}$ of (P) are provided.

The following *aggregated covering constraint* (A) is not restrictive if the formulation contains (P). It provides a lower bound on the number of tasks per maintenance activity that must be scheduled in the qth section.

Lemma 5.1

$$\sum_{t \in Q_q} x_{it} \geq \left\lfloor \frac{|Q_q|}{\pi_i} \right\rfloor \quad \forall i = 1,\ldots,n;\, q = 1,\ldots,m \tag{A}$$

Proof For the sake of brevity, (A) is proven for $Q_q = \{1,\ldots,\tau\}$. Multiply (P) by $\frac{1}{\pi_i}$ and add all $\tau - \pi_i + 1$ constraints to yield

$$\frac{1}{\pi_i} \cdot x_{i1} + \frac{2}{\pi_i} \cdot x_{i2} + \ldots + x_{i\pi_i} + \ldots + x_{i,\tau-\pi_i+1} + \ldots + \frac{2}{\pi_i} \cdot x_{i,\tau-1}$$
$$+ \frac{1}{\pi_i} \cdot x_{i\tau} \geq \frac{\tau - \pi_i + 1}{\pi_i}.$$

Multiply by -1 and round the coefficients as well as the right-hand side down to yield a Chvátal-Gomory inequality.

$$-\sum_{t=1}^{\tau} x_{it} \leq \left\lfloor -\frac{\tau - \pi_i + 1}{\pi_i} \right\rfloor = -\left\lceil \frac{\tau+1}{\pi_i} - 1 \right\rceil = -\left\lceil \frac{\tau+1}{\pi_i} \right\rceil + 1 \tag{5.1}$$

Consider the term $\left\lceil \frac{\tau+1}{\pi_i} \right\rceil$. It holds that $\frac{\tau}{\pi_i} < \frac{\tau+1}{\pi_i}$ because $\tau \geq 0$ and $\pi_i > 0$. $\frac{1}{\pi_i} \leq 1$ holds because $\pi_i \in \mathbb{N}$ implies $\pi_i \geq 1$. It follows $\frac{\tau}{\pi_i} < \frac{\tau+1}{\pi_i} = \frac{\tau}{\pi_i} + \frac{1}{\pi_i} \leq \left\lfloor \frac{\tau}{\pi_i} \right\rfloor + 1$. Thus, $\left\lceil \frac{\tau+1}{\pi_i} \right\rceil = \left\lfloor \frac{\tau}{\pi_i} \right\rfloor + 1$ holds and (5.1) becomes (A). □

The *task upper bound constraint* (U) limits the total number of tasks within a given section Q_q for every maintenance activity. Let $u_{iq} \in \mathbb{N}$ be an upper bound on the number of tasks in section Q_q of an optimal solution to the CPMP.

$$\sum_{t \in Q_q} x_{it} \leq u_{iq} \quad \forall i = 1, \ldots, n; q = 1, \ldots, m \tag{U}$$

A value for u_{iq} is determined in Lemma 5.2 and an Example 5.2 is given afterwards. Let $UB \in \mathbb{N}$ be an upper bound to the CPMP.[4] Let $X_{iq} \in \mathbb{N}$ be a lower bound on the number of open periods in $\{1, \ldots, T\} \setminus Q_q$.[5] Let Y be an upper bound on the total capacity of an optimal solution.[6] Let $a = 2$ if $\pi_i + 1 < |Q_q|$ (otherwise, 0).

Lemma 5.2

$$u_{iq} = \min \left\{ \left|\{t \in Q_q \mid r_i \leq \bar{r}_t\}\right|, UB - X_{iq}, \left\lfloor \frac{Y}{r_i} - \sum_{j=1 \wedge j \neq i}^{n} \frac{r_j}{r_i} \cdot \left\lfloor \frac{T}{\pi_j} \right\rfloor \right\rfloor - X_{iq}, \right.$$
$$\left. 2 + \left\lfloor a \cdot \frac{|Q_q| - 1}{\pi_i + 1} \right\rfloor \right\}$$

with $u_{iq} \leq |Q_q|$ and $u_{iq} \in \mathbb{N}$ holds for all $i = 1, \ldots, n$ and $q = 1, \ldots, m$.

Proof Since all four terms are integer, it holds that $u_{iq} \in \mathbb{N}$. Let (x, y) be a feasible solution to Z. Consider the first term of u_{iq}. Clearly, $\sum_{t \in Q_q} x_{it} \leq \left|\{t \in Q_q \mid r_i \leq \bar{r}_t\}\right| \leq |Q_q|$ holds for each feasible solution, which also proves $u_{iq} \leq |Q_q|$.

---

[4] $UB \in \mathbb{N}$ always exists because of Remark 2.1. If no upper bound UB can be determined, set $UB = T$ because it is an upper bound on any feasible solution if one exists.
[5] A valid value for X_{iq} is obtained with (A) as follows:

$$\sum_{t=1 \wedge t \notin Q_q}^{T} x_{it} = \sum_{t=1}^{Q_q^{min}-1} x_{it} + \sum_{t=Q_q^{max}+1}^{T} x_{it} \geq \left\lfloor \frac{Q_q^{min} - 1}{\pi_i} \right\rfloor + \left\lfloor \frac{T - Q_q^{max}}{\pi_i} \right\rfloor = X_{iq}.$$

[6] A valid value for Y is: Sort the capacity $(\bar{r}_1, \ldots, \bar{r}_T)$ according to non-increasing values and sum up the first UB elements $\bar{r}_{t_1} \geq \ldots \geq \bar{r}_{t_{UB}}$. Clearly, $Y = \sum_{h=1}^{UB} \bar{r}_{t_h} \geq \sum_{t=1}^{T} \bar{r}_t \cdot y_t^*$ holds for an optimal solution (x^*, y^*).

5.1 Valid Inequalities and Polyhedral Properties

Consider the second term of u_{iq}. Sum up (V) for i over all periods and obtain

$$\sum_{t \in Q_q} x_{it} \leq \sum_{t=1}^{T} y_t - \sum_{t=1 \wedge t \notin Q_q}^{T} x_{it} \leq UB - X_{iq}.$$

Consider the third term of u_{iq}. Sum up (C) over all periods and divide by r_i to yield

$$\sum_{t=1}^{T} x_{it} + \sum_{t=1}^{T} \sum_{j=1 \wedge j \neq i}^{n} \frac{r_j}{r_i} \cdot x_{jt} \leq \sum_{t=1}^{T} \frac{\bar{r}_t}{r_i} \cdot y_t$$

$$\sum_{t \in Q_q} x_{it} \leq \frac{1}{r_i} \cdot \sum_{t=1}^{T} \bar{r}_t \cdot y_t - \sum_{j=1 \wedge j \neq i}^{n} \frac{r_j}{r_i} \cdot x_{jt} - \sum_{t=1 \wedge t \notin Q_q}^{T} x_{it} \quad \text{(A), } Y \text{ and } X_{iq} \text{ are bounds}$$

$$\leq \frac{1}{r_i} \cdot Y - \sum_{j=1 \wedge j \neq i}^{n} \frac{r_j}{r_i} \cdot \left\lfloor \frac{T}{\pi_j} \right\rfloor - X_{iq}.$$

Because of the binary variable domain (X), rounding down the right-hand side is feasible.

Consider the fourth term of u_{iq}. For the sake of brevity, let $Q_q = \{1, \ldots, \tau\}$. Assume that $2 \cdot \pi_i + 2 \leq |Q_q|$ holds. Multiply (T) by $\frac{1}{\pi_i + 1}$ and add (T) for $t = 1, \ldots, \tau - \pi_i$ to yield

$$\frac{1}{\pi_i + 1} \cdot x_{i1} + \frac{2}{\pi_i + 1} \cdot x_{i2} + \ldots + x_{i, \pi_i + 1} + \ldots + x_{i, \tau - \pi_i} + \ldots$$
$$+ \frac{2}{\pi_i + 1} \cdot x_{i, \tau - 1} + \frac{1}{\pi_i + 1} \cdot x_{i\tau} \leq 2 \cdot \frac{\tau - \pi_i}{\pi_i + 1}.$$

Multiply (T) by $\frac{1}{\pi_i + 1}$ and add π_i constraints (T) for the first task horizon and π_i constraints (T) for the last task horizon. Therefore, all coefficients are greater or equal to one but strictly smaller than two.

$$x_{i1} + \frac{\pi_i + 2}{\pi_i + 1} \cdot x_{i2} + \ldots + \frac{2 \cdot \pi_i + 1}{\pi_i + 1} \cdot x_{i, \pi_i + 1} + x_{i, \pi_i + 2} + \ldots + x_{i, \tau - \pi_i - 1}$$
$$+ \frac{2 \cdot \pi_i + 1}{\pi_i + 1} \cdot x_{i, \tau - \pi_i} + \ldots + \frac{\pi_i + 2}{\pi_i + 1} \cdot x_{i, \tau - 1} + x_{i\tau} \leq 2 \cdot \frac{\tau + \pi_i}{\pi_i + 1}. \quad (5.2)$$

Round the coefficients and the right-hand side down to yield a Chvátal-Gomory inequality, which gives the result for $a = 2$.

$$\sum_{t \in Q_q} x_{it} \leq \left\lfloor 2 \cdot \frac{\tau + \pi_i}{\pi_i + 1} \right\rfloor = 2 + \left\lfloor 2 \cdot \frac{\tau - 1}{\pi_i + 1} \right\rfloor$$

Table 5.2 Upper bounds on the number of tasks for (U)

i	q	u'_{iq}	u''_{iq}	u'''_{iq}	u''''_{iq}	u_{iq}
1	1	4	3	4	4	3
1	2	1	2	3	2	1
2	1	3	4	2	2	2
2	2	1	3	1	2	1

For $2 \cdot \pi_i + 2 > |Q_q|$, the aggregation that yields (5.2) would yield coefficients larger or equal to two. Hence, (T) is trivially applied. If $2 \cdot \pi_i + 2 > |Q_q| > \pi_i + 1$ holds, apply (T) twice and obtain $\sum_{t \in Q_q} x_{it} \leq 4 = 2 + \left\lfloor 2 \cdot \frac{\tau-1}{\pi_i+1} \right\rfloor$ with $\tau < 2 \cdot \pi_i + 2$, which yields the result for $a = 2$. If $\pi_i + 1 \geq |Q_q|$ holds, then (T) applies once, which gives the result for $a = 0$.

Example 5.2 Let $n = 2$, $T = 6$, $\pi = (2, 4)^T$, $r = (3, 4)^T$ and $\bar{r} = (3, 4, 4, 7, 0, 4)^T$. The optimal solution has $Z = 3$ and $x_{12} = x_{14} = x_{16} = x_{24} = 1$ and $y_2 = y_4 = y_6 = 1$. Example 5.1 shows $m = 2$, $Q_1 = \{1, \ldots, 4\}$, $Q_2 = \{5, 6\}$, $S_1 = \{1, 3, 5\}$ and $S_2 = \{1\}$. Denote the first (second, third, fourth) term of u_{iq} in Lemma 5.2 as u'_{iq} (u''_{iq}, u'''_{iq}, u''''_{iq}). Let $UB = 4$ and it follows $X = \begin{pmatrix} 1 & 2 \\ 0 & 1 \end{pmatrix}$ and $Y = 19$. Table 5.2 demonstrates that there is no dominance between the four upper bounds.

The *non-coupling capacity constraint* (D) is valid since it equals (C) but does not couple the decision variables.

$$\sum_{i=1}^{n} r_i \cdot x_{it} \leq \bar{r}_t \quad \forall t = 1, \ldots, T \qquad (D)$$

Assuming (S) holds, Example 5.3 implies that (C) and (V) together yield a tighter formulation than (C) or (D) and (V) (i.e., $H(C)(V)(S) \subseteq H(C)(S), H(D)(V)(S)$). However, constraint (V) can be aggregated in various ways but these aggregations are not considered further because the problem structure becomes less distinct.[7]

Example 5.3 Let $n = 2$, $T = 1$, $\bar{r}_1 = 1$ and $r = \left(1, \frac{b+c}{a+b}\right)^T$ with $a, b \geq 0$ and $c \in \mathbb{R}$. Let $a + b + c \leq 1$ and $b + c > 0 \wedge a + b > 0$ to yield $r_2 > 0$. Given a solution $x_{11} = a$ and $x_{21} = a + b$. Clearly, $x \in H(C)(S)$ and $x \in H(D)(V)(S)$. It follows $a + b + c \leq y$ in (C) and $\max\{a, a + b\} \leq y$ in (V). Thus, if $c > 0$, (C) is more restrictive than (V). If $c < 0$, (V) is more restrictive than (C).

---

[7]Disaggregated formulations might have several advantages. The aggregated formulation replaces (V) by the linear combination of (V) that is $\sum_{i=1}^{n} x_{it} \leq n \cdot y_t \ \forall t = 1, \ldots, T$. However, the LP relaxation yields a weaker lower bound than the disaggregated formulation with (V) does (cf. the UFLP (Krarup and Pruzan 1983)). Furthermore, disaggregated formulations might yield integral polytopes (cf. the Standardization Problem (Domschke and Wagner 2005)).

5.1 Valid Inequalities and Polyhedral Properties

The *minimal capacity constraint* (C^y) is implied by (P) as well as (C) and stated in following lemma. (C^y) can be used to define a necessary condition for the existence of a feasible solution to the CPMP. Let $r_{n+1} = 0$ and $\pi_{n+1} = T$.

Lemma 5.3

$$\sum_{\tau=t}^{t+\pi_i-1} \bar{r}_\tau \cdot y_\tau \geq \sum_{j=1}^{i} r_j \cdot \left\lfloor \frac{\pi_i}{\pi_j} \right\rfloor \quad \forall i = 1, \ldots, n+1; t = 1, \ldots, T - \pi_i + 1 \quad (C^y)$$

Proof Let $i = 1, \ldots, n+1$ and $t = 1, \ldots, T$. From (C) follows with $x_{n+1,t} = 0$ that

$$\bar{r}_t \cdot y_t \geq \sum_{j=1}^{n+1} r_j \cdot x_{jt} \geq \sum_{j=1}^{i} r_j \cdot x_{jt}$$

holds, which is summed up for all periods $\tau = t, \ldots, t + \pi_i - 1$.

$$\sum_{\tau=t}^{t+\pi_i-1} \bar{r}_\tau \cdot y_\tau \geq \sum_{\tau=t}^{t+\pi_i-1} \sum_{j=1}^{i} r_j \cdot x_{j\tau} = \sum_{j=1}^{i} r_j \cdot \sum_{\tau=t}^{t+\pi_i-1} x_{j\tau} \quad \text{with (A) and } \pi_j \leq \pi_i$$

$$\geq \sum_{j=1}^{i} r_j \cdot \left\lfloor \frac{\pi_i}{\pi_j} \right\rfloor$$

The CPMP possesses all polyhedral properties of the WUPMP and of the UPMP, if it holds that $\bar{r}_t = r^{sum}$ $\forall t$. Therefore, this section assumes that there exists at least one period where the capacity constraint (C) is restrictive. The following simple examples show that even for very restricted instances the LP relaxation of (X) and (Y) yields a lower bound and that the polytope $H(P)(C)(V)(S)$ can contain fractional extreme points.

Example 5.4 Let $n = 2$, $T = 3$, $\pi = (2, 3)^T$, $r = (1, 1)^T$ and $\bar{r} = (2, 1, 1)^T$. Extreme points of the polytope $H(P)(C)(V)(S)$ are the fractional optimal solution $x = \begin{pmatrix} \frac{1}{2} & \frac{1}{2} & \frac{1}{2} \\ \frac{1}{2} & \frac{1}{2} & 0 \end{pmatrix}$, $y = (\frac{1}{2}, 1, \frac{1}{2})^T$ and the optimal integer solution $x_{12} = x_{21} = 1$ and $y_1 = y_2 = 1$ that have an objective function value 2.

Example 5.5 Let $n = 2$, $T = 2$, $\pi = (2, 2)^T$, $r = (1, 1)^T$ and $\bar{r} = (1\frac{1}{2}, 1)^T$. Extreme points of the polytope $H(P)(C)(V)(S)$ are the fractional optimal solution $x = \begin{pmatrix} \frac{1}{2} & \frac{1}{2} \\ 1 & 0 \end{pmatrix}$, $y = (1, \frac{1}{2})^T$, which has an optimal objective function value $1\frac{1}{2}$, and the optimal integer solution $x_{12} = x_{21} = 1$ and $y_1 = y_2 = 1$, which has an objective function value 2.

Since the objective function values of the fractional and integer solution coincide in Example 5.4, Remark 5.1 exemplary proves that the fractional optimal solution is an extreme point of the polytope $H(P)(C)(V)(S)$.

Remark 5.1 The basic matrix A_B to the coefficient matrix of the constraints (P), (C), (V) and (S^y) contains the columns of the basic variables x_{11}, y_1, x_{21}, s_7, x_{22}, x_{23}, x_{12}, s_{12}, x_{13}, y_2, s_{10}, y_3, s_{13}, s_9 and s_{15}. Denote s_7, s_9, s_{10} and s_{12} (s_{13} and s_{15}) as the slack variables of constraint (V) (of constraint (S^y)). The vector of basic variables $x_B \in \mathbb{R}_+^{15}$ is obtained by $x_B = A_B^{-1} \cdot b$ (Chvátal 2002, p. 120). This requires the inverse matrix of A_B that is

$$A_B^{-1} = \begin{pmatrix}
\frac{1}{2} & 0 & \frac{1}{2} & -\frac{1}{2} & \frac{1}{2} & \frac{1}{2} & 0 & 1 & 0 & 0 & -\frac{1}{2} & 0 & 0 & \frac{1}{2} & 0 \\
\frac{1}{2} & 0 & \frac{1}{2} & \frac{1}{2} & \frac{1}{2} & \frac{1}{2} & 0 & 0 & 0 & 0 & -\frac{1}{2} & 0 & 0 & \frac{1}{2} & 0 \\
\frac{1}{2} & 0 & \frac{1}{2} & \frac{1}{2} & \frac{1}{2} & \frac{1}{2} & 0 & -1 & 0 & 0 & -\frac{1}{2} & 0 & 0 & \frac{1}{2} & 0 \\
0 & 0 & 0 & 1 & 0 & 0 & -1 & -1 & 0 & 0 & 0 & 0 & 0 & 0 & 0 \\
-\frac{1}{2} & 0 & \frac{1}{2} & -\frac{1}{2} & -\frac{1}{2} & \frac{1}{2} & 0 & 1 & 0 & 0 & -\frac{1}{2} & 0 & 0 & -\frac{1}{2} & 0 \\
0 & 0 & 0 & 0 & 0 & -1 & 0 & 0 & 0 & 0 & 1 & 0 & 0 & 0 & 0 \\
\frac{1}{2} & 0 & -\frac{1}{2} & \frac{1}{2} & -\frac{1}{2} & -\frac{1}{2} & 0 & -1 & 0 & 0 & \frac{1}{2} & 0 & 0 & -\frac{1}{2} & 0 \\
-\frac{1}{2} & 1 & \frac{1}{2} & -\frac{1}{2} & \frac{1}{2} & \frac{3}{2} & 0 & 1 & 0 & 0 & -\frac{1}{2} & -1 & 0 & \frac{1}{2} & 0 \\
-\frac{1}{2} & 1 & \frac{1}{2} & -\frac{1}{2} & \frac{1}{2} & \frac{1}{2} & 0 & 1 & 0 & 0 & -\frac{1}{2} & 0 & 0 & \frac{1}{2} & 0 \\
0 & 0 & 0 & 0 & 0 & 0 & 0 & 0 & 0 & 0 & 0 & 0 & -1 & 0 & 0 \\
\frac{1}{2} & 0 & -\frac{1}{2} & \frac{1}{2} & -\frac{1}{2} & -\frac{1}{2} & 0 & -1 & 0 & -1 & \frac{1}{2} & 0 & 0 & -\frac{1}{2} & 0 \\
-\frac{1}{2} & 1 & \frac{1}{2} & -\frac{1}{2} & \frac{1}{2} & \frac{1}{2} & 0 & 1 & 0 & 0 & \frac{1}{2} & 0 & 0 & \frac{1}{2} & 0 \\
-\frac{1}{2} & 0 & -\frac{1}{2} & -\frac{1}{2} & -\frac{1}{2} & -\frac{1}{2} & 0 & 0 & 0 & 0 & \frac{1}{2} & 0 & -1 & -\frac{1}{2} & 0 \\
-\frac{1}{2} & 0 & \frac{1}{2} & -\frac{1}{2} & \frac{1}{2} & \frac{1}{2} & 0 & 1 & -1 & 0 & -\frac{1}{2} & 0 & 0 & -\frac{1}{2} & 0 \\
\frac{1}{2} & -1 & -\frac{1}{2} & \frac{1}{2} & -\frac{1}{2} & -\frac{1}{2} & 0 & -1 & 0 & 0 & -\frac{1}{2} & 0 & 0 & -\frac{1}{2} & -1
\end{pmatrix}$$

and $b = (1, 1, 1, 0, 0, 0, 0, 0, 0, 0, 0, 0, -1, -1, -1)^T$, which is the vector of the right-hand side of A_B. Define the non-basic variables $x_N = 0 \in \mathbb{R}_+^9$. The basic feasible solution ($\begin{smallmatrix}x_B\\x_N\end{smallmatrix}$) coincides with the fractional optimal solution of Example 5.4. A basic feasible solution is an extreme point. Consequently, the fractional optimal solution from Example 5.4 is an extreme point of $H(P)(C)(V)(S^y)$ and hence, also of $H(P)(C)(V)(S)$.

The following three lemmata show that the polytope $H(P)(C)(V)(S)$ might be integral or have optimal integer solutions. These polyhedral results prove that the aforementioned Examples 5.4 and 5.5 have the smallest possible dimensions n and T such that the extreme points are fractional. In addition, neither π can be changed in Example 5.4 nor $\bar{r}$ can be set to integer values in Example 5.5.

Lemma 5.4 If $n = 1$, then $\text{Conv}(P)(C)(V)(X)(Y) = H(P)(C)(V)(S)$ holds.

Proof '⇒': Since $H(P)(C)(V)(S)$ is convex, $\text{Conv}(P)(C)(V)(X)(Y) \subseteq H(P)(C)(V)(S)$ holds.

'⇐': Because of the Assumptions (2.3) and (2.4), it holds that $\bar{r}_t \in \{0, r_1\}$ $\forall t$. $\text{Conv}(P)(C)(V)(X)(Y) \supseteq H(P)(C)(V)(S)$ is shown by an argument presented for Lemma 4.8 that uses Theorem 4.1 with $a_{1t} = 1$ $\forall t$ and $b_{1t} = \frac{\bar{r}_t}{r_1}$ $\forall t$.

Lemma 5.5 If $r_i = 1$ $\forall i = 1, \ldots, n \wedge \pi_i = T$ $\forall i = 1, \ldots, n \wedge \bar{r}_t \in \{0, 1\}$ $\forall t = 1, \ldots, T$, then $\text{Conv}(P)(C)(V)(X)(Y) = H(P)(C)(V)(S)$ holds.

Proof Denote A as the coefficient matrix of (P) and (C) that is shown in Table 5.3 for the maintenance activity i. The matrix A is a $\{-1, 0, 1\}$ matrix and it has at most two non-zero elements per column. An asterisk in Table 5.3 represents the assignment

5.1 Valid Inequalities and Polyhedral Properties

Table 5.3 Coefficient matrix of (P) and (C) for $r_i = 1 \; \forall i \wedge \pi_i = T \; \forall i \wedge \bar{r}_t \in \{0, 1\} \; \forall t$

	x_{i1}	...	x_{iT}	y_1	...	y_T	A_1	A_2
(P)	1	...	1				*	
(C)	-1			1			*	
		⋱			⋱		⋮	
			-1			1	*	

of the rows of A to the set A_1. Let $A_2 = \emptyset$. Since A satisfies Lemma 3.7, A is totally unimodular. The coefficient matrix with slack variables and (S) satisfies Lemma 3.8 and thus, $Conv(P)(C)(X)(Y) = H(P)(C)(S)$.

Observe that (V) is redundant because (C) is more restrictive. Therefore, $Conv(P)(C)(V)(X)(Y) = Conv(P)(C)(X)(Y)$ and $H(P)(C)(V)(S) = H(P)(C)(S)$. Using the above result, $Conv(P)(C)(V)(X)(Y) = H(P)(C)(V)(S)$ follows.

Lemma 5.6 If $r_i = 1 \; \forall i = 1, \ldots, n \wedge \pi_i = T \; \forall i = 1, \ldots, n \wedge \bar{r}_t \in \{0, 1, n\} \; \forall t = 1, \ldots, T$, the smallest convex set of all optimal solutions to $H(P)(C)(V)(S)$ has integer extreme points.

Proof If there exists no period t with $\bar{r}_t = n$, the result follows from Lemma 5.5. Otherwise, assume an optimal solution $(\bar{x}, \bar{y}) \in H(P)(C)(V)(S)$ where at least one fractional component $0 < \bar{x}_{it} < 1$ or $0 < \bar{y}_t < 1$. The proof is conducted in three steps.

First, it is proven that $Z = 1$. Because of (P^y), it holds that $Z \geq 1$. Since there exists a period τ with $\bar{r}_\tau = n$, the solution $y_t = 1$ and $x_{it} = 1$ if $t = \tau$ (0, otherwise) $\forall i; t$ is feasible. Thus, the optimal objective function value is $Z = 1$.

Second, it is shown that an optimal solution satisfies

$$x_{it} = y_t \quad \forall i = 1, \ldots, n; t = 1, \ldots, T. \tag{5.3}$$

Assume there exists a pair $i = 1, \ldots, n$ and $\tau = 1, \ldots, T$ with $\bar{x}_{i\tau} < \bar{y}_\tau$. From (P) and (V) follows

$$1 \leq \sum_{t=1}^{T} \bar{x}_{it} < \sum_{t=1 \wedge t \neq \tau}^{T} \bar{x}_{it} + \bar{y}_\tau \leq \sum_{t=1}^{T} \bar{y}_t.$$

The result $\sum_{t=1}^{T} \bar{y}_t > 1$ contradicts optimality because $Z = 1$. Hence, such a pair i and τ does not exist and (5.3) holds for each optimal solution.

Finally, it is proven that there exists a convex combination of K integer solutions that yield $(\bar{x}, \bar{y})$. Therefore, each optimal solution in $H(P)(C)(V)(S)$ is obtained from a convex combination of optimal integer solutions and $(\bar{x}, \bar{y})$ cannot be an extreme point. Let K be the number of non-zeros in $\bar{y}$. Let the function $\tau(k)$ yield the period t where the non-zero $\bar{y}_t > 0$ is the kth non-zero in $\bar{y}$ for a given $k = 1, \ldots, K$.

Table 5.4 The convex combination (5.4) yields the fractional solution $(\bar{x}, \bar{y})$

	t				
	1	2	3	4	5
$\bar{y}_t$		0.2	0.5	0.3	
$\bar{x}_{1t}$		0.2	0.5	0.3	
$\bar{x}_{2t}$		0.2	0.5	0.3	
y_t^1		1			
x_{1t}^1		1			
x_{2t}^1		1			
y_t^2			1		
x_{1t}^2			1		
x_{2t}^2			1		
y_t^3				1	
x_{1t}^3				1	
x_{2t}^3				1	

Since (5.3) holds, construct K integer solutions (x^k, y^k) with $x^k \in \{0, 1\}^{n \times T}$, $y^k \in \{0, 1\}^T$ and a scalar α_k for all $k = 1, \ldots, K$ as follows (see Example 5.6): For all $k = 1, \ldots, K$, set $y_t^k = 1$ if $t = \tau(k)$ (otherwise, 0) $\forall t = 1, \ldots, T$, $x_{it}^k = y_t^k$ $\forall i; t$ and $\alpha_k = \bar{y}_{\tau(k)}$.
K optimal integer solutions $(x^k, y^k) \in H(P)(C)(V)(S)$ $\forall k$ are clearly obtained. Hence,

$$(\bar{x}, \bar{y}) = \sum_{k=1}^{K} \alpha_k \cdot (x^k, y^k) \tag{5.4}$$

is a *strict* convex combination since $\alpha_k > 0$ $\forall k$ and $\sum_{k=1}^{K} \alpha_k = 1$ because (P^y) holds and an optimal solution has $Z = 1$. This contradicts that $(\bar{x}, \bar{y})$ is an extreme point (Nemhauser and Wolsey 1988, p. 93).

Example 5.6 Let $n = 2$, $T = 5$, $\pi = (5, 5)^T$, $r = (1, 1)^T$ and $\bar{r} = (1, 2, 2, 2, 0)^T$. In Table 5.4, K feasible integer solutions (x^k, y^k) are obtained from the feasible fractional solution $(\bar{x}, \bar{y})$. It follows $K = 3$, $\alpha = (0.2, 0.5, 0.3)^T$ and $(\bar{x}, \bar{y})$ is obtained by the convex combination (5.4).

A similar result to Lemma 5.6 with $\bar{r}_t = n$ $\forall t$ is Lemma 4.19 for the UPMP.

5.2 Computational Complexity

Assume that a CPMP instance comprises only integers (Remark 2.2). Table 5.5 summarizes the results about the computational complexity. Note that polynomially solvable problem variants are presented if and only if the approach of a more general

5.2 Computational Complexity

Table 5.5 Computational complexity of capacitated planned maintenance problems

Problem variant	Computational complexity	Proof
$\pi_i = T \ \forall i \wedge \bar{r}_t = \bar{r}_1 \ \forall t$	Strongly $\mathcal{NP}$-hard	Lemma 5.7
General case	$O(\min\{n \cdot \frac{\log T}{\sqrt{T}} \cdot \bar{r}^{max\,T} \cdot 4^T,$	Lemma 5.9
	$n \cdot T^{n+1} \cdot 2^n\})$	
$n = 1$	$O(T)$	Lemma 5.10
$n = k \in O(1)$	$O(T^{k+1})$	Lemma 5.9
$T = 2$	Binary $\mathcal{NP}$-hard	Lemma 5.11
	$O(n \cdot \min\{\bar{r}_1, \bar{r}_2\})$	
$T = k \in O(1)$	$O(n \cdot \bar{r}^{max\,k})$	Lemma 5.9
$T \in O(1) \wedge \bar{r}_t \in O(1) \ \forall t$	$O(n)$	Lemma 5.9
$\pi_i = 2 \ \forall i \wedge \bar{r}_t = \bar{r}_{t+2} \ \forall t$	$O(n \cdot \min\{\bar{r}_1, \bar{r}_2\} + n \cdot T)$	Lemma 5.12
$\pi_i = 2 \ \forall i \wedge \bar{r}_t = \bar{r}_{t+2} \ \forall t \wedge \bar{r}_t \in O(1) \ \forall t$	$O(n \cdot T)$	Lemma 5.12
$\pi_i = T \ \forall i$	$O(\min\{n \cdot T \cdot \log T \cdot \bar{r}^{max\,T},$	Lemma 5.13
	$n \cdot T \cdot 4^n\})$	
$\pi_i = T \ \forall i \wedge n \in O(1)$	$O(T)$	Lemma 5.13
$r_i = r_1 \ \forall i$	Open	
	$O(\min\{n^{T+1} \cdot \frac{\log T}{\sqrt{T}} \cdot 4^T,$	Lemma 5.14
	$\sqrt{n} \cdot T^{n+1} \cdot 2^n\})$	
$r_i = r_1 \ \forall i \wedge T = k \in O(1)$	$O(n^{k+1})$	Lemma 5.14
$r_i = r_1 \ \forall i \wedge \pi_i = \pi_1 \ \forall i$	Open	
	$O(\min\{n^T \cdot T^2, n \cdot T^{n+1}\})$	Lemma 5.15
$r_i = r_1 \ \forall i \wedge \pi_i = \pi_1 \ \forall i \wedge T = k \in O(1)$	$O(n^k)$	Lemma 5.15
$r_i = r_1 \ \forall i \wedge \pi_i = \pi_1 \ \forall i \wedge \pi_1 = k \in O(1)$	$O(n^k \cdot T)$	Lemma 5.15
$r_i = r_1 \ \forall i \wedge \pi_i = T \ \forall i$	$O(n + T \cdot \log T)$	Lemma 5.16
$r_i = r_1 \ \forall i \wedge \bar{r}_t = \bar{r}_1 \ \forall t$	Open	
	$O(\min\{n^{T+1} \cdot \frac{\log T}{\sqrt{T}} \cdot 4^T,$	Lemma 5.14
	$\sqrt{n} \cdot T^{n+1} \cdot 2^n\})$	
$r_i = r_1 \ \forall i \wedge \pi_i = \pi_1 \ \forall i \wedge \bar{r}_t \leq \bar{r}_{t+1} \ \forall t$	$O(\min\{n^2 \cdot T, T^3 + n \cdot T\})$	Lemma 5.17

problem variant is less efficient. The CPMP is strongly $\mathcal{NP}$-hard even if $\pi_i = T \ \forall i$ and $\bar{r}_t = \bar{r}_1 \ \forall t$ holds (Lemma 5.7). Furthermore, the CPMP is binary $\mathcal{NP}$-hard even if $T = 2$ holds (Lemma 5.11) but pseudo-polynomially solvable if $T \in O(1)$ (Lemma 5.9). The CPMP becomes strongly polynomially solvable if either $n \in O(1)$ or if $T \in O(1) \wedge \bar{r}_t \in O(1) \ \forall t$ holds (Lemma 5.9). The most applicable problem variant for real-world applications is the open problem $r_i = r_1 \ \forall i$. Lemma 5.15 states the interesting result that the open problem $r_i = r_1 \ \forall i \wedge \pi_i = \pi_1 \ \forall i$ is already strongly polynomially solvable if $\pi_1 \in O(1)$.[8]

In Sect. 5.1, two cases with integral polytopes are presented, namely Lemma 5.4 and Lemma 5.5. Thus, optimal integer solutions are obtained by solving the LP relaxation with a variant of Tardos' algorithm (Vavasis and Ye 1996) in time $O(n^{3.5} \cdot$

---
[8]Some results are improved in Kuschel and Bock (2016).

$T^{3.5} \cdot L$) (cf. Lemma 4.10). More efficient algorithms solve the case $n = 1$ from Lemma 5.4 in time $O(T)$ (Lemma 5.10) and the case $r_i = 1 \; \forall i \wedge \pi_i = T \; \forall i \wedge \bar{r}_t \in \{0, 1, n\} \; \forall t$ from Lemma 5.6 that generalizes Lemma 5.5 in time $O(n + T \cdot \log T)$ (Lemma 5.16).

Lemma 5.7 The CPMP is strongly $\mathcal{NP}$-hard even if $\pi_i = T \; \forall i = 1, \ldots, n \wedge \bar{r}_t = \bar{r}_1 \; \forall t = 1, \ldots, T$. Furthermore, finding a feasible solution to the CPMP is strongly $\mathcal{NP}$-complete.

Proof A polynomial reduction from the strongly $\mathcal{NP}$-complete 3-PARTITION shows strong $\mathcal{NP}$-hardness of the CPMP that is 3-PARTITION $\leq_p$ CPMP$_{\pi_i=T \wedge \bar{r}_t=\bar{r}_1}$.[9]

3-PARTITION (Garey and Johnson 1979, pp. 96–100, 224; Garey and Johnson 1978): Given $P \in \mathbb{N}$ bins of a capacity $B \in \mathbb{N}$ and $3 \cdot P \in \mathbb{N}$ items with a size $s_i \in \mathbb{N} \; \forall i$ such that $\frac{B}{4} < s_i < \frac{B}{2} \; \forall i$ and $\sum_{i=1}^{3 \cdot P} s_i = P \cdot B$. Exists a partition of all items into disjoint sets $A_1, \ldots, A_P$ such that $\sum_{i \in A_t} s_i = B \; \forall t$? Note that the definition of s_i implies $|A_t| = 3 \; \forall t$.

An instance of the CPMP is obtained in polynomial time with $n = 3 \cdot P$, $T = P$, $\pi_i = P \; \forall i$, $r_i = s_i \; \forall i$ and $\bar{r}_t = B \; \forall t$. The feasibility problem states that the CPMP instance is called solvable if it is feasible and the objective function value is smaller or equal to T. Equivalence of both instances is proven by showing that the instance to 3-PARTITION is solvable if and only if the instance to the CPMP is solvable.

'$\Rightarrow$': Assume the instance to 3-PARTITION is solvable. Hence, there exists a partition $A_1, \ldots, A_P$ that satisfies 3-PARTITION. Let $y_t = 1 \; \forall t$ and $x_{it} = 1$ if $i \in A_t$ (otherwise, 0) $\forall i, t$ be a solution to the CPMP. For every item there exists a set A_t where the item is contained exactly once and this satisfies (P) as an equality. Because of $y_t = 1 \; \forall t$, (V) is satisfied and together with $\sum_{i \in A_t} s_i = B \; \forall t$, (C) holds. Thus, (x, y) is feasible to the CPMP with an objective function value $\sum_{t=1}^{T} y_t = T$.

'$\Leftarrow$': Assume that the instance to the CPMP is solvable. Hence, there exists a feasible solution (x, y) to the CPMP with an objective function value smaller or equal to T. Let $A_t = \{i = 1, \ldots, n \mid x_{it} = 1\} \; \forall t$ be a solution to 3-PARTITION. Since $\sum_{i=1}^{3 \cdot P} s_i = P \cdot B$ and there are $P = T$ periods with $\bar{r}_t = B \; \forall t$, the slack of (C) is zero that is $\sum_{i \in A_t} s_i = B \; \forall t$ and every item is assigned to exactly one set A_t. Thus, $A_1, \ldots, A_P$ is a partition and the solution satisfies 3-PARTITION.

The feasibility problem is a decision problem and in $\mathcal{NP}$ because there exists an algorithm that checks in polynomial time if a given maintenance plan satisfies the feasibility problem. Thus, strong $\mathcal{NP}$-completeness of the feasibility problem holds.

Lemma 5.8 states an upper bound on the binomial coefficient that is used in the proofs below. This upper bound is tighter than other upper bounds such as $\left(\frac{e \cdot n}{k}\right)^k$, which is derived with Stirling's approximation (Cormen et al. 2001, p. 1119).

---

[9] $\mathcal{NP}$-hardness can be derived from the Generalized Bin Packing Problem (see Footnote 34 on page 150).

5.2 Computational Complexity

Lemma 5.8

$$\binom{n}{k} \leq \frac{2^n}{\sqrt{n}} \text{ where } n, k \in \mathbb{N}; k \leq n; n \geq 1$$

Proof It is well-known that $\binom{n}{k} \leq \binom{n}{\lfloor n/2 \rfloor}$ holds. In what follows, an upper bound to the central binomial coefficient $\binom{n}{\lfloor n/2 \rfloor}$ is presented. Assume that n is an even number. It holds that $\binom{2 \cdot h}{h} \leq 2^{2 \cdot h} \cdot (2 \cdot h)^{-1/2}$ with $h \in \mathbb{N} \setminus \{0\}$ (Koshy 2009, p. 50). Therefore, $\binom{n}{\lfloor n/2 \rfloor} = \binom{n}{n/2} \leq 2^n \cdot n^{-1/2}$ follows. Otherwise, n is an odd number and it holds that $\binom{2 \cdot h}{h} \leq 2^{2 \cdot h} \cdot (2 \cdot h + 1)^{-1/2}$ with $h \in \mathbb{N} \setminus \{0\}$ (Koshy 2009, pp. 48–49). It follows

$$\binom{n}{\lfloor \frac{n}{2} \rfloor} = \frac{n}{n - \lfloor \frac{n}{2} \rfloor} \cdot \binom{n-1}{\lfloor \frac{n}{2} \rfloor} \quad \text{because } \binom{n}{k} = \frac{n}{n-k} \cdot \binom{n-1}{k} \text{ (Koshy 2009, p. 5)}$$

$$\leq 2 \cdot \binom{n-1}{\lfloor \frac{n}{2} \rfloor} \quad \text{because } \frac{n}{2} \leq n - \lfloor \frac{n}{2} \rfloor$$

$$= 2 \cdot \binom{n-1}{\frac{n-1}{2}} \quad \text{because } n-1 \text{ is an even number}$$

$$\leq 2 \cdot \frac{2^{n-1}}{\sqrt{n}} = \frac{2^n}{\sqrt{n}}.$$

Lemma 5.9 *The CPMP is solvable in time $O(\min\{n \cdot \frac{\log T}{\sqrt{T}} \cdot \bar{r}^{max \, T} \cdot 4^T, n \cdot T^{n+1} \cdot 2^n\})$. Therefore, if $T = k \in O(1)$, the CPMP is pseudo-polynomially solvable in time $O(n \cdot \bar{r}^{max \, k})$. Additional restrictions state strong polynomial solvability:*

$$n = k \in O(1) \Rightarrow O(T^{k+1})$$

$$T \in O(1) \wedge \bar{r}_t \in O(1) \, \forall t \Rightarrow O(n)$$

Proof Two optimal algorithms for the CPMP are presented.

'$O(n \cdot T^{-\frac{1}{2}} \cdot \log T \cdot \bar{r}^{max \, T} \cdot 4^T)$': The first algorithm solves a reformulation of the CPMP via dynamic programming. Let P_i with $i = 1, \ldots, n$ be the index set of all task plans of the maintenance activity i that satisfy (P). Specifically, the pth task plan with $p \in P_i$ of the maintenance activity i is a vector $(\hat{x}_{i1}^p, \ldots, \hat{x}_{iT}^p) \in \{0, 1\}^T$ that satisfies (P) as

$$\sum_{\tau=t}^{t+\pi_i-1} \hat{x}_{i\tau}^p \geq 1 \quad \forall i = 1, \ldots, n; p \in P_i.$$

Let P_0 be the index set of all slot plans. Since (P^y) holds in the CPMP, it follows that $P_0 = P_1$ and $(\hat{y}_1^p, \ldots, \hat{y}_T^p) = (\hat{x}_{11}^p, \ldots, \hat{x}_{1T}^p)$ $\forall p \in P_0$. The number of open periods f_p is $f_p = \sum_{t=1}^{T} \hat{y}_t^p$ $\forall p \in P_0$. The binary variable z_{ip} equals 1 if the pth task or slot plan with $p \in P_i$ is selected for i with $i = 0, \ldots, n$ (otherwise, 0). The CPMP is reformulated as

$$\min \sum_{p \in P_0} f_p \cdot z_{0p} \qquad \text{subject to} \qquad (5.5)$$

$$\sum_{p \in P_i} z_{ip} = 1 \qquad \forall i = 0, \ldots, n \qquad (5.6)$$

$$\sum_{i=1}^{n} \sum_{p \in P_i} r_i \cdot \hat{x}_{it}^p \cdot z_{ip} \leq \bar{r}_t \cdot \sum_{p \in P_0} \hat{y}_t^p \cdot z_{0p} \qquad \forall t = 1, \ldots, T \qquad (5.7)$$

$$z_{ip} \in \{0, 1\} \qquad \forall i = 0, \ldots, n; p \in P_i. \qquad (5.8)$$

The objective function (5.5) minimizes the number of established maintenance slots. Because of (5.6), exactly one slot plan $(\hat{y}_1^p, \ldots, \hat{y}_T^p)$ and one task plan $(\hat{x}_{i1}^p, \ldots, \hat{x}_{iT}^p)$ for every maintenance activity i is selected. (5.7) represents (C). The variable domain is binary because of (5.8). The variable transformation from an optimal solution z to an optimal solution (x, y) of the original CPMP formulation is

$$x_{it} = \hat{x}_{it}^p \; \forall i = 1, \ldots, n; t = 1, \ldots, T \text{ where } z_{ip} = 1$$
$$y_t = \hat{y}_t^p \; \forall t = 1, \ldots, T \text{ where } z_{0p} = 1.$$

Consider an optimal solution to the reformulated CPMP where the pth task plan with $z_{ip} = 1$ of a maintenance activity $i = 1, \ldots, n$ is removed. Clearly, the remaining partial solution is an optimal solution to the subproblem where the capacity of each period t is adjusted accordingly $\bar{r}_t - r_i \cdot \hat{x}_{it}^p$. Consequently, the reformulated CPMP (5.5)–(5.8) is solvable via dynamic programming for a given selected slot plan with $z_{0p} = 1$. Define the initial states

$$F(0, d_1, \ldots, d_T) = \begin{cases} 0 & \text{if } d_t = 0 \; \forall t \\ 1 & \text{otherwise} \end{cases} \qquad \forall t = 1, \ldots, T; d_t = 0, \ldots, \bar{r}_t \cdot \hat{y}_t^p$$

and solve the recursive equations

$$F(i, d_1, \ldots, d_T) = \min_{q \in P_i} F(i-1, d_1 - r_i \cdot \hat{x}_{i1}^q, \ldots, d_T - r_i \cdot \hat{x}_{iT}^q) \qquad (5.9)$$

for all $\forall i = 1, \ldots, n; t = 1, \ldots, T; d_t = 0, \ldots, \bar{r}_t \cdot \hat{y}_t^p$. Set $F(i, d_1, \ldots, d_T) = 1$ if there exists a $t = 1, \ldots, T$ with $d_t < 0$. If there exists a vector $(d_1, \ldots, d_T)$ with $F(n, d_1, \ldots, d_T) = 0$, a feasible solution (x, y) with f_p open periods exists. The task plans x are obtained from backtracking and the slot plan is determined by the given $p \in P_0$.

5.2 Computational Complexity

The number of configurations of the vector $(d_1, \ldots, d_T)$ is asymptotically upper bounded by $O(\bar{r}^{max\,T})$. Therefore, the number of states $F(i, d_1, \ldots, d_T)$ is asymptotically bounded from above by $O(n \cdot \bar{r}^{max\,T})$. Since the ith index set P_i is a power set, it holds for the cardinality that $|P_i| \leq 2^T$ and each state is evaluated in time $O(2^T)$. Thus, the equations (5.9) are solvable in time $O(n \cdot \bar{r}^{max\,T} \cdot 2^T)$.

In order to solve the CPMP to optimality, the decision problem of the CPMP asks for the existence of a feasible solution where the number of established maintenance slots is smaller or equal to $K \in \mathbb{N}$. Hence, (5.9) has to be solved for each slot plan $p \in P_0$ with $f_p = K$. The number of these slot plans is upper bounded by $\binom{T}{K} \leq \frac{2^T}{\sqrt{T}}$ (Lemma 5.8). A subsequent binary search over the parameter K yields the optimum of the CPMP in additional time $O(\log T)$ (Ottmann and Widmayer 2012, pp. 174–176). Hence, the decision problem is solved at most $O(2^T \cdot T^{-\frac{1}{2}} \cdot \log T)$ times.[10] Constructing the solution (x, y) requires $O(n \cdot T)$. A set P_i and the respective vector can be constructed in time $O(T \cdot 2^T)$. Hence, all index sets $P_0, \ldots, P_n$ and the vectors $\hat{x}$ and $\hat{y}$ can be obtained in time $O(n \cdot T \cdot 2^T)$. Thus, the first algorithm solves the CPMP in time $O(2^T \cdot T^{-\frac{1}{2}} \cdot \log T) \cdot O(n \cdot \bar{r}^{max\,T} \cdot 2^T) + O(n \cdot T) + O(n \cdot T \cdot 2^T) = O(n \cdot T^{-\frac{1}{2}} \cdot \log T \cdot \bar{r}^{max\,T} \cdot 4^T + n \cdot T + n \cdot T \cdot 2^T)$. For the worst case running time, neglect the factor $O(\log T)$ and consider

$$n \cdot T^{-\frac{1}{2}} \cdot \bar{r}^{max\,T} \cdot 4^T + n \cdot T + n \cdot T \cdot 2^T \leq d \cdot n \cdot T^{-\frac{1}{2}} \cdot \bar{r}^{max\,T} \cdot 4^T.$$

There exists a constant $d \geq 1 + \frac{T^{\frac{3}{2}}}{\bar{r}^{max\,T} \cdot 4^T} + \frac{T^{\frac{3}{2}}}{\bar{r}^{max\,T} \cdot 2^T} > 0$ for all $n, T \geq 1$ and $\bar{r}^{max} > 0$ because $\frac{T^{\frac{3}{2}}}{4^T}, \frac{T^{\frac{3}{2}}}{2^T} < 1$.[11] Since $T \geq 2$ (Assumption (2.2)) and consequently, $\log_2 T \geq 1$ holds, it follows for the running time $O(n \cdot T^{-\frac{1}{2}} \cdot \log T \cdot \bar{r}^{max\,T} \cdot 4^T)$.

'$O(n \cdot \bar{r}^{max\,k})$': If $T = k \in O(1)$, the running time and the space requirement $O(n \cdot \bar{r}^{max\,k})$ is pseudo-polynomial.

'$O(n)$': If $T \in O(1)$ and $\bar{r}_t \in O(1)$ $\forall t$, the running time $O(n)$ is strongly polynomial.

'$O(n \cdot T^{n+1} \cdot 2^n)$': In the second algorithm, consider the WUPMP algorithm from Lemma 4.7. Define the set of successors from a given node in layer $t = 0, \ldots, T-1$ as $S^{succ}(t, \alpha_1, \ldots, \alpha_n) = \{(t+1, \beta_1, \ldots, \beta_n) \mid \beta_i \in \{\pi_i, \alpha_i - 1\} \; \forall i \wedge \beta_i \geq 1 \; \forall i \wedge \sum_{i=1}^{n} r_i \cdot \max\{0, \beta_i + 1 - \pi_i\} \leq \bar{r}_{t+1}\}$ to ensure that the nodes in the layers $1, \ldots, t+1$ satisfy (P) and (C). Set $f_t = 1$ $\forall t$ and $c_{it} = 0$ $\forall i; t$. The second algorithm solves the CPMP in time $O(n \cdot |V| + n \cdot |E|) + O(|V| + |E|) + O(n \cdot T) = O(n \cdot |V| + n \cdot |E| + n \cdot T) = O(n \cdot T^{n+1} \cdot 2^n)$.

---

[10] A trivial asymptotic upper bound is that the decision problem has to be solved at most $O(2^T)$ times. Since $\frac{2^T}{\sqrt{T}} \cdot \log_a T \leq 2^T$ holds already for $a \geq e^{\sqrt{4 \cdot e^{-2}}} \approx 2.087$, the applied approach has an improved asymptotic running time.

[11] The function $f(a, b, c) = c^a \cdot b^{-c}$ with $a, b, c \in \mathbb{R}$ has one maximum at $c^* = \frac{a}{\ln b}$. For $f(\frac{3}{2}, 4, c) = c^{\frac{3}{2}} \cdot 4^{-c}$ follows $c^* \approx 1.082$ and $f(\frac{3}{2}, 4, c^*) \approx 0.251$. For $f(\frac{3}{2}, 2, c) = c^{\frac{3}{2}} \cdot 2^{-c}$ holds $c^* \approx 2.164$ and $f(\frac{3}{2}, 2, c^*) \approx 0.710$.

'$O(T^{k+1})$': If $n = k \in O(1)$, the running time $O(T^{k+1})$ is strongly polynomial. Altogether, the CPMP is solvable in time $O(\min\{n \cdot T^{-\frac{1}{2}} \cdot \log T \cdot \bar{r}^{max\ T} \cdot 4^T, n \cdot T^{n+1} \cdot 2^n\})$.

The CPMP with $n = 1$ can be reformulated as a WUPMP and solved in time $O(T^2)$ (Lemma 4.7 or Lemma 4.8). However, Lemma 5.10 yields a lower running time of $O(T)$.

Lemma 5.10 If $n = 1$, the CPMP is strongly polynomially solvable in time $O(T)$.

Proof Because of the Assumptions (2.3) and (2.4), $\bar{r}_t \in \{0, r_1\}$ $\forall t$ holds. Since $n = 1$, the variables y are redundant and the CPMP simplifies to

$$Z = \min \sum_{t=1}^{T} x_{1t} \qquad \text{subject to}$$

$$\sum_{\tau=t}^{t+\pi_1-1} x_{1\tau} \geq 1 \qquad \forall t = 1, \ldots, T - \pi_1 + 1 \qquad v_{1t} \qquad \text{(P)}$$

$$x_{1t} \leq \frac{\bar{r}_t}{r_1} \qquad \forall t = 1, \ldots, T \qquad \mu_t \qquad \text{(C)}$$

$$x_{1t} \in \{0, 1\} \qquad \forall t = 1, \ldots, T. \qquad \text{(X)}$$

The corresponding dual problem of the LP relaxation yields the lower bound

$$ZD = \max \sum_{t=1}^{T-\pi_1+1} v_{1t} - \sum_{t=1}^{T} \frac{\bar{r}_t}{r_1} \cdot \mu_t \qquad \text{subject to}$$

$$\sum_{\tau=\max\{1,t-\pi_1+1\}}^{\min\{t,T-\pi_1+1\}} v_{1\tau} - \mu_t \leq 1 \qquad \forall t = 1, \ldots, T \qquad x_{1t} \qquad (5.10)$$

$$v_{1t} \geq 0 \ \forall t = 1, \ldots, T - \pi_1 + 1 \wedge \mu_t \geq 0 \ \forall t = 1, \ldots, T.$$

Construct a feasible integer primal solution (x, y) and a dual solution (v, μ) in time $O(T)$ as follows: Set $x_{1t} = 0$ $\forall t$ and $v_{1t} = 0$ $\forall t$. Let $a = 1$ and $b = \pi_1$. In an iteration, find the period t with $\bar{r}_t = r_1$ that has the largest index in $a, \ldots, b$. If there exists no period t, terminate because the instance is infeasible. Otherwise, set $x_{1t} = 1$ and $v_{1a} = 1$. Set $a = t+1$ and $b = t+\pi_1$. Commence the next iteration until $a > T - \pi_1 + 1$. Finally, set $y_t = x_{1t}$ $\forall t$ and set $\mu_t = \pi_1$ if $\bar{r}_t = 0$ (otherwise, 0) $\forall t$.

In what follows optimality is proven if the CPMP instance is feasible. Every iteration selects a period t with maximal distance to the selected period τ of the previous iteration such that $t \leq \tau + \pi_1$. This ensures that (P) is satisfied. Since (C) and (V) are clearly satisfied, (x, y) is feasible. However, if $t = \tau + \pi_1$, then constraint (5.10) holds. Otherwise, $t < \tau + \pi_1$ holds and the periods $\theta = t+1, \ldots, \tau + \pi_1$ have $\bar{r}_\theta = 0$. In this case, (5.10) is also satisfied because $\mu_\theta = \pi_1$. Thus, (v, μ) is feasible. In every iteration, it holds that $\sum_{t=1}^{T} x_{1t} = \sum_{t=1}^{T-\pi_1+1} v_{1t}$. Since $\bar{r}^T \cdot \mu = 0$, it follows $Z = ZD$ and hence, optimality because of Theorem 3.3.2.

5.2 Computational Complexity

Lemma 5.11 If $T = 2$, the CPMP is binary $\mathcal{NP}$-hard and pseudo-polynomially solvable in time $O(n \cdot \min\{\bar{r}_1, \bar{r}_2\})$.

Proof A polynomial reduction with the binary $\mathcal{NP}$-complete PARTITION shows binary $\mathcal{NP}$-hardness of the CPMP that is PARTITION $\leq_p$ CPMP$_{T=2}$.

PARTITION (Garey and Johnson 1979, p. 223): Given $I \in \mathbb{N}$ items of a size $s_i \in \mathbb{N}$. Exists a partition of the items into two disjoint sets A_1 and A_2 such that $\sum_{i \in A_1} s_i = \sum_{i \in A_2} s_i$?

Let $B = \sum_{i=1}^{I} \frac{s_i}{2}$. An instance of the CPMP is obtained in polynomial time with $n = I$, $T = 2$, $\pi_i = 2$ $\forall i$ because of Assumption (2.2), $r_i = s_i$ $\forall i$ and $\bar{r} = (B, B)^T$. The CPMP instance is called solvable if it is feasible and the objective function value is smaller or equal to T. Equivalence of both instances is proven by showing that the instance to PARTITION is solvable if and only if the instance to the CPMP is solvable.

'$\Rightarrow$': Assume that the instance to PARTITION is solvable. Hence, there exists a partition A_1 and A_2 that satisfies PARTITION. Let $y = (1, 1)^T$ and $x_{it} = 1$ if $i \in A_t$ (otherwise, 0) $\forall i; t$ be a solution to the CPMP. (V) is satisfied because $y = (1, 1)^T$. Every item is either in A_1 or in A_2, which satisfies (P). Feasibility to PARTITION requires that $\sum_{i \in A_1} s_i = \sum_{i \in A_2} s_i = B$. Together with $y = (1, 1)^T$, (C) is satisfied. Thus, (x, y) is feasible to the CPMP with an objective function value $\sum_{t=1}^{T} y_t = T$.

'$\Leftarrow$': Assume that the instance to the CPMP is solvable. Hence, there exists a feasible solution (x, y) to the CPMP with an objective function value smaller or equal to T. Since $\sum_{i=1}^{I} s_i > B$, it follows $y = (1, 1)^T$ because of (P) and (C). Let $A_1 = \{i = 1, \ldots, n \mid x_{i1} = 1\}$ and $A_2 = \{1, \ldots, n\} \setminus A_1$ be a solution to PARTITION. Because of (P), an item is either in A_1 or in A_2. Since $\sum_{i=1}^{I} s_i = 2 \cdot B$ and $\bar{r} = (B, B)^T$, the slack of (C) is zero and $\sum_{i \in A_1} s_i = \sum_{i \in A_2} s_i$ holds. Every item is assigned to exactly one set that is either A_1 or A_2. Thus, A_1, A_2 is a partition and the solution satisfies PARTITION.

In what follows, an algorithm is developed. If $r^{sum} \leq \bar{r}_2$, obtain a trivial solution $y = (0, 1)^T$ and $x_{it} = 1$ if $t = 2$ (otherwise, 0) $\forall i; t$. If $r^{sum} \leq \bar{r}_1$, obtain a trivial solution $y = (1, 0)^T$ and $x_{it} = 1$ if $t = 1$ (otherwise, 0) $\forall i; t$. Both trivial solutions are clearly feasible, optimal and obtained in time $O(n)$. If at this point $\min\{\bar{r}_1, \bar{r}_2\} = 0$ holds, the instance is infeasible. Otherwise, the CPMP is solved in what follows. Let $t = 1$ be the period with $\bar{r}_1 \leq \bar{r}_2$ and solve the SSP

$$\bar{Z} = \max \sum_{i=1}^{n} r_i \cdot x_{i1} \quad \text{subject to} \quad \sum_{i=1}^{n} r_i \cdot x_{i1} \leq \bar{r}_1 \wedge x_{i1} \in \{0, 1\} \; \forall i = 1, \ldots, n$$

in time $O(n \cdot \bar{r}_1)$ (Martello and Toth 1990, pp. 105–107).[12] Let $y = (1, 1)^T$ as well as $x_{i2} = 1 - x_{i1}$ $\forall i$ and obtain a solution (x, y) in time $O(n)$. It follows $r^{sum} - \bar{Z} \leq \bar{r}_2 + \bar{r}_1 - \bar{Z}$ where $\bar{r}_1 - \bar{Z} \geq 0$.

---

[12]Using a word-RAM implementation, the SSP is even solvable in time $O(\frac{n \cdot \bar{r}_1}{\log \bar{r}_1})$ (Kellerer et al. 2004, pp. 76–79).

If $r^{sum} - \bar{Z} > \bar{r}_2$, the instance is infeasible because (C) is violated for period $t = 2$. This is proven by contradiction. Assume there exists a feasible solution (x', y'). Satisfying (C) requires $r^{sum} - \sum_{i=1}^{n} r_i \cdot x'_{i1} \leq \bar{r}_2$. It follows $\sum_{i=1}^{n} r_i \cdot x'_{i1} > \bar{Z}$ contradicting optimality of $\bar{Z}$.

If $r^{sum} - \bar{Z} \leq \bar{r}_2$, (V) holds because of $y = (1, 1)^T$. Together with $\bar{Z} \leq \bar{r}_1$ and $r^{sum} - \bar{Z} \leq \bar{r}_2$, (C) is satisfied. (P) is satisfied because all maintenance activities are scheduled in either period 1 or 2. Thus, (x, y) is feasible and clearly optimal.

Since computational complexity is determined by the non-trivial solution, the CPMP is solvable in time $O(n \cdot \min\{\bar{r}_1, \bar{r}_2\})$.

Lemma 5.12 If $\pi_i = 2 \; \forall i = 1, \ldots, n \land \bar{r}_t = \bar{r}_{t+2} \; \forall t = 1, \ldots, T-2$, the CPMP is pseudo-polynomially solvable in time $O(n \cdot \min\{\bar{r}_1, \bar{r}_2\} + n \cdot T)$. Therefore, if additionally $\bar{r}_t \in O(1) \; \forall t$, the CPMP is strongly polynomially solvable in time $O(n \cdot T)$.

Proof '$O(n \cdot \min\{\bar{r}_1, \bar{r}_2\} + n \cdot T)$': A feasible solution (x, y) is obtained in two steps. First, yield a partial solution for $t = 1, 2$ in time $O(n \cdot \min\{\bar{r}_1, \bar{r}_2\})$ (Lemma 5.11). Afterwards, set $y_t = y_{t-2} \; \forall t = 3, \ldots, T$ and $x_{it} = x_{i,t-2} \; \forall i = 1, \ldots, n; t = 3, \ldots, T$ in time $O(n \cdot T)$. Because of $\pi_i = 2 \; \forall i$ and $\bar{r}_t = \bar{r}_{t+2} \; \forall t$, the obtained solution (x, y) is feasible. Hence, the CPMP is solvable in time $O(n \cdot \min\{\bar{r}_1, \bar{r}_2\}) + O(n \cdot T) = O(n \cdot \min\{\bar{r}_1, \bar{r}_2\} + n \cdot T)$.

In what follows, optimality of (x, y) is proven. Assume there exists a trivial solution. Lemma 5.11 yields the trivial solution if there exists a $t = 1, 2$ with $r^{sum} \leq \bar{r}_t$. The construction of the partial solution for $T = 2$ first tries to open the period $t = 2$ and afterwards, period $t = 1$. Consequently, if T is odd and $y = (1, 0)^T$ holds, then $Z = \lfloor \frac{T+1}{2} \rfloor$ (cf. (A)). Otherwise, $Z = \lfloor \frac{T}{2} \rfloor$ (cf. (A)). Optimality follows because $Z = Z^{(C)}$ and the UPMP yields a lower bound to the CPMP (Lemma 4.17).

Assume there exists no trivial solution. The non-trivial solution solves one SSP and yields $Z = T$. Assume there exists a feasible non-trivial solution (x', y') with $Z < T$. Hence, there exists a period t with $y'_t = 1$ and $y'_{t+1} = 0$. Since $r^{sum} > \bar{r}_t$, there exists a maintenance activity i with $x'_{it} = 0$. Together with $x'_{i,t+1} = 0$, (P) is violated and this contradicts feasibility. Hence, a non-trivial solution has $Z = T$ that is optimal.

'$O(n \cdot T)$': If $\bar{r}_t \in O(1) \; \forall t$, the running time $O(n \cdot T)$ is strongly polynomial. □

Lemma 5.13 If $\pi_i = T \; \forall i = 1, \ldots, n$, the CPMP is solvable in time $O(\min\{n \cdot T \cdot \log T \cdot \bar{r}^{max\,T}, n \cdot T \cdot 4^n\})$. Therefore, if additionally $n \in O(1)$, the CPMP is strongly polynomially solvable in time $O(T)$.

Proof In what follows, two optimal algorithms are presented.

'$O(n \cdot T \cdot \log T \cdot \bar{r}^{max\,T})$': The first algorithm from Lemma 5.9 can be simplified because (P) reduces to an equality and a solution is dominated by a solution with the same objective function value where periods with a higher capacity are used. This holds because swapping a non-empty schedule of an open period t with an empty schedule of a closed period τ where $\bar{r}_t \leq \bar{r}_\tau$ holds, yields a solution with the same number of established maintenance slots if t is closed and τ is opened. Therefore,

5.2 Computational Complexity

assume that $\bar{r}_1, \ldots, \bar{r}_T$ is sorted according to non-increasing values $\bar{r}_t \geq \bar{r}_{t+1}$ $\forall t$. The decision problem tries to find a feasible solution where at most K maintenance slots are established in the first K periods $t_1, \ldots, t_K$ with $1 \leq K \leq T$. The initial states are

$$F(0, d_1, \ldots, d_K) = \begin{cases} 0 & \text{if } d_t = 0 \; \forall t \\ 1 & \text{otherwise} \end{cases} \quad \forall t = 1, \ldots, K; d_t = 0, \ldots, \bar{r}_t.$$

The dynamic programming approach solves the recursive equations

$$F(i, d_1, \ldots, d_K) = \min_{\tau=1,\ldots,K} F(i-1, d_1, \ldots, d_\tau - r_i, \ldots, d_K)$$

for all $i = 1, \ldots, n; t = 1, \ldots, K; d_t = 0, \ldots, \bar{r}_t$. Set $F(i, d_1, \ldots, d_K) = 1$ if there exists a $t = 1, \ldots, K$ with $d_t < 0$. If there exists a vector $(d_1, \ldots, d_K)$ with $F(n, d_1, \ldots, d_K) = 0$, a feasible solution (x, y) with K open periods exists. The task plans x can be obtained from backtracking and the slot plan y has $y_t = 1$ if $t \leq K$ (otherwise, 0) $\forall t$. Compared to solving (5.9), the running time reduces significantly. Since each state is evaluated in time $O(T)$, the decision problem is solved in time $O(n \cdot T \cdot \bar{r}^{max \, T})$. Furthermore, one slot plan needs to be evaluated for a given parameter K. Using binary search over the parameter K, the first algorithm solves CPMP in time $O(n \cdot T \cdot \log T \cdot \bar{r}^{max \, T})$.

'$O(n \cdot T \cdot 4^n)$': The second algorithm of the CPMP from Lemma 5.9 is simplified because (P) reduces to an equality. Let an entry α_i of a partial solution $(t, \alpha_1, \ldots, \alpha_n)$ equal 1 if a maintenance activity i is performed in period t (otherwise, 0). Therefore, each layer $t = 1, \ldots, T$ has at most 2^n nodes and the number of nodes $|V|$ is asymptotically bounded from above by $O(T \cdot 2^n)$. The set of successors from a given node in the layer $t = 0, \ldots, T-1$ becomes $S^{succ}(t, \alpha_1, \ldots, \alpha_n) = \{(t+1, \beta_1, \ldots, \beta_n) \mid \beta_i \in \{0, 1\} \; \forall i \wedge \sum_{i=1}^{n} r_i \cdot \beta_i \leq \bar{r}_{t+1}\}$. The maintenance vector evaluates to $m_i(\beta) = \beta_i \; \forall i$. The number of arcs $|E|$ is asymptotically upper bounded by $O(|V| \cdot 2^n) = O(T \cdot 2^{2 \cdot n}) = O(T \cdot 4^n)$. Thus, the CPMP is solved in time $O(n \cdot |V| + n \cdot |E| + n \cdot T) = O(n \cdot T \cdot 4^n)$.

'$O(T)$': If $n \in O(1)$, the running time $O(T)$ is strongly polynomial.
Altogether, the CPMP is solved in time $O(\min\{n \cdot T \cdot \log T \cdot \bar{r}^{max \, T}, n \cdot T \cdot 4^n\})$.

Remark 5.2 allows to interpret the capacity constraint (C) as an upper bound on the number of tasks, which can be scheduled in a period.

Remark 5.2 An instance to the CPMP with $r_i = r_1 \; \forall i = 1, \ldots, n$ is reformulated as $r_i = 1 \; \forall i = 1, \ldots, n$ and $\bar{r}_t = \lfloor \frac{\bar{r}_t}{r_1} \rfloor \; \forall t = 1, \ldots, T$. An integer number of maintenance activities is scheduled per period because of (X) and any fractional capacity cannot be used.

Lemma 5.14 If $r_i = r_1 \; \forall i = 1, \ldots, n$, the CPMP is solvable in time $O(\min\{n^{T+1} \cdot \frac{\log T}{\sqrt{T}} \cdot 4^T, \sqrt{n} \cdot T^{n+1} \cdot 2^n\})$. Therefore, if additionally $T = k \in O(1)$, the CPMP is strongly polynomially solvable in time $O(n^{k+1})$.

Proof Reformulate the instance with Remark 5.2. Two algorithms are presented.

'$O(n^{T+1} \cdot \frac{\log T}{\sqrt{T}} \cdot 4^T)$': Since $\bar{r}_t \leq n \ \forall t$, the first algorithm from Lemma 5.9 has a running time of $O(n^{T+1} \cdot \frac{\log T}{\sqrt{T}} \cdot 4^T)$.

'$O(n^{k+1})$': If $T = k \in O(1)$, the running time $O(n^{k+1})$ is strongly polynomial.

'$O(\sqrt{n} \cdot T^{n+1} \cdot 2^n)$': Consider the second algorithm from Lemma 5.9. Since the variable costs are zero, it is clearly advantageous to schedule as many maintenance activities as possible in every period that is $\bar{r}_{t+1}$ maintenance activities. Hence, the number of successors in layer $t+1$ from a given node in layer $t = 0, \ldots, T-1$ is $1 + \binom{n}{\bar{r}_{t+1}}$ because no maintenance activity and $\bar{r}_{t+1}$ maintenance activities are scheduled. Since $\bar{r}_{t+1} \leq n$ holds, $\binom{n}{\bar{r}_{t+1}} \leq 2^n \cdot n^{-1/2}$ (Lemma 5.8) and the number of arcs $|E|$ is asymptotically upper bounded by $O(|V| \cdot 2^n \cdot n^{-1/2}) = O(T^{n+1} \cdot 2^n \cdot n^{-1/2})$. Considering Lemma 5.9, an optimal solution is obtained in time $O(n \cdot |V| + n \cdot |E| + n \cdot T) = O(n \cdot T^{n+1} + \sqrt{n} \cdot T^{n+1} \cdot 2^n + n \cdot T)$. For the worst case running time, consider

$$n \cdot T^{n+1} + \sqrt{n} \cdot T^{n+1} \cdot 2^n + n \cdot T \leq d \cdot \sqrt{n} \cdot T^{n+1} \cdot 2^n.$$

There exists a constant $d \geq \frac{\sqrt{n}}{2^n} + 1 + \frac{\sqrt{n}}{T^n \cdot 2^n} > 0$ for all $n, T \geq 1$ because $\frac{\sqrt{n}}{2^n} < 1$.[13] Therefore, the second algorithm solves the CPMP in time $O(\sqrt{n} \cdot T^{n+1} \cdot 2^n)$. Altogether, the CPMP is solvable in time $O(\min\{n^{T+1} \cdot \frac{\log T}{\sqrt{T}} \cdot 4^T, \sqrt{n} \cdot T^{n+1} \cdot 2^n\})$.

Lemma 5.14 solves the open problem $r_i = r_1 \ \forall i \land \pi_i = \pi_1 \ \forall i$ in time $O(n^{k+1})$ if $T = k \in O(1)$. The following Lemma 5.15 reduces the running time to $O(n^k)$. It is particularly interesting that this open problem is strongly polynomially solvable if $\pi_1 \in O(1)$. Note that π_1 is bounded from above by T.

Lemma 5.15 *If $r_i = r_1 \ \forall i = 1, \ldots, n \land \pi_i = \pi_1 \ \forall i = 1, \ldots, n$, the CPMP is solvable in time $O(\min\{n^T \cdot T^2, n \cdot T^{n+1}\})$. Additional restrictions state strong polynomial solvability:*

$$T = k \in O(1) \Rightarrow O(n^k)$$

$$\pi_1 = k \in O(1) \Rightarrow O(n^k \cdot T)$$

Proof Reformulate the instance with Remark 5.2. Two algorithms are presented.

'$O(n^T \cdot T^2)$': In the first algorithm, observe that all maintenance activities have an identical coverage and maintenance time. Hence, the maintenance activities are indistinguishable, which is taken into account. The algorithm is based on the second algorithm from Lemma 5.9 and solves the CPMP as a SPP. However, the definition of a partial solution and the enumeration scheme is different. Each node is a partial solution with $(t, \gamma_0, \ldots, \gamma_{\pi_1})$ with $1 \leq \gamma_i \leq n \ \forall i = 0, \ldots, \pi_1$ and

---

[13] Using the result from Footnote 11 on page 117, it follows for $f(\frac{1}{2}, 2, c) = \sqrt{c} \cdot 2^{-c}$ that $c^* \approx 0.721$ and $f(\frac{1}{2}, 2, c^*) \approx 0.515$.

5.2 Computational Complexity

$t = 1, \ldots, T$. An entry γ_i defines the number of maintenance activities with a remaining coverage i that is measured in periods at period t. The entry γ_0 is used to indicate feasibility of a partial solution if $\gamma_0 = 0$. Introduce an initial node $(t, \gamma_0^{init} = 0, \ldots, \gamma_{\pi_1-1}^{init} = 0, \gamma_{\pi_1}^{init} = n)$ with zero costs $F(t, \gamma_0^{init}, \ldots, \gamma_{\pi_1}^{init}) = 0$ and a final node $(T + 1, \gamma_0^{final}, \ldots, \gamma_{\pi_1}^{final})$.

Consider the number of arcs $|E|$. Each node $(t, \gamma_0, \ldots, \gamma_{\pi_1})$ in layer $t = 0, \ldots, T-1$ has at most two successors that represent the feasible succeeding solutions. In particular, both succeeding nodes are obtained as follows:

If no maintenance activity is conducted, the remaining coverage reduces by one for each maintenance activity and this results in a left shift of the variables γ_i. The node $(t, \gamma_0, \ldots, \gamma_{\pi_1})$ has the node $(t+1, \beta_0 = \gamma_1, \ldots, \beta_{\pi_1-1} = \gamma_{\pi_1}, \beta_{\pi_1} = 0)$ as a successor if $\beta_0 = 0$. Note that if $\beta_0 \geq 1$ at least one maintenance activity should have been scheduled in an earlier period and (P) would be violated. The arc weight is $C(t+1, \beta) = 0$.

Otherwise, at least one maintenance activity is scheduled. Since the variable costs are zero, it is clearly advantageous to schedule as many maintenance activities as possible in every period. These are $\bar{r}_{t+1}$ maintenance activities with the smallest remaining coverage. The node $(t, \gamma_0, \ldots, \gamma_{\pi_1})$ has the node $(t+1, \beta_0, \ldots, \beta_{\pi_1})$ as a successor if $\bar{r}_{t+1} > 0$ and $\beta_0 = 0$.

$$\beta_i = \begin{cases} \max\left\{0, \gamma_{i+1} - \max\left\{0, \bar{r}_{t+1} - \sum_{j=1}^{i} \gamma_j\right\}\right\} & \text{if } i < \pi_1 \\ \bar{r}_{t+1} & \text{otherwise} \end{cases} \quad \forall i = 0, \ldots, \pi_1$$

holds and the arc weight is $C(t+1, \beta) = 1$.

Moreover, each node $(T, \gamma_0, \ldots, \gamma_{\pi_1})$ in the layer T defines a complete solution and is connected by an arc with zero weight to the sink node $(T+1, \gamma_0^{final}, \ldots, \gamma_{\pi_1}^{final})$. Therefore, the number of arcs $|E|$ is asymptotically upper bounded by $O(|V| \cdot 2) = O(|V|)$.

Consider the number of nodes $|V|$. In every node, the sum over all non-zeros amounts to $\sum_{i=0}^{\pi_1} \gamma_i = n$. Therefore, there are at most $\binom{n+\pi_1-1}{n}$ nodes per layer $t = 1, \ldots, T$.[14] Hence, $|V|$ could be asymptotically upper bounded by $O(T \cdot \binom{n+\pi_1-1}{n})$. However, the graph $G(V, E)$ contains only feasible solutions and hence, it holds that $(t, 0, \gamma_1, \ldots, \gamma_{\pi_1})$. Since $1 \leq \gamma_i \leq n \; \forall i = 1, \ldots, \pi_1$, there are at most n^{π_1} nodes per layer $t = 0, \ldots, T-1$. Consequently, $|V|$ is asymptotically upper bounded by $O(T \cdot n^{\pi_1})$.

The computational effort to obtain a node and a single successor from a given node is $O(\pi_1)$. Consequently, the entire graph $G(V, E)$ is constructed in time

---

[14] In the well-known and so called "stars and bars" principle, n indistinguishable balls are assigned to π_1 different boxes. An assignment is encoded as a binary string comprising n digits with a value of "1" for a ball and $\pi_1 - 1$ digits with a value of "0" representing a box. Hence, there are $\binom{n+\pi_1-1}{n}$ combinations.

$O(\pi_1 \cdot |V| + \pi_1 \cdot |E|)$. The CPMP becomes solvable as a SPP in time $O(\pi_1 \cdot |V| + \pi_1 \cdot |E| + n \cdot T) = O(\pi_1 \cdot n^{\pi_1} \cdot T)$ that is asymptotically upper bounded by $O(n^T \cdot T^2)$ since $\pi_1 \leq T$.

'$O(n^k)$': If $T = k \in O(1)$, strong polynomial solvability in time $O(n^T \cdot T^2) = O(n^k)$ follows.

'$O(n^k \cdot T)$': If $\pi_1 = k \in O(1)$, the running time $O(\pi_1 \cdot n^{\pi_1} \cdot T) = O(n^k \cdot T)$ is strongly polynomial.

'$O(n \cdot T^{n+1})$': Consider the second algorithm from Lemma 5.9. Observe that a node $(t, \alpha_1, \ldots, \alpha_n)$ is easily transformed into a node $(t, \gamma_0, \ldots, \gamma_{\pi_1})$ and vice versa. Consequently, the entries $(\alpha_1, \ldots, \alpha_n)$ can be sorted such that $\alpha_i \geq \alpha_{i+1} \; \forall i$ holds. By arguments presented for the first algorithm, $|V|$ is asymptotically upper bounded by $O(T \cdot \binom{n+\pi_1-1}{n})$ and $|E|$ is asymptotically upper bounded by $O(|V|)$ because every node has at most two successors. Note that $|V|$ is asymptotically bounded from above by $O(T^{n+1})$. Thus, the CPMP is solvable in time $O(n \cdot |V| + n \cdot |E| + n \cdot T) = O(n \cdot |V| + n \cdot T) = O(n \cdot T^{n+1})$.

All in all, the CPMP is solvable in time $O(\min\{n^T \cdot T^2, n \cdot T^{n+1}\})$.

Lemma 5.13 suggests exponential algorithms to solve the CPMP with $r_i = r_1 \; \forall i \wedge \pi_i = T \; \forall i$. However, Lemma 5.16 states polynomial solvability of this problem variant.

Lemma 5.16 If $r_i = r_1 \; \forall i = 1, \ldots, n \wedge \pi_i = T \; \forall i = 1, \ldots, n$, the CPMP is strongly polynomially solvable in time $O(n + T \cdot \log T)$.

Proof Construct a lower bound as follows: Because of $\pi_i = T \; \forall i$, there exists an optimal solution where (P) is satisfied as

$$\sum_{t=1}^{T} x_{it} = 1 \quad \forall i = 1, \ldots, n. \tag{5.11}$$

Apply Remark 5.2, sum up (C) over all periods, substitute with (5.11) and obtain

$$\sum_{t=1}^{T} \bar{r}_t \cdot y_t \geq \sum_{i=1}^{n} \sum_{t=1}^{T} x_{it} = n. \tag{5.12}$$

The lower bound is calculated via the Minimization KP

$$\min \sum_{t=1}^{T} y_t \quad \text{subject to } (5.12) \wedge (Y).$$

that is solvable as follows: Sort $\bar{r}_t$ according to non-increasing values $\bar{r}_{t_1} \geq \ldots \geq \bar{r}_{t_T}$ in time $O(T \cdot \log T)$ (Ottmann and Widmayer 2012, pp. 106–112). Find the period t_m with $\sum_{p=1}^{m} \bar{r}_{t_p} \geq n$ and $\sum_{p=1}^{m-1} \bar{r}_{t_p} < n$ in time $O(\log T)$ (Dietrich et al. 1993). Set $y_t = 1$ if $t = t_1, \ldots, t_m$ (otherwise, 0) $\forall t$ in time $O(T)$. The values for y are optimal to the lower bound because closing a period $t_1, \ldots, t_m$ and opening a

5.2 Computational Complexity

period $t_{m+1}, \ldots, t_T$ cannot yield a feasible solution with a strictly lower objective function value.

If $\sum_{t=1}^{T} \bar{r}_t < n$, the instance is infeasible. Otherwise, the feasible solution to the lower bound provides enough capacity to schedule all n maintenance activities. Any assignment of maintenance activities to open periods with at most $\bar{r}_t$ tasks per period t yields a feasible integer solution (x, y). Since the aforementioned construction of y provides all open periods, this step requires the time $O(n)$. Optimality of (x, y) follows because the objective function value is identical to the lower bound. Therefore, an optimal integer solution (x, y) is obtained in time $O(T \cdot \log T) + O(\log T) + O(T) + O(n) = O(n + T \cdot \log T)$. Note that $T \cdot \log_2 T \geq T$ holds for $T \geq 2$, which applies because of $\pi_i \geq 2$ $\forall i$ (Assumption (2.2)).

Lemma 5.17 If $r_i = r_1$ $\forall i = 1, \ldots, n \land \pi_i = \pi_1$ $\forall i = 1, \ldots, n \land \bar{r}_t \leq \bar{r}_{t+1}$ $\forall t = 1, \ldots, T-1$, the CPMP is strongly polynomially solvable in time $O(\min\{n^2 \cdot T, T^3 + n \cdot T\})$.

Proof Reformulate the instance with Remark 5.2.

'$O(\min\{n^2 \cdot T, n \cdot T^2\})$': Consider the second algorithm from Lemma 5.15. In what follows, the structure of the graph is discussed. Consider a given node in the layer t and the successors in the layers $t+1, t+2$. It follows

$$(t, \alpha_1, \ldots, \alpha_n) \xrightarrow{0} (t+1, \alpha_1 - 1, \ldots, \alpha_n - 1) \xrightarrow{} \begin{matrix} \nearrow^{0} \bar{\bar{d}} \\ \xrightarrow{1} \bar{d} \end{matrix}$$
$$\searrow^{1} (t+1, \pi_1, \ldots, \pi_1, \alpha_1 - 1, \ldots, \alpha_{n-\bar{r}_{t+1}} - 1) \xrightarrow{0} \bar{\bar{d}}$$
$$\searrow^{1}$$

with the nodes

$$\bar{\bar{d}} = \left(t+2, \bar{\bar{\alpha}}_1 = \pi_1, \ldots, \bar{\bar{\alpha}}_{\bar{r}_{t+2}} = \pi_1, \bar{\bar{\alpha}}_{1+\bar{r}_{t+2}} = \alpha_1 - 2, \ldots, \bar{\bar{\alpha}}_n = \alpha_{n-\bar{r}_{t+2}} - 2\right)$$

$$\bar{d} = \left(t+2, \bar{\alpha}_1 = \pi_1 - 1, \ldots, \bar{\alpha}_{\bar{r}_{t+1}} = \pi_1 - 1, \bar{\alpha}_{1+\bar{r}_{t+1}} = \alpha_1 - 2, \ldots, \bar{\alpha}_n = \alpha_{n-\bar{r}_{t+1}} - 2\right).$$

Since $\bar{r}_{t+1} \leq \bar{r}_{t+2}$ and $\pi_1 \geq \alpha_i \geq \alpha_{i+1}$ $\forall i$, it holds for the conducted maintenance activities

$$\bar{\alpha}_i = \pi_1 > \begin{cases} \bar{\bar{\alpha}}_i = \pi_1 - 1 & \forall i = 1, \ldots, \bar{r}_{t+1} \\ \bar{\bar{\alpha}}_i = \alpha_{i-\bar{r}_{t+1}} - 2 & \forall i = 1 + \bar{r}_{t+1}, \ldots, \bar{r}_{t+2} \end{cases}$$

and for the remaining maintenance activities

$$\bar{\alpha}_i = \alpha_{i-\bar{r}_{t+2}} - 2 \geq \bar{\bar{\alpha}}_i = \alpha_{i-\bar{r}_{t+1}} - 2 \quad \forall i = 1 + \bar{r}_{t+2}, \ldots, n.$$

Therefore, it holds that $\bar{\bar{\alpha}}_i \geq \bar{\alpha}_i \ \forall i = 1,\ldots,n$. Together with the fact that both nodes $\bar{d}$ and $\bar{\bar{d}}$ are evaluated the same costs that is $F(\bar{d}) = F(\bar{\bar{d}})$, the node $\bar{\bar{d}}$ can be pruned because $\bar{d}$ has more maintenance activities with a higher remaining coverage than $\bar{\bar{d}}$. Thus, to each objective function value exists exactly one non-dominated node per layer.

In what follows, this fact is used to asymptotically upper bound the number of nodes per layer. From the cognition above follows that the graph $G(V,E)$ can be constructed such that there are at most $t+1$ nodes in layer $t = 0,\ldots,T-1$. This gives the asymptotic upper bound $O(T)$ on the number of nodes per layer.

Another upper bound on the number of nodes is derived as follows. If there exists a feasible solution it has at least one but at most n open periods per task horizon. Since there exists exactly one non-dominated node per objective function value per layer, the number of nodes per layer is asymptotically upper bounded by $O(n)$. For this purpose, the layers are evaluated in the order $0,\ldots,t$. In order to evaluate and to prune successors, consider the nodes in the tth layer in the order of increasing objective function values. This implies that per node in layer t a successor in layer $t+1$ that has a closed period is evaluated prior to the successor that leads to an open period. Because of this enumeration and assuming there exists a feasible solution, there exists a node $(\alpha_1,\ldots,\alpha_n)$ in layer $t+1$ with an objective function value $F(t+1,\alpha_1,\ldots,\alpha_n)$ that represents a feasible partial solution and all nodes with a lower (larger) objective function value represent an infeasible (feasible) partial solution. When this node has become reachable, it is the first node in layer $t+1$. Since the objective function value of this node can increase by at most $O(n)$, there are at most $O(n)$ nodes in layer $t+1$.

Altogether, the number of nodes $|V|$ is asymptotically upper bounded by $O(\min\{n,T\} \cdot T)$. Note that the graph is constructed in time $O(n \cdot |V| + n \cdot |E|) = O(n \cdot |V|)$. Using the second algorithm from Lemma 5.15, the CPMP is solvable in time $O(n \cdot |V| + n \cdot T) = O(\min\{n^2 \cdot T, n \cdot T^2\})$.

'$O(\min\{n \cdot T^2, T^3 + n \cdot T\})$': Consider the first algorithm from Lemma 5.15. Since a node $(t, \alpha_1, \ldots, \alpha_n)$ is easily transformed into a node $(t, \gamma_0, \ldots, \gamma_{\pi_1})$ and vice versa, the number of nodes $|V|$ is asymptotically upper bounded by $O(\min\{n,T\} \cdot T)$. Because of $\pi_1 \leq T$, the CPMP is solvable in time $O(T \cdot |V| + n \cdot T) = O(\min\{n \cdot T^2, T^3 + n \cdot T\})$.

'$O(\min\{n^2 \cdot T, T^3 + n \cdot T\})$': The overall computational effort to solve the CPMP is $O(\min\{n^2 \cdot T, T^3 + n \cdot T, n \cdot T^2\})$. However, if $n \leq T$, it becomes $O(n^2 \cdot T)$. Otherwise, $n > T$ holds and it follows that $O(T^3 + n \cdot T)$. Therefore, the CPMP is solvable in time $O(\min\{n^2 \cdot T, T^3 + n \cdot T\})$.

Using ideas from Lemma 5.17, the CPMP with $r_i = r_1 \ \forall i \land \pi_i = \pi_1 \ \forall i \land \bar{r}_t = \bar{r}_1 \ \forall t$ becomes solvable in time $O(n \cdot T)$ and an optimal solution can be constructed by opening the last $\lceil \frac{n}{r_1} \rceil$ periods per consecutive task horizon.

5.3 Lower Bounds

Section 5.3.1 evaluates 99 lower bounds that are obtained from neglecting constraints completely as well as from Lagrangean relaxation. In Sect. 5.3.2, Theorem 5.22 presents the relative strengths of the relevant lower bounds and provides the respective computational complexity. Insights into the problem structure are given in Sect. 5.3.3. The mathematical formulations of all Lagrangean duals are presented in Sect. 5.3.4. Consider Notation 3.14 and recall that the CPMP Z is stated as

$$Z = \min \sum_{t=1}^{T} y_t \qquad \text{subject to} \tag{Z}$$

$$\sum_{\tau=t}^{t+\pi_i-1} x_{i\tau} \geq 1 \qquad \forall i = 1, \ldots, n; t = 1, \ldots, T - \pi_i + 1 \tag{P}$$

$$\sum_{i=1}^{n} r_i \cdot x_{it} \leq \bar{r}_t \cdot y_t \qquad \forall t = 1, \ldots, T \tag{C}$$

$$x_{it} \leq y_t \qquad \forall i = 1, \ldots, n; t = 1, \ldots, T \tag{V}$$

$$0 \leq x_{it}, y_t \leq 1 \qquad \forall i = 1, \ldots, n; t = 1, \ldots, T \tag{S}$$

$$x_{it} \in \{0, 1\} \qquad \forall i = 1, \ldots, n; t = 1, \ldots, T \tag{X}$$

$$y_t \in \{0, 1\} \qquad \forall t = 1, \ldots, T \tag{Y}$$

5.3.1 Considered Lower Bounds

The number of combinations to neglect at least one of the constraints (P), (C), (V) is $\sum_{i=1}^{3} \binom{3}{i} = 7$. For these relaxations, the number of possibilities to neglect the integrality constraints (X) and (Y) is $\sum_{i=0}^{2} \binom{2}{i} = 4$. Furthermore, there are $\sum_{i=1}^{2} \binom{2}{i} = 3$ LP relaxations of the CPMP. Since a variant of the Simplex algorithm implicitly considers upper bounds on variables (Domschke and Drexl 2005, pp. 48–51), the constraint (S) is not relaxed. The total number of relaxations with neglected constraints is $7 \cdot 4 + 3 = 31$ but only five of them are relevant.

Lemma 5.18 The relaxations with neglected constraints are $Z^{(X)}$, $Z^{(X)(Y)}$, $Z^{(V)(Y)}$, $Z^{(V)(X)(Y)}$, $Z^{(C)}$ and

$$Z^{(P)} = Z^{(P)(C)} = Z^{(P)(V)} = Z^{(P)(Y)} = Z^{(P)(X)} = Z^{(P)(C)(V)} = Z^{(P)(C)(Y)} = Z^{(P)(C)(X)}$$
$$= Z^{(P)(V)(Y)} = Z^{(P)(V)(X)} = Z^{(P)(X)(Y)} = Z^{(P)(C)(V)(Y)} = Z^{(P)(C)(V)(X)}$$
$$= Z^{(P)(C)(X)(Y)} = Z^{(P)(V)(X)(Y)} = Z^{(P)(C)(V)(X)(Y)} = 0 \qquad (5.13)$$

$$Z^{(C)(V)} = Z^{(C)(V)(Y)} = Z^{(C)(V)(X)} = Z^{(C)(V)(X)(Y)} = 0 \qquad (5.14)$$

$$Z^{(C)} = Z^{(C)(Y)} = Z^{(C)(X)} = Z^{(C)(X)(Y)} \qquad (5.15)$$

$$Z^{(X)} = Z^{(V)(X)} \wedge Z = Z^{(V)} \qquad (5.16)$$

$$Z = Z^{(Y)}. \qquad (5.17)$$

Proof '(5.13)' and '(5.14)': The optimal objective function value is zero if either (P) is relaxed because there is no lower bound on x or if (C) and (V) are relaxed because there exists no coupling between x and y.

'(5.15)': $Conv(P)(V)(X)(Y) \subseteq Conv(P)(V)(S)(X)$, $Conv(P)(V)(S)(Y) \subseteq H(P)(V)(S)$ yields $Z^{(C)} \leq Z^{(C)(X)}, Z^{(C)(Y)} \leq Z^{(C)(X)(Y)}$. (5.15) follows from $Z^{(C)} = Z^{(C)(X)(Y)}$ (Lemma 4.17).

'(5.16)': (V) is redundant because of (C) and (Y).

'(5.17)': y is binary because of (V) and (X).

The total number of combinations to relax at least one of the constraints (P), (C), (V) in a Lagrangean fashion is $\sum_{i=1}^{3} \binom{3}{i}$. Having relaxed i constraints, at most $3 - i$ constraints can be neglected completely, which yields $\sum_{j=0}^{3-i} \binom{3-i}{j} = 2^{3-i}$ combinations. The only interesting Lagrangean LP relaxations are those, which are stronger than the LP relaxation $Z^{(X)(Y)}$. This implies that (X) and (Y) are not simultaneously relaxed because of the integrality property of Corollary 3.12 and thus, $\sum_{i=1}^{2} \binom{2}{i} = 3$ additional combinations are obtained for each Lagrangean relaxation. Assuming few computational advantages, (S) is not relaxed at all.[15] The total number of Lagrangean relaxations is $\sum_{i=1}^{3} \binom{3}{i} \cdot 2^{3-i} \cdot 3 = 57$, which are characterized in Lemma 5.19. However, only five Lagrangean duals are relevant.

Lemma 5.19 The Lagrangean duals are $Z_{(P)}$, $Z_{(C)}$, $Z_{(V)}^{(Y)}$, $Z_{(P)}^{(V)(Y)}$, $Z_{(P)(V)}^{(Y)}$ and

$$Z_{(C)} = Z_{(C)}^{(X)} \qquad (5.18)$$

$$Z^{(X)(Y)} = Z_{(P)}^{(X)} = Z_{(P)(V)}^{(X)} \wedge Z^{(V)(X)(Y)} = Z_{(P)}^{(V)(X)} \qquad (5.19)$$

---

[15]This assumption is analogous to Cornuejols et al. (1991), who investigate relaxations of the related Capacitated Facility Location Problem.

5.3 Lower Bounds

$$Z_{(P)} = Z_{(P)}^{(Y)} \wedge Z_{(C)} = Z_{(C)}^{(Y)} \wedge Z_{(P)(C)} = Z_{(P)(C)}^{(Y)} \tag{5.20}$$

$$Z_{(P)} = Z_{(P)}^{(V)} \wedge Z = Z_{(V)} \wedge Z_{(P)} = Z_{(P)(V)} \wedge Z^{(X)} = Z_{(V)}^{(X)} \tag{5.21}$$

$$Z^{(X)(Y)} = Z_{(P)(C)} = Z_{(P)(C)}^{(X)} \tag{5.22}$$

$$Z^{(X)(Y)} = Z_{(C)(V)} = Z_{(C)(V)}^{(X)} = Z_{(C)(V)}^{(Y)} \tag{5.23}$$

$$Z^{(X)(Y)} = Z_{(P)(C)(V)} = Z_{(P)(C)(V)}^{(X)} = Z_{(P)(C)(V)}^{(Y)} \tag{5.24}$$

$$Z^{(V)(X)(Y)} = Z_{(C)}^{(V)} = Z_{(C)}^{(V)(X)} = Z_{(C)}^{(V)(Y)} = Z_{(P)(C)}^{(V)} = Z_{(P)(C)}^{(V)(X)} = Z_{(P)(C)}^{(V)(Y)} \tag{5.25}$$

$$Z^{(C)} = Z_{(P)}^{(C)} = Z_{(P)}^{(C)(Y)} = Z_{(P)}^{(C)(X)} = Z_{(V)}^{(C)} = Z_{(V)}^{(C)(Y)} = Z_{(V)}^{(C)(X)}$$
$$= Z_{(P)(V)}^{(C)} = Z_{(P)(V)}^{(C)(Y)} = Z_{(P)(V)}^{(C)(X)} \tag{5.26}$$

$$Z_{(C)}^{(P)} = Z_{(C)}^{(P)(V)} = Z_{(C)}^{(P)(Y)} = Z_{(C)}^{(P)(X)} = Z_{(C)}^{(P)(V)(Y)} = Z_{(C)}^{(P)(V)(X)}$$
$$= Z_{(V)}^{(P)} = Z_{(V)}^{(P)(X)} = Z_{(V)}^{(P)(Y)} = Z_{(V)}^{(P)(C)(X)} = Z_{(V)}^{(P)(C)(Y)} = Z_{(V)}^{(P)(C)} \tag{5.27}$$
$$= Z_{(C)(V)}^{(P)} = Z_{(C)(V)}^{(P)(Y)} = Z_{(C)(V)}^{(P)(X)} = Z_{(P)}^{(C)(V)} = Z_{(P)}^{(C)(V)(Y)} = Z_{(P)}^{(C)(V)(X)} = 0.$$

Proof For the sake of brevity, a reference to neither Lemma 3.7, Lemma 3.8 nor Geoffrion's Theorem 3.11 is given.

'(5.18)': This result follows from the single-assignment property of Lemma 4.3.
'(5.19)': $Z_{(P)}^{(X)} = Z_{(P)(V)}^{(X)}$ because $H(P) \cap Conv(C)(V)(S)(Y) = H(P)(V) \cap Conv(C)(S)(Y)$, where (V) is redundant because of (C) and (Y). With the polyhedral result from Lemma 5.20, which is given below, it follows $H(P)(V) \cap Conv(C)(S)(Y) = H(P)(C)(V)(S)$. The second equality is analogously proven.
'(5.20)': y is binary because of (V) and (X).
'(5.21)': (V) is redundant because of (C) and (Y).
'(5.22), (5.23) and (5.24)': In (5.22), $Z_{(P)(C)}$ has $H(P)(C) \cap Conv(V)(X)(Y) = H(P)(C)(V)(S)$ because of a totally unimodular coefficient matrix of (V) and (S) (Rhys 1970).[16] Hence, $Z^{(X)(Y)} = Z_{(P)(C)}$ holds. In (5.23), $Z_{(C)(V)}$ has $H(C)(V) \cap Conv(P)(X)(Y)$ and the set of constraints divides into a problem with $Conv(Y)$ and a problem with $Conv(P)(X)$. The first problem is a selection problem with $Conv(Y) = H(S^y)$ because of a totally unimodular coefficient matrix. The second problem has $Conv(P)(X) = H(P)(S^x)$ because of Theorem 4.1 with $a_{it} = 1 \ \forall i; t$ and $b_{it} = 1 \ \forall i; t$.[17] Hence, $H(C)(V) \cap Conv(P)(X)(Y) = H(P)(C)(V)(S)$ and $Z^{(X)(Y)} = Z_{(C)(V)}$ holds. In (5.24), $Z_{(P)(C)(V)}$ has $H(P)(C)(V) \cap Conv(X)(Y)$. A selection problem with $Conv(X)(Y) = H(S)$ is obtained where (S) has a totally unimodular coefficient matrix. Hence, $H(P)(C)(V) \cap Conv(X)(Y) = H(P)(C)(V)(S)$

---

[16] In $Z_{(P)(C)}$, the Lagrangean relaxation is a Shared Fixed-Cost Problem (Rhys 1970).
[17] In $Z_{(C)(V)}$, the Lagrangean relaxation can be reformulated with Remark 4.2 to yield n SPPs.

and $Z^{(X)(Y)} = Z_{(P)(C)(V)}$ holds. A relaxation of (X) or (Y) yields a Lagrangean dual of the same strength because of the integrality property of Corollary 3.12, which proves the remaining equalities.

'(5.25), (5.26)': The proof is analogous to (5.22), (5.23) and (5.24) for $Z^{(C)(X)(Y)}$. Note that $Z^{(C)} = Z^{(C)(X)(Y)}$ holds by Lemma 4.17.

'(5.27)': The proof is analogous to (5.13) and (5.14).

The result $Z^{(X)(Y)} = Z^{(X)}_{(P)}$ from (5.19) is surprising. $ZL^{(X)}_{(P)}$ decomposes in the same way as $ZL_{(P)}$ (5.44) that is presented in Lemma 5.23 but it has T Continuous KPs and one trivial selection problem. Intuitively, $ZL^{(X)}_{(P)}$ has the integrality property from which $Z^{(X)(Y)} = Z^{(X)}_{(P)}$ follows.[18] Lemma 5.19 requires the following polyhedral result that is found in a very short version in Cornuejols et al. (1991).

Lemma 5.20 $Conv(C)(S)(Y) = H(C)(S)$

Proof '$\Rightarrow$': $H(C)(S)$ is convex and hence, $Conv(C)(S)(Y) \subseteq H(C)(S)$ holds.
'$\Leftarrow$': Assume a feasible solution $(\bar{x}, \bar{y}) \in H(C)(S)$ where at least one component $0 < \bar{y}_t < 1$ is fractional. $Conv(C)(S)(Y) \supseteq H(C)(S)$ is shown by the existence of a convex combination of K mixed integer solutions that yield $(\bar{x}, \bar{y})$. Therefore, $(\bar{x}, \bar{y})$ cannot be an extreme point and it follows that $(\bar{x}, \bar{y}) \in Conv(C)(S)(Y)$.

Construct K mixed integer solutions (x^k, y^k) with $x^k \in \mathbb{R}_+^{n \times T} \wedge y^k \in \{0, 1\}^T$ and a scalar α as follows (see Example 5.7): Using the approach from Theorem 4.1, represent all non-zeros in $\bar{y}$ as fractions,[19] let K be the least common denominator among of all values of $\bar{y}$ and let $\alpha = \frac{1}{K}$ be a quant. Obtain K integer vectors y^k as follows: Initialize $y_t^k = 0$ $\forall t = 1, \ldots, T; k = 1, \ldots, K$ and set $k = 1, t = 0, d = 0$. Repeat the following steps per iteration: While $d > 0$ holds, set $y_t^k = 1$, $d = d - \alpha$ and $k = k + 1$. If $d = 0$ holds, select the next period $t = t + 1$ and set $d = \bar{y}_t, k = 1$. Until $t > T$ holds, commence with the next iteration. It follows that $\bar{y} = \sum_{k=1}^{K} \alpha \cdot y^k$. Set

$$x_{it}^k = \begin{cases} \dfrac{\bar{x}_{it} \cdot y_t^k}{\bar{y}_t} & \text{if } \bar{y}_t > 0 \\ 0 & \text{otherwise} \end{cases} \quad \forall i = 1, \ldots, n; t = 1, \ldots, T; k = 1, \ldots, K.$$

(5.28)

K feasible mixed integer solutions $(x^k, y^k) \in Conv(C)(S)(Y)$ $\forall k$ are obtained. Consider constraint (S). Constraint (C) implies $\bar{x}_{it} \leq \bar{y}_t \leq 1$ and together with (5.28),

---

[18]This intuitive approach that does not hold in general. One counter example is a production scheduling problem from a tile factory presented in Guignard (2003). A Lagrangean relaxation of the demand constraint leads to a Lagrangean relaxation that decomposes because of the integer linearization property. The subproblems consist of continuous KPs and SPPs. Although all subproblems have the integrality property, the Lagrangean relaxation does not have it and the obtained lower bound is tighter than the LP relaxation.

[19]See Footnote 4 on page 74.

5.3 Lower Bounds

$x_{it}^k \leq 1$ holds. Together with the non-negativity of $\bar{x}$ and $\bar{y}$, (S) is satisfied. Constraint (C) is clearly satisfied if $y_t^k = 0$. Assume $y_t^k = 1$. Substitute with (5.28) in (C) and obtain

$$\sum_{i=1}^n r_i \cdot x_{it}^k = \sum_{i=1}^n r_i \cdot \frac{\bar{x}_{it}}{\bar{y}_t} \leq \bar{r}_t,$$

which is satisfied because $(\bar{x}, \bar{y}) \in H(C)(S)$ holds. Furthermore,

$$\forall i; t : \forall \bar{y}_t > 0 : \sum_{k=1}^K \alpha \cdot x_{it}^k = \bar{x}_{it} \cdot \sum_{k=1}^K \alpha \cdot \frac{y_t^k}{\bar{y}_t} \quad \text{from (5.28)}$$

$$= \bar{x}_{it} \qquad \text{because } \bar{y}_t = \sum_{k=1}^K \alpha \cdot y_t^k,$$

and

$$\forall i; t : \forall \bar{y}_t = 0 : \sum_{k=1}^K \alpha \cdot x_{it}^k = 0 = \bar{x}_{it} \quad \text{because } x_{it}^k = 0 \ \forall k \land \bar{x}_{it} = 0 \text{ from (C)}$$

holds. Hence,

$$(\bar{x}, \bar{y}) = \sum_{k=1}^K \alpha \cdot (x^k, y^k) \tag{5.29}$$

is a *strict* convex combination since $\alpha > 0$, $K \cdot \alpha = 1$ and this contradicts that $(\bar{x}, \bar{y})$ is an extreme point (Nemhauser and Wolsey 1988, p. 93).

Example 5.7 Let $n = 2$, $T = 3$ and $\pi = (3, 3)^T$. In Table 5.6, K feasible mixed integer solutions (x^k, y^k) are obtained from a given feasible fractional solution $(\bar{x}, \bar{y})$ with $K = 3$ and $\alpha = \frac{1}{3}$. $(\bar{x}, \bar{y})$ is obtained by the convex combination (5.29).

Table 5.6 The convex combination (5.29) yields the fractional solution $(\bar{x}, \bar{y})$

	t		
	1	2	3
$\bar{y}_t$		$\frac{2}{3}$	1
$\bar{x}_{1t}$		$\frac{2}{3}$	$\frac{1}{3}$
$\bar{x}_{2t}$		$\frac{1}{3}$	$\frac{1}{3}$
y_t^1, y_t^2		1	1
x_{1t}^1, x_{1t}^2		1	$\frac{1}{3}$
x_{2t}^1, x_{2t}^2		$\frac{1}{2}$	$\frac{1}{3}$
y_t^3			1
x_{1t}^3			$\frac{1}{3}$
x_{2t}^3			$\frac{1}{3}$

Some additional relaxations are investigated. A proper Lagrangean decomposition is obtained, if the variable splitting is established between the constraints (P) and (C). Otherwise, one Lagrangean relaxation is the CPMP itself and the Lagrangean dual coincides with Z (see Theorem 3.13). The corresponding Lagrangean LP relaxations are omitted because at least one subproblem has the integrality property and thus, the lower bound coincides with at least one Lagrangean dual from Lemma 5.19. Thus, only four Lagrangean decompositions are relevant, which are

$$Z_{(P)/(C)}, Z_{(P)(V)/(C)}, Z_{(P)/(C)(V)}, Z_{(P)/(C)}^{(V)}.$$

Leading to well-known and insightful structures, the substitution $H(P) = H(Q) \cap H(R)$ is additionally considered for $Z_{(P)}$. The non-overlapping, consecutive task horizons of (Q) are kept in the Lagrangean relaxation, whereas the constraints (R) are relaxed. This yields $Z_{(R)}^{(P)+(Q)}$ and $Z_{(R)}^{(P)+(Q)}$. (Y) and (X) are not relaxed simultaneously because of the integrality property of Corollary 3.12. This yields seven lower bounds

$$Z_{(R)}^{(P)+(Q)}, Z_{(R)}^{(P)(X)+(Q)}, Z_{(R)}^{(P)(Y)+(Q)}, Z^{(P)+(Q)}, Z^{(P)(X)+(Q)}, Z^{(P)(Y)+(Q)}, Z^{(P)(X)(Y)+(Q)}.$$

Equivalence of some of these eleven lower bounds is stated in Lemma 5.21 and only six of them are relevant.

Lemma 5.21 The additional relaxations are $Z_{(P)/(C)}$, $Z_{(R)}^{(P)+(Q)}$, $Z_{(R)}^{(P)(X)+(Q)}$, $Z^{(P)+(Q)}$, $Z^{(P)(X)+(Q)}$, $Z^{(P)(X)(Y)+(Q)}$ and

$$Z_{(P)/(C)} = Z_{(P)(V)/(C)} \wedge Z_{(P)} = Z_{(P)/(C)(V)} = Z_{(P)/(C)}^{(V)} \qquad (5.30)$$

$$Z_{(R)}^{(P)+(Q)} = Z_{(R)}^{(P)(Y)+(Q)} \wedge Z^{(P)+(Q)} = Z^{(P)(Y)+(Q)}. \qquad (5.31)$$

Proof In what follows, there is no reference to Geoffrion's Theorem 3.11 or Theorem 3.13.

'(5.30)': $Z_{(P)/(C)} = Z_{(P)(V)/(C)}$ holds because $Conv(P)(V)(X)(Y) \cap Conv(C)(V)(X)(Y) = Conv(P)(V)(X)(Y) \cap Conv(C)(X)(Y)$ since (V) is redundant. $Z_{(P)/(C)(V)} = Z_{(P)}$ holds because $Conv(P)(X)(Y) \cap Conv(C)(V)(X)(Y) = H(P)(S) \cap Conv(C)(V)(X)(Y) = H(P) \cap Conv(C)(V)(X)(Y)$ holds from an analogous argument that proves $Z^{(X)(Y)} = Z_{(C)(V)}$ in (5.23). $Z_{(P)/(C)}^{(V)} = Z_{(P)/(C)(V)}$ holds because $Conv(P)(X)(Y) \cap Conv(C)(X)(Y) = Conv(P)(X)(Y) \cap Conv(C)(V)(X)(Y)$ since (V) is redundant.
'(5.31)': y is binary because of (V) and (X).

5.3 Lower Bounds

Table 5.7 Computational complexity of the relevant lower bounds from Theorem 5.22

Problem variant	Computational complexity	Proof
$Z^{(X)}$	Open	
$Z_{(P)}$	Binary $\mathcal{NP}$-hard, $O(n \cdot T + n \cdot \bar{r}^{sum})$	Lemma 5.23
$Z_{(P)}$ with $\bar{r}^{sum} \in O(1)$	$O(n \cdot T)$	Lemma 5.23
$Z_{(C)}$	Open	
$Z_{(P)/(C)}$	Strongly $\mathcal{NP}$-hard	Lemma 5.24
$Z_{(V)}^{(Y)}$	Strongly $\mathcal{NP}$-hard	Lemma 5.25
$Z^{(V)(Y)}$	Strongly $\mathcal{NP}$-hard	Lemma 5.25
$Z_{(R)}^{(P)+(Q)}$	Strongly $\mathcal{NP}$-hard	Lemma 5.26
$Z_{(R)}^{(P)(X)+(Q)}$	Binary $\mathcal{NP}$-hard	Lemma 5.27
$Z^{(P)+(Q)}$	Strongly $\mathcal{NP}$-hard	Lemma 5.26
$Z^{(P)(X)+(Q)}$	Open	
$Z^{(P)(X)(Y)+(Q)}$	$O(n^{3.5} \cdot T^{3.5} \cdot L)$	Lemma 5.28
$Z^{(X)(Y)}$	$O(n^{3.5} \cdot T^{3.5} \cdot L)$	Lemma 5.28
$Z^{(V)(X)(Y)}$	$O(n^{3.5} \cdot T^{3.5} \cdot L)$	Lemma 5.28
$Z^{(C)}$	$O(n \cdot T)$	Lemma 4.13

5.3.2 Relative Strengths and Computational Complexity

From all 99 lower bounds eight relaxations with neglected constraints and six Lagrangean duals are examined, namely

$$Z^{(X)}, Z^{(V)(Y)}, Z^{(X)(Y)}, Z^{(V)(X)(Y)}, Z^{(C)}, Z^{(P)+(Q)}, Z^{(P)(X)+(Q)}, Z^{(P)(X)(Y)+(Q)},$$

$$Z_{(P)/(C)}, Z_{(P)}, Z_{(C)}, Z_{(V)}^{(Y)}, Z_{(R)}^{(P)+(Q)}, Z_{(R)}^{(P)(X)+(Q)}.$$

The lower bounds (5.13), (5.14) and (5.27) are not considered because any feasible solution to the CPMP has an objective function value that is strictly larger than 0 because of the Assumptions (2.2) and (2.5). Furthermore, $Z_{(P)(V)}^{(Y)}$ and $Z_{(P)}^{(V)(Y)}$ are not investigated because they have the same computational complexity as $Z_{(P)}$ that yields a stronger lower bound (Lemma 5.23).

Table 5.7 summarizes the computational complexity. Let L be the input length of n, T, π, r, $\bar{r}$ and the coefficient matrix.

Theorem 5.22 extensively compares the relative strength of all lower bounds.[20] In Sect. 7.2, a computational study investigates the absolute strength.

---

[20]Note that Theorem 5.22 shows that there is no dominance between $Z_{(C)}$ and $Z_{(P)}$. This result is important for the Lagrangean Hybrid Heuristic that is presented in Sect. 6.3. Some results correspond to similar results for the related *Single Source Capacitated Facility Location Problem (SSCFLP)* (Klose and Drexl 2005; Klose 2001, pp. 218–222).

Theorem 5.22

$$Z \geq Z^{(X)}, Z_{(P)/(C)}, Z_{(R)}^{(P)+(Q)}, Z_{(R)}^{(P)(X)+(Q)} \geq Z^{(X)(Y)} \tag{5.32}$$

$$Z_{(P)/(C)} \geq Z_{(P)}, Z_{(C)} \tag{5.33}$$

$$Z^{(X)} \geq Z_{(C)}, Z_{(R)}^{(P)(X)+(Q)} \tag{5.34}$$

$$Z_{(R)}^{(P)+(Q)} \geq Z_{(P)}, Z^{(P)+(Q)}, Z_{(R)}^{(P)(X)+(Q)} \tag{5.35}$$

$$Z^{(P)+(Q)}, Z_{(R)}^{(P)(X)+(Q)} \geq Z^{(P)(X)+(Q)} \geq Z^{(P)(X)(Y)+(Q)} \tag{5.36}$$

$$Z^{(X)(Y)} \geq Z^{(C)}, Z^{(V)(X)(Y)}, Z^{(P)(X)(Y)+(Q)} \tag{5.37}$$

$$Z_{(V)}^{(Y)} \geq Z^{(V)(Y)} \geq Z^{(V)(X)(Y)} \tag{5.38}$$

Every inequality is strict for at least one instance of the CPMP. No other inequality exists, except those obtained from transitivity.

Proof For the sake of brevity a reference to neither Geoffrion's Theorem 3.11, Theorem 3.13 nor $H(P) = H(Q) \cap H(R)$ is given.

'(5.32)': All relaxations provide lower bounds to Z. The solution space of Z is included in those of the Lagrangean duals that are included in the solution space of $Z^{(X)(Y)}$. Inclusion yields $Z^{(X)} \geq Z^{(X)(Y)}$.

'(5.33)': $Conv(P)(V)(X)(Y) \cap Conv(C)(V)(X)(Y) \subseteq H(P) \cap Conv(C)(V)(X)(Y)$ and $Conv(P)(V)(X)(Y) \cap Conv(C)(V)(X)(Y) \subseteq H(C) \cap Conv(P)(V)(X)(Y)$ yield the result.

'(5.34)': Because of equality (5.18), $Z_{(C)} = Z_{(C)}^{(X)}$ holds. $Z^{(X)} \geq Z_{(C)}^{(X)}$ since $Conv(P)(C)(V)(S)(Y) \subseteq H(C) \cap Conv(P)(V)(S)(Y)$. $Z^{(X)} \geq Z_{(R)}^{(P)(X)+(Q)}$ holds because of $Conv(P)(C)(V)(S)(Y) \subseteq H(R) \cap Conv(Q)(C)(V)(S)(Y)$.

'(5.35)': $Z_{(R)}^{(P)+(Q)} \geq Z_{(P)}$ holds because of $H(R) \cap Conv(Q)(C)(V)(X)(Y) \subseteq H(P) \cap Conv(C)(V)(X)(Y)$. $Z_{(R)}^{(P)+(Q)} \geq Z^{(P)+(Q)}$ holds because of $H(R) \cap Conv(Q)(C)(V)(X)(Y) \subseteq Conv(Q)(C)(V)(X)(Y)$. Since it is a LP relaxation, $Z_{(R)}^{(P)+(Q)} \geq Z_{(R)}^{(P)(X)+(Q)}$ holds.

'(5.36)': $Z_{(R)}^{(P)(X)+(Q)} \geq Z^{(P)(X)+(Q)}$ holds because of $H(R) \cap Conv(Q)(C)(V)(S)(Y) \subseteq Conv(Q)(C)(V)(S)(Y)$. Since a LP relaxation provides a lower bound, $Z^{(P)+(Q)} \geq Z^{(P)(X)+(Q)} \geq Z^{(P)(X)(Y)+(Q)}$ holds.

'(5.37)': Lemma 4.17 proves $Z^{(X)(Y)} \geq Z^{(C)}$. The remaining lower bounds are LP relaxations and the inclusion of solution spaces proves that they are lower bounds to $Z^{(X)(Y)}$.

'(5.38)': $H(V) \cap Conv(P)(C)(S)(X) \subseteq Conv(P)(C)(S)(X) \subseteq H(P)(C)(S)$ holds.

Showing that all inequalities are strict and that no other inequality exists is done by a comparison of optimal objective function values of two lower bounds of the same instance. Table 5.8 presents the instances and the optimal objective function values of all lower bounds are provided in Table 5.9. The mathematical formulations of the Lagrangean duals are given in Sect. 5.3.4 and each Lagrangean dual is solved to optimality as presented in Sect. 7.2. Table 5.10 yields results where the inequalities are strict.

5.3 Lower Bounds

Table 5.8 Relative strength: instances

Instance	n	T	$\pi_1,\ldots,\pi_n$	$r_1,\ldots,r_n$	$\bar{r}_1,\ldots,\bar{r}_T$
1	10	20	5,5,5,5,5,5,5,5,5,5	7,2,9,7,1,10,4,7,9,1	10,12,14,12,18,18,20,14,18,18,20,19,17,21,11,20,19,10,17,18
2	5	20	5,5,7,7,9	2,3,1,1,4	5,5,4,5,4,4,5,4,0,0,4,4,0,0,5,0,5,5,5,0
3	5	20	7,8,9,9,9	1,4,5,3,2	8,8,10,5,0,15,15,10,15,15,0,12,10,5,0,15,7,0,0,14
4	5	20	4,4,5,6,8	5,3,5,1,1	5,7,5,7,7,7,5,5,5,7,6,7,6,6,7,5,6,7
5	2	10	3,4	1,1	2,2,2,2,2,2,2,2,2,2
6	2	4	2,4	1,10	10,10,10,10
7	2	4	3,3	2,2	2,1,3,2

Table 5.9 Relative strength: optimal objective function values for Table 5.8

Instance	Z	$Z^{(X)}$	$Z_{(P)/(C)}$	$Z_{(P)}$	$Z_{(C)}$	$Z^{(Y)}_{(V)}$	$Z^{(V)(Y)}$	$Z^{(P+Q)}_{t(R)}$	$Z^{(P)(X)+(Q)}_{(R)}$	$Z^{(P+Q)}$	$Z^{(P)(X)+(Q)}$	$Z^{(X)(Y)}$	$Z^{(V)(X)(Y)}$	$Z^{(C)}$
1	16	16	13.0915	13.0915	13.0915	13.2122	13.1565	15.6667	15.6667	15	12.9977	13.0915	13.0915	4
2	10	9	8.2500	8.2500	7.5900	8.4389	8.3500	8.2500	7.8167	6	4.7000	7.5900	7.5500	4
3	3	3	3	2.6667	2.8182	2.5882	2.2000	2.6667	2.6471	2	2	2.5882	2.2000	2
4	14	11	14	14	9.7879	10.1781	10.0762	14	10.1096	10	7.9286	9.7874	9.7833	5
5	3	3	3	3	3	3	2.5000	3	3	2	2	3	2.5000	3
6	3	2	3	3	2	2	1.2000	3	2	2	2	2	1.2000	2
7	3	2	3	3	2	2.6667	2.6667	3	2	2	1.5000	2	2	1

Table 5.10 Relative strength: comparison of Table 5.9

	Z	$Z^{(X)}$	$Z_{(P)/(C)}$	$Z_{(P)}$	$Z_{(C)}$	$Z^{(Y)}_{(V)}$	$Z^{(V)(Y)}$	$Z^{(P+Q)}_{(R)}$	$Z^{(P)(X+Q)}_{(R)}$	$Z^{(P+Q)}$	$Z^{(P)(X+Q)}$	$Z^{(P)(X)(Y+Q)}$	$Z^{(X)(Y)}$	$Z^{(V)(X)(Y)}$	$Z^{(C)}$
$Z^{(X)}$	2	–	4	4	–	7	7	4	–	4	–	–	–	–	–
$Z_{(P)/(C)}$	2	2	–	–	–	2	2	1	1	1	1	–	–	–	–
$Z_{(P)}$	2	2	3	–	3	2	2	1	1	1	1	–	–	–	–
$Z_{(C)}$	2	2	3	4	–	2	2	1	1	1	1	–	–	–	–
$Z^{(Y)}_{(V)}$	2	2	3	4	3	–	–	4	1	1	1	–	–	–	5
$Z^{(V)(Y)}$	2	2	3	4	3	2	–	4	1	1	1	6	3	–	–
$Z^{(P+Q)}_{(R)}$	2	2	3	–	3	2	2	–	–	–	–	–	–	–	–
$Z^{(P)(X+Q)}_{(R)}$	2	2	3	4	3	2	2	4	–	4	–	–	–	–	–
$Z^{(P+Q)}$	2	2	3	4	3	2	2	1	1	–	–	–	3	3	5
$Z^{(P)(X+Q)}$	2	2	3	4	3	2	2	1	1	4	–	–	3	3	5
$Z^{(P)(X)(Y+Q)}$	2	2	3	4	3	2	2	1	1	1	1	–	3	3	5
$Z^{(X)(Y)}$	2	2	3	4	3	2	2	1	1	1	1	–	–	–	–
$Z^{(V)(X)(Y)}$	2	2	3	4	3	2	2	1	1	1	1	6	3	–	5
$Z^{(C)}$	2	2	3	4	3	2	2	1	1	1	1	6	3	3	–

A number represents an instance where the objective function value of a problem in the row is strictly smaller than the objective function value of a problem in the column

5.3 Lower Bounds

Lemma 5.23 $Z_{(P)}$, $Z_{(P)(V)}^{(Y)}$, $Z_{(P)}^{(V)(Y)}$ are binary $\mathcal{NP}$-hard and the Lagrangean relaxations are pseudo-polynomially solvable in time $O(n \cdot T + n \cdot \bar{r}^{sum})$. If $\bar{r}^{sum} \in O(1)$, then $ZL_{(P)}$, $ZL_{(P)(V)}^{(Y)}$ and $ZL_{(P)}^{(V)(Y)}$ are strongly polynomially solvable in time $O(n \cdot T)$. Furthermore, it holds that

$$Z_{(P)} \geq Z_{(P)(V)}^{(Y)} \geq Z_{(P)}^{(V)(Y)}. \tag{5.39}$$

Proof For the sake of brevity neither reference to Geoffrion's Theorem 3.11 nor to Remark 3.1 is given.

'(5.39)': The relative strength (5.39) holds because of $H(P) \cap Conv(C)(V)(X)(Y) \subseteq H(P)(V) \cap Conv(C)(S)(X) \subseteq H(P) \cap Conv(C)(S)(X)$.

Consider $Z_{(P)}$ with $Z_{(P)} = Z_{(P)}^{(V)}$ from (5.21). Associate the Lagrangean multiplier $v_{it} \geq 0$ with the tth constraint (P) of a maintenance activity i. The *rearrangement equality*

$$\sum_{t=1}^{T-\pi_i+1} v_{it} \cdot \sum_{\tau=t}^{t+\pi_i-1} x_{i\tau} = \sum_{t=1}^{T} x_{it} \cdot \sum_{\tau=\max\{1,t-\pi_i+1\}}^{\min\{t,T-\pi_i+1\}} v_{i\tau} \quad \forall i=1,\ldots,n \tag{5.40}$$

is applied to obtain (5.42). $ZL_{(P)}$ becomes

$$ZL_{(P)}(v) = \min \sum_{t=1}^{T} y_t - \sum_{i=1}^{n} \sum_{t=1}^{T-\pi_i+1} v_{it} \cdot \left(\sum_{\tau=t}^{t+\pi_i-1} x_{i\tau} - 1 \right) \tag{5.41}$$

subject to $(C) \wedge (X) \wedge (Y)$

$$= \min \sum_{t=1}^{T} y_t - \sum_{i=1}^{n} \sum_{t=1}^{T} x_{it} \cdot \sum_{\tau=\max\{1,t-\pi_i+1\}}^{\min\{t,T-\pi_i+1\}} v_{i\tau} + \sum_{i=1}^{n} \sum_{t=1}^{T-\pi_i+1} v_{it} \tag{5.42}$$

subject to $(C) \wedge (X) \wedge (Y)$.

Using the *integer linearization property* (Guignard 2003), (5.42) decomposes into T subproblems z_t and one trivial selection problem that is $y_t = 1$ if $1 - z_t < 0$ (otherwise, 0) $\forall t$.[21]

---

[21] In this problem, the variable y_t is coupled via (C) with the variables x_{it} such that the value of y_t decides on an upper bound of all variables x_{it} of the same index t. Specifically, y_t is binary. If $y_t = 0$, then $x_{it} = 0$ for all i. Conversely, $y_t = 1$ and $(x_{1t}, \ldots, x_{nt})$ represents the solution of a subproblem. The optimal objective function value z_t of the subproblem z_t is the contribution of $y_t = 1$ to the objective function value and the objective function is reformulated with $z_t \cdot y_t$.

138   5 Analyzing the Solvability of the Capacitated Planned Maintenance Problem

$$ZL_{(P)}(\upsilon) = \min \sum_{t=1}^{T}\left(1 - \sum_{i=1}^{n} x_{it} \cdot \sum_{\tau=\max\{1,t-\pi_i+1\}}^{\min\{t,T-\pi_i+1\}} \upsilon_{i\tau}\right) \cdot y_t + \sum_{i=1}^{n}\sum_{t=1}^{T-\pi_i+1} \upsilon_{it} \tag{5.43}$$

subject to (C)∧(X)∧(Y)

$$= \sum_{t=1}^{T} \min\{0, 1 - z_t\} + \sum_{i=1}^{n}\sum_{t=1}^{T-\pi_i+1} \upsilon_{it} \tag{5.44}$$

with

$$z_t = \max \sum_{i=1}^{n} x_{it} \cdot \sum_{\tau=\max\{1,t-\pi_i+1\}}^{\min\{t,T-\pi_i+1\}} \upsilon_{i\tau} \quad \text{subject to} \sum_{i=1}^{n} r_i \cdot x_{it} \leq \bar{r}_t \wedge x_{it} \in \{0,1\} \; \forall i$$

A polynomial reduction from the binary $\mathcal{NP}$-complete KNAPSACK shows binary $\mathcal{NP}$-hardness of $ZL_{(P)}$ that is KNAPSACK $\leq_p ZL_{(P)}$. This result proves binary $\mathcal{NP}$-hardness of $Z_{(P)}$.

KNAPSACK (Garey and Johnson 1979, pp. 65, 247): Given a constant $K \geq 0$, a capacity $c \in \mathbb{N}$ and $J \in \mathbb{N}$ items with a weight $w_j \in \mathbb{N}$ and a profit $p_j \in \mathbb{N}$. Exists a subset $A \subseteq \{1,\ldots,J\}$ with $\sum_{j \in A} p_j \geq K$ such that $\sum_{j \in A} w_j \leq c$?

An instance of $ZL_{(P)}$ is obtained in polynomial time with $n = J$, $T = 2$, $\pi_i = 2 \; \forall i$, $r_i = w_i \; \forall i$, $\bar{r}_1 = c$, $\bar{r}_2 = 0$ and $\upsilon_{i1} = p_i \; \forall i$. It follows $x_{i2} = 0 \; \forall i$ because of (C) and only the subproblem z_1 is relevant. The $ZL_{(P)}$ instance is called solvable if it is feasible and the objective function value of z_1 is greater or equal to K. Equivalence of both instances is proven by showing that the instance to KNAPSACK is solvable if and only if the instance to z_1 is solvable.

'⇒': Assume that the instance to KNAPSACK is solvable. Hence, there exists a set A that satisfies KNAPSACK with $\sum_{j \in A} p_j \geq K$. The solution $x_{i1} = 1$ if $i \in A$ (otherwise, 0) $\forall i$ to z_1 is clearly feasible and has an objective function value $\sum_{j \in A} p_j \geq K$.

'⇐': Assume that the instance to z_1 is solvable. Hence, there exists a feasible solution x to z_1 with an objective function value greater or equal to K. Obviously, the set $A = \{i = 1,\ldots,n \,|\, x_{i1} = 1\}$ is a feasible solution to KNAPSACK with $\sum_{j \in A} p_j \geq K$.

'$O(n \cdot T + n \cdot \bar{r}^{sum})$': This suggests an algorithm to solve $ZL_{(P)}$. Let p_i^t be objective function coefficients of an item i of the tth KP. Set $p_i^0 = 0$, $\upsilon_{i0} = 0$ and $\upsilon_{i,T-\pi_i+2} = 0$. Calculate the profits per period $p_i^t = p_i^{t-1} - \upsilon_{i,\max\{0,t-\pi_i\}} + \upsilon_{i,\min\{t,T-\pi_i+2\}} \; \forall i; t$ in time $O(n \cdot T)$. The sum of p_i^t over all i and t yields the constant of (5.44) in time $O(n \cdot T)$. Let c^t be the capacity of the tth KP. A non-trivial solution has $\bar{r}^{sum} > 0$. For all subproblems z_t, yield an instance to the tth KP with $J = n$, $w_i = r_i \; \forall i$,

5.3 Lower Bounds

$c^t = \bar{r}_t$, $(p_1^t, \ldots, p_n^t)$ and solve it in time $O(n \cdot \bar{r}_t)$ (Martello and Toth 1990, pp. 36–39).[22] Set $y_t = 1$ if $z_t > 0$ (otherwise, 0) and $x_{it} = 1$ if the item i is part of the tth KP solution (otherwise, 0) $\forall i$ in time $O(n)$. Solve all T problems z_t and obtain an optimal solution (x, y) in time $2 \cdot O(n \cdot T) + \sum_{t=1}^{T} O(n \cdot \bar{r}_t + n) = O(n \cdot \bar{r}^{sum} + 3 \cdot n \cdot T)$. For the worst case running time it holds that

$$n \cdot \bar{r}^{sum} + 3 \cdot n \cdot T \leq d \cdot (n \cdot T + n \cdot \bar{r}^{sum})$$

with the constant $d \geq 1$ for all $n, T \geq 1$ and $\bar{r}^{sum} > 0$. Thus, $ZL_{(P)}$ is solvable in time $O(n \cdot \bar{r}^{sum} + 3 \cdot n \cdot T) = O(n \cdot T + n \cdot \bar{r}^{sum})$.

Consider $Z_{(P)}^{(V)(Y)}$. Since the objective function coefficients assigned to y are non-negative and y is continuous, there always exists an optimal solution that satisfies (C) an equality. A substitution with

$$y_t = \sum_{i=1}^{n} \frac{r_i}{\bar{r}_t} \cdot x_{it} \ \forall t \tag{5.45}$$

from (C) yields (5.46). Note that constraint (S) becomes (D). Apply the transformation presented for $ZL_{(P)}$ to yield

$$ZL_{(P)}^{(V)(Y)}(\upsilon) = \min \sum_{t=1}^{T} y_t - \sum_{i=1}^{n}\sum_{t=1}^{T} x_{it} \cdot \sum_{\tau=\max\{1,t-\pi_i+1\}}^{\min\{t,T-\pi_i+1\}} \upsilon_{i\tau} + \sum_{i=1}^{n}\sum_{t=1}^{T-\pi_i+1} \upsilon_{it}$$

subject to $(C) \wedge (S) \wedge (X)$

$$= \min \sum_{i=1}^{n}\sum_{t=1}^{T} \left(\frac{r_i}{\bar{r}_t} - \sum_{\tau=\max\{1,t-\pi_i+1\}}^{\min\{t,T-\pi_i+1\}} \upsilon_{i\tau} \right) \cdot x_{it} + \sum_{i=1}^{n}\sum_{t=1}^{T-\pi_i+1} \upsilon_{it} \tag{5.46}$$

subject to $(D) \wedge (X)$.

An instance of $ZL_{(P)}^{(V)(Y)}$ is obtained in polynomial time with $T = 2$, $\pi_i = 2 \ \forall i$ and $\bar{r}_2 = 0$. It follows $x_{i2} = 0 \ \forall i$ because of (D). A transformation yields $ZL_{(P)}^{(V)(Y)}(\upsilon) = \sum_{i=1}^{n} \upsilon_{i1} - \bar{Z}$ with a subproblem

$$\bar{Z} = \max \sum_{i=1}^{n} \left(-\frac{r_i}{\bar{r}_1} + \upsilon_{i1} \right) \cdot x_{i1} \text{ subject to } \sum_{i=1}^{n} r_i \cdot x_{i1} \leq \bar{r}_1 \wedge x_{i1} \in \{0, 1\} \ \forall i.$$

A polynomial reduction shows binary $\mathcal{NP}$-hardness of $ZL_{(P)}^{(V)(Y)}$ via KNAPSACK $\leq_p ZL_{(P)}^{(V)(Y)}$. This result proves binary $\mathcal{NP}$-hardness of $Z_{(P)}^{(V)(Y)}$. An instance of

---

[22] Using a word-RAM implementation, the tth KP is even solvable in time $O(\frac{n \cdot m}{\log m})$ with $m = \max\{\bar{r}_t, z_t\}$ (Kellerer et al. 2004, pp. 131–136).

$ZL_{(P)}^{(V)(Y)}$ is obtained in polynomial time with $n = J$, $r_i = w_i$ $\forall i$, $\bar{r}_1 = c$ and $v_{i1} = \frac{r_i}{r_1} + p_i$ $\forall i$. Equivalence of both instances $ZL_{(P)}^{(V)(Y)}$ and KNAPSACK is shown with the aforementioned polynomial reduction for $ZL_{(P)}$. Using the approach presented for $ZL_{(P)}$, an optimal solution to $ZL_{(P)}^{(V)(Y)}$ is obtained in time $O(n \cdot T + n \cdot \bar{r}^{sum})$ but set y according to (5.45) in time $O(n)$.

Consider $Z_{(P)(V)}^{(Y)}$. Let $u \in \mathbb{R}_+^{n \times T}$ be the Lagrangean multiplier of (V). It follows

$$ZL_{(P)(V)}^{(Y)}(v, u) = \min \sum_{i=1}^{n} \sum_{t=1}^{T} \left(\frac{r_i}{\bar{r}_t} - \sum_{\tau=\max\{1, t-\pi_i+1\}}^{\min\{t, T-\pi_i+1\}} v_{i\tau} - \sum_{h=1}^{n} u_{ht} + u_{it} \right) \cdot x_{it}$$

$$+ \sum_{i=1}^{n} \sum_{t=1}^{T-\pi_i+1} v_{it}$$

subject to $(D) \wedge (X)$

that is solvable in a similar way as $ZL_{(P)}^{(V)(Y)}$ in time $O(n \cdot T + n \cdot \bar{r}^{sum})$. Setting $u_{it} = 0$ $\forall i; t$, $Z_{(P)(V)}^{(Y)}$ becomes the binary $\mathcal{NP}$-hard $Z_{(P)}^{(V)(Y)}$.

Lemma 5.24 $Z_{(P)/(C)}$ is strongly $\mathcal{NP}$-hard.

Proof $ZL_{(P)/(C)}$ is derived in Sect. 5.3.4 as $ZL_{(P)/(C)}(\alpha, \beta) = ZL'(\alpha, \beta) + ZL''(\alpha, \beta)$ with two subproblems

$$ZL'(\alpha, \beta) = \min \sum_{t=1}^{T} \beta_t \cdot y_t + \sum_{i=1}^{n} \sum_{t=1}^{T} \alpha_{it} \cdot x_{it} \quad \text{subject to } (P) \wedge (V) \wedge (S) \wedge (Y) \quad (5.88)$$

$$ZL''(\alpha, \beta) = \sum_{t=1}^{T} \min\{0, 1 - \beta_t - \bar{z}_t\} \quad (5.89)$$

$$\forall t : \bar{z}_t = \max\left\{ \sum_{i=1}^{n} \alpha_{it} \cdot x_{it} \middle| \sum_{i=1}^{n} r_i \cdot x_{it} \leq \bar{r}_t \wedge x_{it} \in \{0, 1\} \; \forall i \right\}.$$

ZL' is the strongly $\mathcal{NP}$-hard WUPMP (Lemma 4.6) and hence, strong $\mathcal{NP}$-hardness holds for $ZL_{(P)/(C)}$ and $Z_{(P)/(C)}$ (Remark 3.1).

Lemma 5.25 $Z^{(V)(Y)}$ and $Z_{(V)}^{(Y)}$ are strongly $\mathcal{NP}$-hard.

Proof In $Z^{(V)(Y)}$, the objective function coefficients of y are non-negative and y is continuous. Hence, there exists an optimal solution that satisfies (C) an equality. A substitution with $y_t = \sum_{i=1}^{n} \frac{r_i}{\bar{r}_t} \cdot x_{it}$ $\forall t$ yields

$$Z^{(V)(Y)} = \min \sum_{i=1}^{n} \sum_{t=1}^{T} \frac{r_i}{\bar{r}_t} \cdot x_{it} \quad \text{subject to } (P) \wedge (D) \wedge (X).$$

5.3 Lower Bounds

The polynomial reduction of Lemma 5.7 applies to show strong $\mathcal{NP}$-hardness of $Z^{(V)(Y)}$ that is 3-PARTITION $\leq_p Z^{(V)(Y)}$. However, the objective function value becomes

$$\sum_{i=1}^{n} \frac{r_i}{B} \cdot \sum_{t=1}^{T} x_{it} \quad \text{(P) can only be satisfied as an equality}$$

$$= \frac{1}{B} \cdot \sum_{i=1}^{n} r_i \quad \text{since } \sum_{i=1}^{n} r_i = P \cdot B \text{ from 3-PARTITION and } T = P$$

$$= T.$$

Consider $Z_{(V)}^{(Y)}$ and set the Lagrangean multiplier assigned to (V) to zero. Thus, $Z_{(V)}^{(Y)}$ becomes the strongly $\mathcal{NP}$-hard $Z^{(V)(Y)}$.

Lemma 5.26 $Z_{(R)}^{(P)+(Q)}$ and $Z^{(P)+(Q)}$ are strongly $\mathcal{NP}$-hard.

Proof The polynomial reduction of Lemma 5.7 applies to show strong $\mathcal{NP}$-hardness of $Z^{(P)+(Q)}$ that is 3-PARTITION $\leq_p Z^{(P)+(Q)}$. Setting the Lagrangean multiplier associated with (R) to zero, $Z_{(R)}^{(P)+(Q)}$ becomes the strongly $\mathcal{NP}$-hard $Z^{(P)+(Q)}$.

Lemma 5.27 $Z_{(R)}^{(P)(X)+(Q)}$ is binary $\mathcal{NP}$-hard.

Proof A polynomial reduction from the binary $\mathcal{NP}$-complete SUBSET SUM shows binary $\mathcal{NP}$-hardness of $ZL_{(R)}^{(P)(X)+(Q)}$ that is SUBSET SUM $\leq_p ZL_{(R)}^{(P)(X)+(Q)}$. This result proves binary $\mathcal{NP}$-hardness for $Z_{(R)}^{(P)(X)+(Q)}$ (Remark 3.1).

SUBSET SUM (Garey and Johnson 1979, p. 223; Karp 1972): Given a positive integer K and $J \in \mathbb{N}$ items with a weight $w_j \in \mathbb{N}$. Exists a subset $A \subseteq \{1, \ldots, J\}$ such that $\sum_{j \in A} w_j = K$? W.l.o.g. assume $w_j \leq K \; \forall j$ and $w_1 \geq \ldots \geq w_J > 0$.

Let $I = \{2, \ldots, J+1\}$ be the set of items of SUBSET SUM. Denote the objective function coefficients as $c_{1t} = \sum_{\tau=\max\{t-\pi_1+1,1\} \wedge \tau \notin S_1}^{\min\{t, T-\pi_1+1\}} \upsilon_{1\tau} \; \forall t$. Omitting the redundant constraint (V), $ZL_{(R)}^{(P)(X)+(Q)}$, which is stated by (5.64) in Sect. 5.3.4, becomes

$$ZL_{(R)}^{(P)(X)+(Q)}(\upsilon) = \min \sum_{t=1}^{T} y_t - \sum_{t=1}^{T} c_{1t} \cdot x_{1t} + \sum_{t=1 \wedge t \notin S_1}^{T-\pi_1+1} \upsilon_{1t}$$

subject to $(Q) \wedge (C) \wedge (S^x) \wedge (Y)$.

An instance of $ZL_{(R)}^{(P)(X)+(Q)}$ is obtained in polynomial time as follows: Set $n = 1$, $T = 2 \cdot J + 1$, $\pi_1 = J + 1$, $r_1 = K$ and $\bar{r}_t = w_{t-1}$ if $t \in I$ (otherwise, 0) $\forall t$. Setting $m = 1$, $Q_1 = \{1, \ldots, T\}$ and $S_1 = \{1\}$, (R) comprises the Lagrangean relaxed task horizons in I. Set $\upsilon_{1t} = \frac{K}{w_1} - K \cdot f$ if $t = 2$ with a scaling factor $f = 10^{-(\lfloor \log_{10} g \rfloor + 1)}$ and $g = \max\{K, \max_{j=1,\ldots,J} w_j\}$ (otherwise, $\frac{K}{w_{t-1}} - \frac{K}{w_{t-2}}$) $\forall t \in I$. Note that the

Table 5.11 Coefficient matrix of (Q) and (R) for Lemma 5.27

				t			
		1	2	...	$J+1$	$J+2$	...	$2 \cdot J+1$
(Q)		1	1	...	1			
(R)	v_{12}		1	...	1	1		
⋮				⋱			⋱	
	$v_{1,J+1}$				1	1	...	1

non-negativity of the Lagrangean multiplier $v_{1t} \geq 0 \ \forall t$ holds. Table 5.11 shows the coefficient matrix of (Q) and (R). The periods $t \in I$ are $2,\ldots,J+1$.

The definitions simplify $\text{ZL}_{(R)}^{(P)(X)+(Q)}$ significantly. Since $\bar{r}_t \leq r_1 \ \forall t$ holds, constraint (C) ensures $x_{1t} \leq 1 \ \forall t$ and thus, (S^x) is satisfied. Furthermore, $y_t = 0 \ \forall t \notin I$ holds since $\bar{r}_t = 0$ because of (C). Moreover, $c_{1t} = \frac{K}{w_{t-1}} - K \cdot f \ \forall t \in I$ and $c_{1t} \geq 0 \ \forall t \in I$ are obtained. Hence, (Q) and (C) become (5.47) and (5.48). The simplified problem is

$$\text{ZL}_{(R)}^{(P)(X)+(Q)}(v) = \min \sum_{t \in I} y_t - \sum_{t \in I} c_{1t} \cdot x_{1t} + \frac{K}{w_J} - K \cdot f \quad \text{subject to}$$

$$\sum_{t \in I} x_{1t} \geq 1 \tag{5.47}$$

$$K \cdot x_{1t} \leq w_{t-1} \cdot y_t \qquad \forall t \in I \tag{5.48}$$

$x_{1t} \geq 0 \ \forall t \in I \wedge y_t \in \{0,1\} \ \forall t \in I$.

The $\text{ZL}_{(R)}^{(P)(X)+(Q)}$ instance is called solvable if it is feasible and the objective function value is smaller or equal to $\frac{K}{w_J}$. In what follows, equivalence of both instances is proven by showing that the instance to SUBSET SUM is solvable if and only if the instance to $\text{ZL}_{(R)}^{(P)(X)+(Q)}$ is solvable.

'$\Rightarrow$': Assume the instance to SUBSET SUM is solvable. Hence, there exists a set A that satisfies SUBSET SUM with $\sum_{j \in A} w_j = K$. Let (x,y) be a solution to $\text{ZL}_{(R)}^{(P)(X)+(Q)}$ with $x_{1t} = \frac{w_{t-1}}{K}$ and $y_t = 1$ if $\{t-1\} \in A$ (otherwise, 0) $\forall t$. (5.47) is satisfied because $\sum_{t \in I} x_{1t} = \sum_{j \in A} \frac{w_j}{K} = 1$. (5.48) is clearly satisfied. Thus, (x,y) is feasible to $\text{ZL}_{(R)}^{(P)(X)+(Q)}$ and the objective function value becomes

$$\sum_{t \in A} y_{t+1} - \sum_{t \in A} c_{1,t+1} \cdot x_{1,t+1} + \frac{K}{w_J} - K \cdot f \quad \text{since } \forall t \in A : x_{1,t+1} = \frac{w_t}{K} \wedge y_{t+1} = 1$$

$$= f \cdot \sum_{j \in A} w_j + \frac{K}{w_J} - K \cdot f \qquad \text{with } \sum_{j \in A} w_j = K$$

$$= \frac{K}{w_J}.$$

5.3 Lower Bounds

'⇐': Assume that the instance to $\text{ZL}_{(R)}^{(P)(X)+(Q)}$ is solvable. Hence, there exists a feasible solution (x, y) to $\text{ZL}_{(R)}^{(P)(X)+(Q)}$ with an objective function value smaller or equal to $\frac{K}{w_J}$. Let $A = \{t = 1, ..., T \mid y_{t-1} = 1\}$ be a solution to SUBSET SUM. Introduce a slack variable $\Delta_t \geq 0\ \forall t$ and constraint (5.48) becomes $x_{1t} = \frac{w_{t-1}}{K} \cdot y_t - \Delta_t\ \forall t \in I$. Hence, the contribution of a period $t \in I$ to the objective function value is

$$y_t - c_{1t} \cdot x_{1t} = f \cdot w_{t-1} \cdot y_t + c_{1t} \cdot \Delta_t = \begin{cases} f \cdot w_{t-1} + c_{1t} \cdot \Delta_t & \text{if } x_{1t} > 0 \\ 0 & \text{otherwise.} \end{cases}$$

Feasibility of (5.47) requires that there exists at least one period $t \in A$ with $x_{1,t+1} > 0$. Hence, it holds for the objective function value that

$$\frac{K}{w_J} \geq \sum_{t \in A} y_{t+1} - \sum_{t \in A} c_{1,t+1} \cdot x_{1,t+1} + \frac{K}{w_J} - K \cdot f$$

$$= \sum_{t \in A} f \cdot w_t + \sum_{t \in A} c_{1,t+1} \cdot \Delta_{t+1} + \frac{K}{w_J} - K \cdot f$$

Due to the fact that $c_{1t}, \Delta_t \geq 0\ \forall t$ and $f > 0$ hold, a rearrangement yields

$$K \geq \sum_{t \in A} w_t + \sum_{t \in A} \frac{c_{1,t+1}}{f} \cdot \Delta_{t+1} \geq \sum_{t \in A} w_t.$$

Rewrite (5.48) as $\frac{w_{t-1}}{K} \cdot y_t \geq x_{1t}\ \forall t \in I$ and sum up over all periods $t \in A$ to obtain

$$\sum_{t \in A} \frac{w_t}{K} \cdot y_{t+1} \geq \sum_{t \in A} x_{1,t+1} \geq 1 \quad \text{with (5.47) and } x_{1,t+1} = 0\ \forall t \notin A \text{ (5.48)}$$

$$\sum_{t \in A} w_t \geq K \quad \text{because of } y_{t+1} = 1\ \forall t \in A.$$

Altogether, $\sum_{j \in A} w_j = K$ holds and the set A satisfies SUBSET SUM.

Lemma 5.28 $Z^{(P)(X)(Y)+(Q)}$, $Z^{(X)(Y)}$ and $Z^{(V)(X)(Y)}$ are polynomially solvable in time $O(n^{3.5} \cdot T^{3.5} \cdot L)$ with L as the input length of n, T, π, r, $\bar{r}$ and the coefficient matrix.

Proof By an argument of Lemma 4.10, the LPs are solvable in time $O(n^{3.5} \cdot T^{3.5} \cdot L)$ by Karmarkar's algorithm (Karmarkar 1984).

Being a specific WUPMP, $ZL_{(C)}$ (5.51) is closely related to the UFLP. Neither polynomial reductions with UFLP, SET COVER and *VERTEX COVER*[23] could have been established nor polynomially solvable cases of the UFLP could have been applied such as Cornuejols et al. (1990), Krarup and Pruzan (1983), Kolen and Tamir (1990), Kolen (1983), Jones et al. (1995) and Billionnet and Costa (1994). It is assumed that $ZL_{(C)}$ is $\mathcal{NP}$-hard. From a polyhedral point of view the strongest evidence is the absence of the integrality property of Corollary 3.12 because Theorem 5.22 proves $Z_{(C)} > Z^{(X)(Y)}$.[24] Two problems are related to the *Capacitated Facility Location Problem (CFLP)*,[25] namely $Z^{(X)}$ and $Z^{(P)(X)+(Q)}$, which has m subproblems that are similar to ZL_q (5.64). However, because of the objective function, polynomial reductions of the CFLP, *3-DIMENSIONAL MATCHING*,[26] SET COVER and VERTEX COVER could not have been established.

Constraint (U) with Lemma 5.2 potentially tightens $Z_{(P)}$, $Z_{(R)}^{(P)+(Q)}$ and $Z_{(R)}^{(P)(X)+(Q)}$. Constraint (P^y) tightens $Z_{(P)}$ such that T KPs and one SPP is obtained from the decomposition presented in Lemma 5.23.

5.3.3 Transformations of the Lower Bounds

The lower bounds (5.49) are special cases of network flow problems, whereas the lower bounds (5.50) are interpreted as extended facility location problems.[27]

$$Z, Z^{(X)}, Z_{(C)}, Z_{(V)}^{(Y)}, Z^{(V)(Y)}, Z_{(R)}^{(P)+(Q)}, Z^{(P)+(Q)}, Z_{(R)}^{(P)(X)+(Q)}, Z^{(P)(X)+(Q)} \qquad (5.49)$$

$$Z_{(R)}^{(P)+(Q)}, Z^{(P)+(Q)}, Z_{(R)}^{(P)(X)+(Q)}, Z^{(P)(X)+(Q)} \qquad (5.50)$$

---

[23]Given a graph $G(V, E)$ and a positive integer $K \leq |V|$. Exists a node cover of size K or less for $G(V, E)$ that is a subset $V' \subseteq V$ with $|V'| \leq K$ such that for each edge $(a, b) \in E$ at least one node a or b belongs to V'? VERTEX COVER is strongly $\mathcal{NP}$-complete (Garey and Johnson 1979, p. 190).

[24]This is also shown by Example 6.2. Note that $ZL_{(C)}$ is a WUPMP with a specific cost structure and not a trivial problem as the one presented in Footnote 18 on page 130.

[25]Given a set of facilities with fixed costs and a maximal supply. Given a set of customers with transportation costs and a demand. Serve the customers from the facilities such that the demand of all customers is satisfied and the capacity of each facility is not exceeded. The CFLP is strongly $\mathcal{NP}$-hard (Cornuejols et al. 1991), which is proven with a polynomial reduction with the strongly $\mathcal{NP}$-complete 3-DIMENSIONAL MATCHING (Cornuejols et al. 1991).

[26]Given a set $M \subseteq W \times X \times Y$ with W, X and Y being disjoint and consisting of an equal number of m elements. Does M contain a matching, that is a subset $M' \subseteq M$ such that $|M'| = m$ and no two elements of M' agree in any coordinate? 3-DIMENSIONAL MATCHING is strongly $\mathcal{NP}$-complete (Garey and Johnson 1979, p. 221).

[27]The following discussion does not include $Z^{(X)(Y)}$, $Z^{(V)(X)(Y)}$, $Z^{(P)(X)(Y)+(Q)}$, which are LPs, $Z_{(P)}$ because it is equivalent to solving T KPs (see Lemma 5.23), $Z_{(P)/(C)}$ because it is solvable with $Z_{(P)}$ as well as $Z_{(C)}$ and $Z^{(C)}$ because it is a WUPMP with $c_{it} = 0 \ \forall i; t$, which is a particular UFLP (Sect. 4.3).

5.3.3.1 Network Flow Problems

The general idea of the *network-transformation* that has been introduced in Remark 4.3 is to reformulate (P) with Remark 4.2 as a path within a directed graph. Afterwards, each node t is split into two nodes a, b such that the ingoing (outgoing) arcs of t are the ingoing (outgoing) arcs of a (b). In order to model a maintenance slot, insert a new arc (a, b) with zero transportation cost but with respective fixed costs. All other arcs have transportation costs from the objective function coefficients of x and fixed costs that are zero. A task plan for maintenance activity i is obtained with the flow from the source to the sink.

In what follows, $ZL_{(C)}$ is exemplary transformed. Because of Lemma 4.3, $ZL_{(C)}$ has the single-assignment property with $ZL_{(C)}(\mu) = ZL_{(C)}^{(X)}(\mu)$ and becomes

$$ZL_{(C)}(\mu) = \min \sum_{t=1}^{T} (1 - \mu_t \cdot \bar{r}_t) \cdot y_t + \sum_{i=1}^{n} \sum_{t=1}^{T} r_i \cdot \mu_t \cdot x_{it} \tag{5.51}$$

subject to (P)∧(V)∧(S)∧(Y).

Let $V = \{0, \ldots, 2 \cdot T + 1\}$ be the node set with a source node 0 and a sink node $2 \cdot T + 1$.

$E' = \{(a, b) \in V \times V \mid a \text{ is odd} \land b = a + 1\}$
$E_i'' = \{(a, b) \in V \times V \mid a \text{ is even} \land b = a + 1, a + 3, \ldots, a + 2 \cdot \pi_i - 1\} \quad \forall i = 1, \ldots, n$

are two arc sets. If a feasible solution contains an arc $(a, b) \in E'$, a maintenance slot is established in period $\frac{b}{2}$. The arc set E_i'' is used to reformulate (P). If a feasible solution has a flow on an arc $(a, b) \in E_i''$, a maintenance activity is scheduled in period $\frac{a}{2}$. $E_i'' \subseteq E_{i+1}'' \ \forall i = 1, \ldots, n-1$ holds because of Assumption (2.1). Let $E = E' \cup E_n''$ be the arc set of the directed graph $G(V, E)$. The fixed costs $F_{ab} = 1 - \mu_{\frac{b}{2}} \cdot \bar{r}_{\frac{b}{2}}$ arise if $(a, b) \in E'$ (otherwise, 0) $\forall (a, b) \in E$. Flow costs C_{iab} are defined as

$$C_{iab} = \begin{cases} 0 & \text{if } (a, b) \in E' \\ r_i \cdot \mu_{\frac{a}{2}} & \text{if } (a, b) \in E_i'' \quad \forall i = 1, \ldots, n; (a, b) \in E. \\ \infty & \text{otherwise} \end{cases}$$

The variable z_{iab} represents the flow of maintenance activity i along each arc $(a, b) \in E$ and the binary variable ζ_{ab} equals 1 if the arc $(a, b) \in E$ is used (otherwise, 0). The network-transformed problem $ZL_{(C)}$ becomes the UNDP (Balakrishnan et al. 1989).

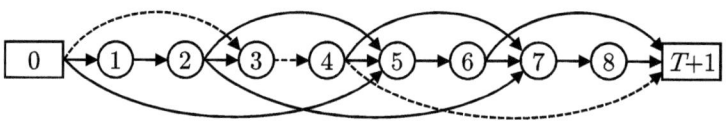

Fig. 5.1 Network-transformation: directed graph

$$ZL_{(C)}(\mu) = \min \sum_{(a,b) \in E} F_{ab} \cdot \zeta_{ab} + \sum_{i=1}^{n} \sum_{(a,b) \in E} C_{iab} \cdot z_{iab} \quad \text{subject to} \tag{5.52}$$

$$\sum_{(a,b) \in E} z_{iab} - \sum_{(b,a) \in E} z_{iba} = \begin{cases} 1 & \forall i = 1, \ldots, n; a = 0 \\ 0 & \forall i = 1, \ldots, n; a = 1, \ldots, 2 \cdot T \\ -1 & \forall i = 1, \ldots, n; a = 2 \cdot T + 1 \end{cases} \tag{5.53}$$

$$z_{iab} \leq \zeta_{ab} \quad \forall i = 1, \ldots, n; (a,b) \in E \tag{5.54}$$

$$z_{iab} \geq 0 \; \forall i = 1, \ldots, n; (a,b) \in E \wedge \zeta_{ab} \in \{0,1\} \; \forall (a,b) \in E \tag{5.55}$$

The objective function (5.52) minimizes the total costs. The flow balance equations (5.53) represent (P). Constraint (5.54) corresponds to (V). The variable domains (5.55) are binary for ζ and the flow z is continuous. The following variable transformation is applied

$$x_{it} = \sum_{(2 \cdot t, \tau) \in E_i''} z_{i,2 \cdot t,\tau} \; \forall i = 1, \ldots, n; t = 1, \ldots, T \wedge y_t = \zeta_{2 \cdot t-1, 2 \cdot t} \; \forall t = 1, \ldots, T.$$

Example 5.8 Let $n = 1$, $T = 4$ and $\pi_1 = 3$. It follows $E' = \{(1,2),(3,4),(5,6),(7,8)\}$ and the remaining arcs in Fig. 5.1 define E_1''. Dashed lines represent a flow from the sink to the source in the directed graph $G(V, E)$. The solution $z_{103} = z_{134} = z_{149} = 1$ and $\zeta_{03} = \zeta_{34} = \zeta_{49} = 1$ corresponds to $x_{12} = 1$ and $y_2 = 1$.

Section A.2 of the appendix presents the transformation of (C). $ZL_{(V)}^{(Y)}$ and $Z^{(V)(Y)}$ become special cases of the *Origin-Destination Integer Multi-Commodity Network Flow Problem (MCNFP)* (Ahuja et al. 1993, pp. 649–650; Barnhart et al. 2000).[28] $Z^{(X)}$ and the remaining problems are special cases of the *Fixed-Charge Capacitated Network Design Problem (FCNDP)* (Rodríguez-Martín and

---

[28]Using the definitions of the UNDP, let an upper bound restrict the total flow of all commodities on an arc $(a,b) \in E$ and remove the variables for the fixed costs. Find a flow of minimal cost for all commodities through the graph such that the flow is balanced in every node, the arc capacity is never exceeded and the flow is split (i.e., integral). Using a polynomial reduction with the strongly $\mathcal{NP}$-complete 3-PARTITION, the MCNFP is strongly $\mathcal{NP}$-hard.

5.3 Lower Bounds

Salazar-González, 2010; Herrmann et al. 1996; Gendron 2002).[29] Integrality on the flow variables must be imposed for Z, $ZL_{(R)}^{(P)+(Q)}$ and $ZL^{(P)+(Q)}$. Reformulate (Q) comparable to Remark 4.2 and apply the node splitting to transform $ZL_{(R)}^{(P)+(Q)}$, $Z^{(P)+(Q)}$, $ZL_{(R)}^{(P)(X)+(Q)}$ and $Z^{(P)(X)+(Q)}$.

5.3.3.2 Facility Location Problems

The *facility-transformation* interprets periods as facilities and task horizons as customers.[30] Associate the periods $1, \ldots, T$ with *regular facilities*. Consider a maintenance activity i and the qth subproblem from (5.64). The minimal requirement to satisfy (Q) is to schedule maintenance activity i once per task horizon. Interpret such a task horizon $t, \ldots, t + \pi_i - 1$ with $t \in S_i \cap Q_q$ as a *regular customer*. Setting the transportation costs from a regular customer j to the set of regular facilities $Q_q \setminus \{t, \ldots, t + \pi_i - 1\}$ to infinity, the customer's unit demand must be served from the regular facilities $t, \ldots, t + \pi_i - 1$. A set of *dummy customers* and one *dummy facility* $T + q$ are introduced because of the slack of (Q).[31]

In what follows, $ZL_{(R)}^{(P)+(Q)}$ is exemplary transformed. Add the valid inequality (U) to (5.64) and set $c_{it} = \sum_{\tau=\max\{t-\pi_i+1,1\} \wedge \tau \notin S_i}^{\min\{t, T-\pi_i+1\}} \upsilon_{i\tau}$ $\forall i; t$. The subproblems for $ZL_{(R)}^{(P)+(Q)(U)}$ become

$$\forall q : ZL_q(\upsilon) = \min \sum_{t \in Q_m} y_t - \sum_{i=1}^{n} \sum_{t \in Q_m} c_{it} \cdot x_{it}$$

subject to $(U) \wedge (Q) \wedge (C) \wedge (V) \wedge (X) \wedge (Y)$.

The maximal number of customers per maintenance activity and subproblem q is $u_i^{max} = \max_{r=1,\ldots,m} u_{ir}$ $\forall i$ and the set $J = \{1, \ldots, \sum_{h=1}^{n} u_h^{max}\}$ contains all customers. In order to map customers to maintenance activities, let $U_i = \{1 + \sum_{h=1}^{i-1} u_h^{max}, \ldots, \sum_{h=1}^{i} u_h^{max}\}$ be the set of all customers that are assigned to a maintenance activity i. Per set U_i the customers $j_1, \ldots, j_{|S_i \cap Q_q|}$ are regular customers, the customers $j_{1+|S_i \cap Q_q|}, \ldots, j_{u_{iq}}$ are dummy customers and the customers $j_{1+u_{iq}}, \ldots, j_{|U_i|}$ are not considered. Task horizons are mapped as follows: Assume the

---

[29] Using the definitions of the UNDP, let an upper bound restrict the total flow of all commodities on an arc $(a, b) \in E$. Find a flow of minimal cost for all commodities through the graph such that the flow is balanced in every node and the arc capacity is never exceeded. Using a polynomial reduction with the strongly $\mathcal{NP}$-complete STEINER TREE IN GRAPHS (Garey and Johnson 1979, p. 208), the FCNDP is strongly $\mathcal{NP}$-hard (Magnanti and Wong 1984).

[30] Note that this transformation differs from the transformation of the WUPMP to the extended UFLP (4.15)–(4.23) in Sect. 4.2.

[31] In this transformation, a so called *customer* (*facility*) is either a regular or a dummy customer (facility).

set S_i is ordered to increasing values. For regular customers it holds that if customer j is the pth regular customer of a maintenance activity i, then let $d(j, i, q)$ yield the pth period in $S_i \cap Q_q$. Otherwise, $d(j, i, q)$ is a big number. Let C_{jt} be the transportation costs defined for all $i = 1, \ldots, n; j \in U_i; q = 1, \ldots, m; t \in Q_q \cup \{T + q\}$ as

$$C_{jt} = \begin{cases} -c_{it} & \text{if } j \text{ is a regular customer for } i \text{ and} \\ & t \in \{d(j,i,q), \ldots, d(j,i,q) + \pi_i - 1\} \cap Q_q \\ -c_{it} & \text{if } j \text{ is a dummy customer for } i \text{ and } t \in Q_q \\ 0 & \text{if } j \text{ is a dummy customer for } i \text{ and } t = T + q \\ \infty & \text{otherwise.} \end{cases}$$

A regular customer is served by the set of regular facilities that represent one task horizon of (Q) at costs $-c_{it}$. A dummy customer can be served by any regular facility at costs $-c_{it}$ or by the dummy facility at zero costs. (Q) is satisfied for a given task horizon if the regular customer is served by a regular facility in the task horizon, which must always hold. (Q) is satisfied as a strict inequality for a given task horizon if one regular customer and at least one dummy customer are served by a regular facility in that task horizon. No more than u_{iq} customers with finite transportation cost exist, which represents (U). The fixed costs for all facilities are $F_t = 1$ if $t \leq T$ (otherwise, 0) $\forall t = 1, \ldots, T + m$. The customer demand is $R_j = r_i \; \forall i = 1, \ldots, n; j \in U_i$ and the facility capacity is $\bar{R}_t = \bar{r}_t$ if $t \leq T$ (otherwise, ∞) $\forall t = 1, \ldots, T + m$. The binary variable ζ_t equals 1 if the a facility $t \in Q_q \cup \{T + q\}$ is established (otherwise, 0) and the binary variable z_{jt} equals 1 if the customer $j \in J$ is served from the facility $t \in Q_q \cup \{T + q\}$ (otherwise, 0).

$$ZL_q(\upsilon) = \min \sum_{t \in Q_q \cup \{T+q\}} F_t \cdot \zeta_t + \sum_{j \in J} \sum_{t \in Q_q \cup \{T+q\}} C_{jt} \cdot z_{jt} \quad \text{subject to} \tag{5.56}$$

$$\sum_{t \in Q_q \cup \{T+q\}} z_{jt} = 1 \qquad \forall j \in J \tag{5.57}$$

$$\sum_{j \in J} R_j \cdot z_{jt} \leq \bar{R}_t \cdot \zeta_t \qquad \forall t \in Q_q \cup \{T+q\} \tag{5.58}$$

$$\sum_{j \in U_i} z_{jt} \leq 1 \qquad \forall i = 1, \ldots, n; t \in Q_q \tag{5.59}$$

$$z_{jt} \leq \zeta_t \qquad \forall j \in J; t \in Q_q \cup \{T+q\} \tag{5.60}$$

$$\zeta_t \in \{0, 1\} \qquad \forall t \in Q_q \cup \{T+q\} \tag{5.61}$$

$$z_{jt} \in \{0, 1\} \qquad \forall j \in J; t \in Q_q \cup \{T+q\} \tag{5.62}$$

is the facility-transformed subproblem ZL_q. The objective function (5.56) minimizes the total costs. Because of constraint (5.57), every customer must be assigned to

5.3 Lower Bounds

exactly one facility. Together with the structure of C this represents (Q) and (U). The capacity constraint (5.58) represents (C). The clique inequality (5.59) ensures that a regular facility t serves at most one customer $j \in U_i$. If constraint (5.59) was relaxed, the variable x would be integer and not binary because a regular facility t could serve more than one customer $j \in U_i$. Constraint (5.60) is the corresponding valid inequality (V). The variable domains (5.61) and (5.62) are binary and the variable transformation is

$$x_{it} = \sum_{j \in U_i} z_{jt} \ \forall i = 1, \ldots, n; t = 1, \ldots, T \wedge y_t = \zeta_t \ \forall t = 1, \ldots, T.$$

Example 5.9 Let $n = 2$, $T = 6$, $\pi = (2, 4)^T$, $r = (3, 4)^T$, $\bar{r} = (3, 7, 7, 7, 0, 4)^T$ and $UB = 3$. Let $m = 2$, $Q_1 = \{1, \ldots, 4\}$, $Q_2 = \{5, 6\}$, $S_1 = \{1, 3, 5\}$, $S_2 = \{1\}$ and $u = \binom{4}{3}\binom{1}{1}$ from $u = u'$ (Example 5.2). It follows $J = \{1, \ldots, 6\}$, $U_1 = \{1, 2, 3, 4\}$ and $U_2 = \{5, 6, 7\}$. In the subproblem $q = 1$, regular customers are 1, 2 for $i = 1$ and 5 for $i = 2$. Dummy customer are 3, 4 for $i = 1$ and 6, 7 for $i = 2$. In the subproblem $q = 2$, there is one regular customer 1 for $i = 1$ and one dummy customer 5 for $i = 2$. The customers 2, 3, 4, 6 and 7 are not considered for $q = 2$. Regular facilities are 1, $\ldots$, 6 and dummy facilities are 7 for $q = 1$ and 8 for $q = 2$. It follows $d(1, 1, 1) = 1$, $d(2, 1, 1) = 3$, $d(5, 2, 1) = 1$ and $d(1, 1, 2) = 5$. Table 5.12 provides the transportation costs for $v_{it} = 1 \ \forall i; t$ and the fixed costs are $F = (1, 1, 1, 1, 1, 1, 0, 0)^T$. If an entry (j, t) is given in italic, $z_{jt} = 1$ (otherwise, 0). This yields the optimal solutions $\zeta_2 = \zeta_3 = \zeta_4 = \zeta_7 = 1$ with $ZL_1(v) = -5$ for $q = 1$ and $\zeta_5 = \zeta_8 = 1$ with $ZL_2(v) = -2$ for $q = 2$ that correspond to $x_{i2} = x_{i3} = x_{i4} = x_{i5} = 1 \ \forall i$ and $y_2 = y_3 = y_4 = y_5 = 1$.

The valid inequality (U) tightens the lower bound as demonstrated in Example 5.10.

Example 5.10 Consider the Example 5.9 where (U) is tightened by $u = \binom{2}{3}\binom{1}{1}$ with $u_{iq} = \min\{u'_{iq}, u'''_{iq}\} \ \forall i; q$ (Example 5.2). Hence, there exists no dummy customer for maintenance activity $i = 1$ and the optimal solution for ZL_1 improves to $ZL_1(v) = -4$.

Table 5.12 Facility-transformation: transportation costs C_{jt}

j \\ t	1	2	3	4	5	6	7	8
1	0	−1	∞	∞	−1	0	∞	∞
2	∞	∞	−1	−1	∞	∞	∞	0
3	0	−1	−1	−1	∞	∞	0	0
4	0	−1	−1	−1	∞	∞	0	0
5	0	−1	−2	−2	−2	−1	∞	0
6	0	−1	−2	−2	∞	∞	0	0
7	0	−1	−2	−2	∞	∞	0	0

Relaxing the clique inequality (5.59) completely, the facility-transformed subproblem ZL_q becomes a SSCFLP[32] with negative transportation costs that aims to establish as many assignments as possible between customers and periods in $1, \ldots, T$.[33] The structure of the clique inequalities (5.59) allows to adapt heuristics and exact solution approaches for the SSCFLP to solve ZL_q such as Klose and Drexl (2005), Cortinhal and Captivo (2003), Sridharan (1993). Note that clique inequalities are valid inequalities to the SSCLFP. The facility-transformation of $ZL_{(R)}^{(P)(X)+(Q)}$ yields a CFLP with negative transportation costs, $Z^{(P)(X)+(Q)}$ a CFLP and $Z^{(P)+(Q)}$ a *Generalized Bin Packing Problem*.[34]

5.3.4 Mathematical Formulations of the Lagrangean Duals

This section gives the mathematical formulations of all Lagrangean duals, which are relevant for Theorem 5.22, as RMPs (3.23)–(3.26) from Sect. 3.3.1. Section 7.2 presents the solution approach.

5.3.4.1 Lagrangean Duals $Z_{(R)}^{(P)+(Q)}$ and $Z_{(R)}^{(P)(X)+(Q)}$

Consider $Z_{(R)}^{(P)+(Q)}$. The rearrangement equality (5.40) is adapted to yield

$$\sum_{t=1 \wedge t \notin S_i}^{T-\pi_i+1} v_{it} \cdot \sum_{\tau \in \{t,\ldots,t+\pi_i-1\} \cap Q_q} x_{i\tau} = \sum_{t \in Q_q} x_{it} \cdot \sum_{\tau=\max\{t-\pi_i+1,1\} \wedge \tau \notin S_i}^{\min\{t,T-\pi_i+1\}} v_{i\tau} \quad (5.63)$$

for all $q = 1, \ldots, m$ and $i = 1, \ldots, n$.

Example 5.11 Let $\pi_i = 3$, $T = 5$ and $m = 2$. Assume $Q_1 = \{1, 2, 3\}$ and $Q_2 = \{4, 5\}$. Table 5.13 shows the coefficient matrix. Both equalities of (5.63) are

---

[32]Given a set of facilities with fixed costs and a maximal supply. Given a set of customers with transportation costs and a demand. Find the cost minimal assignment of customers to facilities such that the demand of all customers is satisfied, the capacity of each facility is not exceeded and each customer is assigned to exactly one facility. A polynomial reduction with the strongly $\mathcal{NP}$-complete 3-PARTITION shows that SSCFLP is strongly $\mathcal{NP}$-hard (Klose and Drexl 2005) with Martello and Toth (1990, p. 8).

[33]Cf. Anti-CFLP presented in Klose (2001, pp. 238–239).

[34]The Generalized Bin Packing Problem (Hung and Brown 1978; Lewis and Parker 1982) is a special case of the SSCFLP with $f_t = 1$ $\forall t$ and $c_{it} = 0$ $\forall i; t$. Strong $\mathcal{NP}$-hardness is shown by a polynomial reduction of the strongly $\mathcal{NP}$-complete BIN PACKING (Garey and Johnson 1979, p. 226).

5.3 Lower Bounds

Table 5.13 Coefficient matrix of Example 5.11

			v_{i1}	v_{i2}	v_{i3}
λ_1	Q_1	x_{i1}	1		
		x_{i2}	1	1	
		x_{i3}	1	1	1
λ_2	Q_2	x_{i4}		1	1
		x_{i5}			1

given below.

$$v_{i1} \cdot (x_{i1} + x_{i2} + x_{i3}) + v_{i2} \cdot (x_{i2} + x_{i3}) + v_{i3} \cdot x_{i3} = x_{i1} \cdot v_{i1} + x_{i2} \cdot (v_{i1} + v_{i2}) +$$
$$x_{i3} \cdot (v_{i1} + v_{i2} + v_{i3})$$

$$v_{i2} \cdot x_{i4} + v_{i3} \cdot (x_{i4} + x_{i5}) = x_{i4} \cdot (v_{i2} + v_{i3}) + x_{i5} \cdot v_{i3}$$

Since no two task horizons of (Q) overlap, the constraints are partitioned into m independent sets and consequently, the Lagrangean relaxation $\text{ZL}_{(R)}^{(P)+(Q)}$ decomposes into m subproblems ZL_q. Let $c_{it} = \sum_{\tau=\max\{t-\pi_i+1,1\} \wedge \tau \notin S_i}^{\min\{t,T-\pi_i+1\}} v_{i\tau}$ $\forall i; t$.

$$\text{ZL}_{(R)}^{(P)+(Q)}(v) = \min \sum_{t=1}^{T} y_t - \sum_{i=1}^{n} \sum_{t=1 \wedge t \notin S_i}^{T-\pi_i+1} v_{it} \cdot \left(\sum_{\tau=t}^{t+\pi_i-1} x_{i\tau} - 1 \right)$$

$$= \sum_{q=1}^{m} \min \left(\sum_{t \in Q_q} y_t - \sum_{i=1}^{n} \sum_{t=1 \wedge t \notin S_i}^{T-\pi_i+1} v_{it} \cdot \sum_{\tau \in \{t,\ldots,t+\pi_i-1\} \cap Q_q} x_{i\tau} \right)$$

$$+ \sum_{i=1}^{n} \sum_{t=1 \wedge t \notin S_i}^{T-\pi_i+1} v_{it}$$

$$= \sum_{q=1}^{m} \min \left(\sum_{t \in Q_q} y_t - \sum_{i=1}^{n} \sum_{t \in Q_q} c_{it} \cdot x_{it} \right) + \sum_{i=1}^{n} \sum_{t=1 \wedge t \notin S_i}^{T-\pi_i+1} v_{it} \quad \text{with (5.63)}$$

$$= \sum_{q=1}^{m} \text{ZL}_q(v) + \sum_{i=1}^{n} \sum_{t=1 \wedge t \notin S_i}^{T-\pi_i+1} v_{it}$$

with m subproblems

$$\forall q: \text{ZL}_q(v) = \min \sum_{t \in Q_q} y_t - \sum_{i=1}^{n} \sum_{t \in Q_q} c_{it} \cdot x_{it} \quad (5.64)$$

subject to $(Q) \wedge (C) \wedge (V) \wedge (X) \wedge (Y)$.

Let P_q (R_q) be the index set for all extreme points for y_t^p and x_{it}^p (extreme rays for y_t^r and x_{it}^r) of $Conv(Q)(C)(V)(Y)(X)$ restricted to the periods $t \in Q_q$ in section $q = 1, \ldots, m$. Since the planning horizon is partitioned, a period $t \in Q_q$ occurs only in P_q and in no other set P_s with $s \neq q$. Hence, an additional index $q = 1, \ldots, m$ (e.g., y_{qt}^p) is superfluous. A convexification yields

$$ZL_q(\upsilon) \leq \begin{cases} -\infty & \text{if } \exists r \in R_q : \sum_{t \in Q_q} y_t^r - \sum_{i=1}^n \sum_{t \in Q_q} c_{it} \cdot x_{it}^r < 0 \\ \sum_{t \in Q_q} y_t^p - \sum_{i=1}^n \sum_{t \in Q_q} c_{it} \cdot x_{it}^p & \text{otherwise; } \forall p \in P_q. \end{cases}$$

The Lagrangean dual $Z_{(R)}^{(P)+(Q)}$ becomes

$$Z_{(R)}^{(P)+(Q)} = \max_{\upsilon \geq 0} ZL_{(R)}^{(P)+(Q)}(\upsilon)$$

$$= \max_{\upsilon \geq 0} \sum_{q=1}^m \min_{p \in P_q} \sum_{t \in Q_q} y_t^p - \sum_{i=1}^n \sum_{t=1 \wedge t \notin S_i}^{T-\pi_i+1} \upsilon_{it} \cdot \sum_{\tau \in \{t, \ldots, t+\pi_i-1\} \cap Q_q} x_{i\tau}^p$$

$$+ \sum_{i=1}^n \sum_{t=1 \wedge t \notin S_i}^{T-\pi_i+1} \upsilon_{it}.$$

A linearization as a LP yields

$$Z_{(R)}^{(P)+(Q)} = \max \sum_{q=1}^m \eta_q + \sum_{i=1}^n \sum_{t=1 \wedge t \notin S_i}^{T-\pi_i+1} \upsilon_{it} \qquad \text{subject to}$$

$$\eta_q + \sum_{i=1}^n \sum_{t=1 \wedge t \notin S_i}^{T-\pi_i+1} \upsilon_{it} \cdot \sum_{\tau \in \{t, \ldots, t+\pi_i-1\} \cap Q_q} x_{i\tau}^p \leq \sum_{t \in Q_q} y_t^p \qquad \begin{array}{l} \forall q = 1, \ldots, m; \\ p \in P_q \end{array} \qquad \lambda_{qp}$$

(5.65)

$$\sum_{i=1}^n \sum_{t=1 \wedge t \notin S_i}^{T-\pi_i+1} \upsilon_{it} \cdot \sum_{\tau \in \{t, \ldots, t+\pi_i-1\} \cap Q_q} x_{i\tau}^r \leq \sum_{t \in Q_q} y_t^r \qquad \begin{array}{l} \forall q = 1, \ldots, m; \\ r \in R_q \end{array} \qquad \lambda_{qr}$$

(5.66)

$\upsilon_{it} \geq 0 \; \forall i = 1, \ldots, n; t = 1, \ldots, T - \pi_i + 1 \wedge \eta_q \in \mathbb{R} \; \forall q = 1, \ldots, m.$

5.3 Lower Bounds

A dualization yields the primal problem with $A_i = \{1, \ldots, T - \pi_i + 1\} \setminus S_i$ for all i as

$$Z_{(R)}^{(P)+(Q)} = \min \sum_{q=1}^{m} \sum_{t \in Q_q} \left(\sum_{p \in P_q} y_t^p \cdot \lambda_{qp} + \sum_{r \in R_q} y_t^r \cdot \lambda_{qr} \right) \quad \text{subject to}$$

$$\sum_{p \in P_q} \lambda_{qp} = 1 \qquad \forall q = 1, \ldots, m \qquad \eta_q$$

$$\sum_{q=1}^{m} \sum_{\tau \in \{t, \ldots, t+\pi_i-1\} \cap Q_q} \left(\sum_{p \in P_q} x_{i\tau}^p \cdot \lambda_{qp} + \sum_{r \in R_q} x_{i\tau}^r \cdot \lambda_{qr} \right) \geq 1 \quad \begin{array}{l} \forall i = 1, \ldots, n; \\ t \in A_i \end{array} \quad \upsilon_{it}$$

(5.67)

$$\lambda_{qp} \geq 0 \; \forall q = 1, \ldots, m; p \in P_q \wedge \lambda_{qr} \geq 0 \; \forall q = 1, \ldots, m; r \in R_q$$

The following example demonstrates the dualization of the constraints (5.65) and (5.66) that yield the primal constraint (5.67).

Example 5.12 Consider Example 5.11 and constraint (5.65) with $P = \{1\}$. The dualization yields (5.67) as $\lambda_1 \cdot (x_{i1} + x_{i2} + x_{i3})$ for υ_{i1}, $\lambda_1 \cdot (x_{i2} + x_{i3}) + \lambda_2 \cdot x_{i4}$ for υ_{i2} and $\lambda_2 \cdot (x_{i3} + x_{i4} + x_{i5})$ for υ_{i3}.

Omit all extreme rays y_t^r and $x_{i\tau}^r$. For a given q, the extreme point $p \in P_q$ with $(x_{i\tau}^p)_{i \in \{1, \ldots, n\} \wedge t \in Q_q}$ and $(y_t^p)_{t \in Q_q}$ yields a maintenance subplan, which is not necessarily feasible to (P). The variable λ_{qp} equals 1, if all maintenance slots and tasks of the periods Q_q are determined by the pth maintenance subplan (otherwise, 0). The RMP is stated as

$$Z_{(R)}^{(P)+(Q)} = \min \sum_{q=1}^{m} \sum_{t \in Q_q} \sum_{p \in P_q} y_t^p \cdot \lambda_{qp} \quad \text{subject to}$$

$$\sum_{p \in P_q} \lambda_{qp} = 1 \qquad \forall q = 1, \ldots, m \qquad \eta_q$$

$$\sum_{q=1}^{m} \sum_{\tau \in \{t, \ldots, t+\pi_i-1\} \cap Q_q} \sum_{p \in P_q} x_{i\tau}^p \cdot \lambda_{qp} \geq 1 \quad \forall i = 1, \ldots, n; t = 1, \ldots, T - \pi_i + 1; t \notin S_i \quad \upsilon_{it}$$

$$\lambda_{qp} \geq 0 \qquad \forall q = 1, \ldots, m; p \in P_q.$$

Constraint (5.65) defines the reduced cost $\bar{c}_q^p$ of an extreme point $p \in P_q$ for a given $q = 1, \ldots, m$ as

$$\bar{c}_q^p = \min \sum_{t \in Q_q} y_t - \sum_{i=1}^{n} \sum_{t \in Q_q} c_{it} \cdot x_{it} - \eta_q \quad \text{subject to } (Q) \wedge (C) \wedge (V) \wedge (X) \wedge (Y).$$

The RMP of $Z_{(R)}^{(P)(X)+(Q)}$ is identical but the reduced costs are

$$\bar{c}_q^p = \min \sum_{t \in Q_q} y_t - \sum_{i=1}^{n} \sum_{t \in Q_q} c_{it} \cdot x_{it} - \eta_q \quad \text{subject to } (Q) \wedge (C) \wedge (V) \wedge (S) \wedge (Y).$$

5.3.4.2 Lagrangean Dual $Z_{(P)}$

The Lagrangean relaxation $ZL_{(P)}$ is stated in Lemma 5.23 as

$$ZL_{(P)}(\upsilon) = \sum_{t=1}^{T} \min\{0, 1-z_t\} + \sum_{i=1}^{n} \sum_{t=1}^{T-\pi_i+1} \upsilon_{it} \tag{5.44}$$

$$\forall t : z_t = \max \left\{ \sum_{i=1}^{n} x_{it} \cdot \sum_{\tau=\max\{1,t-\pi_i+1\}}^{\min\{t,T-\pi_i+1\}} \upsilon_{i\tau} \,\middle|\, \sum_{i=1}^{n} r_i \cdot x_{it} \leq \bar{r}_t \wedge x_{it} \in \{0,1\} \,\forall i \right\}$$

and a linearization with the definition of z_t yields

$$ZL_{(P)}(\upsilon) = \max \sum_{i=1}^{n} \sum_{t=1}^{T-\pi_i+1} \upsilon_{it} - \sum_{t=1}^{T} \eta_t \quad \text{subject to} \tag{5.68}$$

$\eta_t + 1 \geq z_t \,\forall t = 1, \ldots, T \wedge \eta_t \geq 0 \,\forall t = 1, \ldots, T$.

For each $t = 1, \ldots, T$, let P_t (R_t) be the index set for all extreme points x_{it}^p (extreme rays x_{it}^r) of $Conv(D)(X)$ that is restricted to t. A convexification yields

$$z_t \geq \begin{cases} \infty & \text{if } \exists r \in R_t : \sum_{i=1}^{n} x_{it}^r \cdot \sum_{\tau=\max\{t-\pi_i+1,1\}}^{\min\{t,T-\pi_i+1\}} \upsilon_{i\tau} > 0 \\ \sum_{i=1}^{n} x_{it}^p \cdot \sum_{\tau=\max\{t-\pi_i+1,1\}}^{\min\{t,T-\pi_i+1\}} \upsilon_{i\tau} & \text{otherwise}; \forall p \in P_t. \end{cases} \tag{5.69}$$

The following LP is yielded for the Lagrangean dual $Z_{(P)} = \max_{\upsilon \geq 0} ZL_{(P)}(\upsilon)$ of the linearized and decomposed Lagrangean relaxation (5.68).

$$Z_{(P)} = \max \sum_{i=1}^{n} \sum_{t=1}^{T-\pi_i+1} \upsilon_{it} - \sum_{t=1}^{T} \eta_t \quad \text{subject to} \tag{5.70}$$

$$-\eta_t + z_t \leq 1 \qquad \forall t = 1, \ldots, T \qquad \lambda_t^0 \tag{5.71}$$

$$\sum_{i=1}^{n} x_{it}^p \cdot \sum_{\tau=\max\{t-\pi_i+1,1\}}^{\min\{t,T-\pi_i+1\}} \upsilon_{i\tau} - z_t \leq 0 \qquad \forall t = 1, \ldots, T; p \in P_t \qquad \lambda_{tp} \tag{5.72}$$

5.3 Lower Bounds

$$\sum_{i=1}^{n} x_{it}^{r} \cdot \sum_{\tau=\max\{t-\pi_i+1,1\}}^{\min\{t,T-\pi_i+1\}} v_{i\tau} \leq 0 \qquad \forall t = 1,\ldots,T; r \in R_t \qquad \lambda_{tr} \qquad (5.73)$$

$$v_{it} \geq 0 \ \forall i = 1,\ldots,n; t = 1,\ldots,T-\pi_i+1 \qquad (5.74)$$

$$z_t \in \mathbb{R} \ \forall t = 1,\ldots,T \wedge \eta_t \geq 0 \ \forall t = 1,\ldots,T \qquad (5.75)$$

Dualization yields the primal problem of the Lagrangean dual as

$$Z_{(P)} = \min \sum_{t=1}^{T} \lambda_t^0 \qquad \text{subject to}$$

$$-\lambda_t^0 \geq -1 \qquad \forall t = 1,\ldots,T \qquad \eta_t$$

$$\lambda_t^0 - \sum_{p \in P_t} \lambda_{tp} = 0 \qquad \forall t = 1,\ldots,T \qquad z_t \qquad (5.76)$$

$$\sum_{\tau=t}^{t+\pi_i-1} \left(\sum_{p \in P_\tau} x_{i\tau}^p \cdot \lambda_{\tau p} + \sum_{r \in R_\tau} x_{i\tau}^r \cdot \lambda_{\tau p} \right) \geq 1 \qquad \begin{array}{l} \forall i = 1,\ldots,n; \\ t = 1,\ldots,T-\pi_i+1 \end{array} \qquad v_{it}$$

$$\lambda_t^0 \geq 0 \ \forall t = 1,\ldots,T \wedge \lambda_{tp} \geq 0 \ \forall t = 1,\ldots,T; p \in P_t \wedge \lambda_{tr} \geq 0 \ \forall t = 1,\ldots,T; r \in R_t.$$

The variable λ_t^0 is substituted with (5.76). Omit all extreme rays x_{it}^r. The variable λ_{tp} equals 1, if the pth plan $(x_{1t}^p,\ldots,x_{nt}^p)$ is realized in period t (otherwise, 0). The RMP is

$$Z_{(P)} = \min \sum_{t=1}^{T} \sum_{p \in P_t} \lambda_{tp} \qquad \text{subject to} \qquad (5.77)$$

$$-\sum_{p \in P_t} \lambda_{tp} \geq -1 \qquad \forall t = 1,\ldots,T \qquad \eta_t \qquad (5.78)$$

$$\sum_{\tau=t}^{t+\pi_i-1} \sum_{p \in P_\tau} x_{i\tau}^p \cdot \lambda_{\tau p} \geq 1 \qquad \forall i = 1,\ldots,n; t = 1,\ldots,T-\pi_i+1 \qquad v_{it} \qquad (5.79)$$

$$\lambda_{tp} \geq 0 \qquad \forall t = 1,\ldots,T; p \in P_t. \qquad (5.80)$$

Constraint (5.72) defines the reduced cost $\bar{c}_t^p$ of the extreme point $p \in P_t$ for a given t. Since λ_t^0 is eliminated from the formulation, (5.71) and (5.72) yield

$$\sum_{i=1}^{n} x_{it}^p \cdot \sum_{\tau=\max\{t-\pi_i+1,1\}}^{\min\{t,T-\pi_i+1\}} v_{i\tau} \leq 1 + \eta_t \quad \forall t = 1,\ldots,T; p \in P_t.$$

Maximizing the slack, T subproblems are obtained and the reduced costs for all $t = 1, ..., T$ are

$$\bar{c}_t^p = 1 + \eta_t - \max \left\{ \sum_{i=1}^{n} x_{it} \cdot \sum_{\tau=\max\{1,t-\pi_i+1\}}^{\min\{t,T-\pi_i+1\}} v_{i\tau} \,\middle|\, \sum_{i=1}^{n} r_i \cdot x_{it} \leq \bar{r}_t \wedge x_{it} \in \{0,1\} \; \forall i \right\}.$$

5.3.4.3 Lagrangean Dual $Z_{(C)}$

The Lagrangean relaxation $ZL_{(C)}$

$$ZL_{(C)}(\mu) = \min \sum_{t=1}^{T} (1 - \mu_t \cdot \bar{r}_t) \cdot y_t + \sum_{i=1}^{n} \sum_{t=1}^{T} r_i \cdot \mu_t \cdot x_{it} \tag{5.51}$$

subject to $(P) \wedge (V) \wedge (X) \wedge (Y)$.

is a special WUPMP and has the single-assignment property of Lemma 4.3. Let P (R) be the index set for all extreme points for y_t^p and x_{it}^p (extreme rays for y_t^r and x_{it}^r) of $Conv(P)(V)(S)(Y)$. A convexification yields

$$ZL_{(C)}(\mu) \leq \begin{cases} -\infty & \text{if } \exists r \in R: \sum_{t=1}^{T}(1 - \mu_t \cdot \bar{r}_t) \cdot y_t^r + \sum_{t=1}^{T}\sum_{i=1}^{n} \mu_t \cdot r_i \cdot x_{it}^r < 0 \\ \sum_{t=1}^{T}(1 - \mu_t \cdot \bar{r}_t) \cdot y_t^p + \sum_{i=1}^{n}\sum_{t=1}^{T} \mu_t \cdot r_i \cdot x_{it}^p & \text{otherwise;} \; \forall p \in P. \end{cases}$$

The Lagrangean dual $Z_{(C)}$ becomes

$$Z_{(C)} = \max_{\mu \geq 0} ZL_{(C)}(\mu)$$

$$= \max_{\mu \geq 0} \min_{p \in P} \sum_{t=1}^{T} y_t^p + \sum_{t=1}^{T} \left(\sum_{i=1}^{n} r_i \cdot x_{it}^p - \bar{r}_t \cdot y_t^p \right) \cdot \mu_t.$$

A linearization as a LP yields

$$Z_{(C)} = \max \eta \qquad \text{subject to}$$

$$\eta + \sum_{t=1}^{T} \left(\bar{r}_t \cdot y_t^p - \sum_{i=1}^{n} r_i \cdot x_{it}^p \right) \cdot \mu_t \leq \sum_{t=1}^{T} y_t^p \qquad \forall p \in P \qquad \lambda_p \tag{5.81}$$

$$\sum_{t=1}^{T} \left(\bar{r}_t \cdot y_t^r - \sum_{i=1}^{n} r_i \cdot x_{it}^r \right) \cdot \mu_t \leq \sum_{t=1}^{T} y_t^r \qquad \forall r \in R \qquad \lambda_r$$

$$\mu_t \geq 0 \; \forall t = 1, \ldots, T \wedge \eta \in \mathbb{R}.$$

5.3 Lower Bounds

Dualization yields the primal problem of the Lagrangean dual as

$$Z_{(C)} = \min \sum_{p \in P} \sum_{t=1}^{T} y_t^p \cdot \lambda_p + \sum_{r \in R} \sum_{t=1}^{T} y_t^r \cdot \lambda_r \qquad \text{subject to}$$

$$\sum_{p \in P} \lambda_p = 1 \qquad \qquad \eta$$

$$\sum_{p \in P} \lambda_p \cdot \left(\bar{r}_t \cdot y_t^p - \sum_{i=1}^{n} r_i \cdot x_{it}^p \right) + \sum_{r \in R} \lambda_r \cdot \left(\bar{r}_t \cdot y_t^r - \sum_{i=1}^{n} r_i \cdot x_{it}^r \right) \geq 0 \quad \forall t = 1, \ldots, T \quad \mu_t$$

$$\lambda_p \geq 0 \ \forall p \in P \wedge \lambda_r \geq 0 \ \forall r \in R.$$

Omit all extreme rays y_t^r and x_{it}^r. The variable λ_p equals 1, if the pth maintenance plan (x^p, y^p) is realized (otherwise, 0). The RMP is

$$Z_{(C)} = Z_{(C)}^{(X)} = \min \sum_{p \in P} \sum_{t=1}^{T} y_t^p \cdot \lambda_p \qquad \text{subject to} \qquad (5.82)$$

$$\sum_{p \in P} \lambda_p = 1 \qquad \qquad \eta \qquad (5.83)$$

$$\sum_{p \in P} \lambda_p \cdot \left(\bar{r}_t \cdot y_t^p - \sum_{i=1}^{n} r_i \cdot x_{it}^p \right) \geq 0 \qquad \forall t = 1, \ldots, T \qquad \mu_t \qquad (5.84)$$

$$\lambda_p \geq 0 \qquad \forall p \in P. \qquad (5.85)$$

The reduced cost $\bar{c}^p$ of the extreme point $p \in P$ is defined by (5.81) as the subproblem

$$\bar{c}^p = \min \sum_{t=1}^{T} (1 - \bar{r}_t \cdot \mu_t) \cdot y_t + \sum_{i=1}^{n} \sum_{t=1}^{T} r_i \cdot \mu_t \cdot x_{it} - \eta \quad \text{subject to (P)} \wedge \text{(V)} \wedge \text{(S)} \wedge \text{(Y)}.$$

5.3.4.4 Lagrangean Dual $Z_{(P)/(C)}$

A variable splitting yields an equivalent formulation of the CPMP. Let $\alpha \in \mathbb{R}^{n \times T}$ and $\beta \in \mathbb{R}^T$ be the Lagrangean multiplier of (5.86) and (5.87).

$$Z = \min \sum_{t=1}^{T} y_t'' \qquad \text{subject to} \qquad (Z)$$

$$\sum_{\tau=t}^{t+\pi_i-1} x_{i\tau}' \geq 1 \qquad \forall i = 1, \ldots, n\,; t = 1, \ldots, T - \pi_i + 1 \qquad (P)$$

$$-x''_{it} + y'_t \geq 0 \qquad \forall i = 1, \ldots, n; t = 1, \ldots, T \qquad \text{(V)}$$

$$-\sum_{i=1}^{n} r_i \cdot x''_{it} + \bar{r}_t \cdot y''_t \geq 0 \qquad \forall t = 1, \ldots, T \qquad \text{(C)}$$

$$x''_{it} - x'_{it} = 0 \qquad \forall i = 1, \ldots, n; t = 1, \ldots, T \qquad \alpha_{it} \quad (5.86)$$

$$y''_t - y'_t = 0 \qquad \forall t = 1, \ldots, T \qquad \beta_t \quad (5.87)$$

$$y'_t, y''_t \in \{0, 1\} \qquad \forall t = 1, \ldots, T \qquad \text{(X)}$$

$$x'_{it}, x''_{it} \in \{0, 1\} \qquad \forall i = 1, \ldots, n; t = 1, \ldots, T \qquad \text{(Y)}$$

Using $Z_{(P)/(C)} = Z_{(P)(V)/(C)}$ from (5.30), the Lagrangean relaxation of (5.86) and (5.87) yields a Lagrangean dual that is simplified with the single-assignment property of Lemma 4.3.

$$Z_{(P)/(C)} = \min\{1^T \cdot y \mid (x, y) \in \text{Conv}(P)(V)(X)(Y) \cap \text{Conv}(C)(V)(X)(Y)\}$$
$$= \min\{1^T \cdot y \mid (x, y) \in \text{Conv}(P)(V)(S)(Y) \cap \text{Conv}(C)(X)(Y)\}$$

The Lagrangean relaxation

$$ZL_{(P)/(C)}(\alpha, \beta) = \min \sum_{t=1}^{T} y''_t - \sum_{i=1}^{n}\sum_{t=1}^{T} \alpha_{it} \cdot (x''_{it} - x'_{it}) - \sum_{t=1}^{T} \beta_t \cdot (y''_t - y'_t)$$
$$= ZL'(\alpha, \beta) + ZL''(\alpha, \beta)$$

has two subproblems

$$ZL'(\alpha, \beta) = \min \sum_{t=1}^{T} \beta_t \cdot y_t + \sum_{i=1}^{n}\sum_{t=1}^{T} \alpha_{it} \cdot x_{it} \qquad (5.88)$$

subject to $(P) \wedge (V) \wedge (S) \wedge (Y)$

$$ZL''(\alpha, \beta) = \sum_{t=1}^{T} \min\{0, 1 - \beta_t - \bar{z}_t\} \qquad (5.89)$$

$$\forall t : \bar{z}_t = \max\left\{\sum_{i=1}^{n} \alpha_{it} \cdot x_{it} \middle| \sum_{i=1}^{n} r_i \cdot x_{it} \leq \bar{r}_t \wedge x_{it} \in \{0, 1\} \; \forall i\right\}.$$

In order to obtain (5.89), observe that ZL'' is almost identical to $ZL_{(P)}$ (5.44) from Lemma 5.23 and it analogously decomposes into T subproblems $\bar{z}_t$. Let P' (R') be the index set for all extreme points containing y'^p_t and x'^p_{it} (extreme rays of y'^r_t and x'^r_{it}) of $\text{Conv}(P)(V)(S)(Y)$. For each $t = 1, \ldots, T$, let P''_t (R''_t) be the index set for all extreme points x''^p_{it} (extreme rays x''^r_{it}) of $\text{Conv}(D)(X)$ that is restricted to t. A

5.3 Lower Bounds

convexification yields

$$ZL'(\alpha, \beta) \leq \begin{cases} -\infty & \text{if } \exists r \in R' : \sum_{t=1}^{T} \beta_t \cdot y_t^{\prime r} + \sum_{i=1}^{n}\sum_{t=1}^{T} \alpha_{it} \cdot x_{it}^{\prime r} < 0 \\ \sum_{t=1}^{T} \beta_t \cdot y_t^{\prime p} + \sum_{i=1}^{n}\sum_{t=1}^{T} \alpha_{it} \cdot x_{it}^{\prime p} & \text{otherwise;} \ \forall p \in P' \end{cases}$$

$$\bar{z}_t \geq \begin{cases} \infty & \text{if } \exists r \in R_t'' : \sum_{i=1}^{n} x_{it}^r \cdot \sum_{\tau=\max\{t-\pi_i+1,1\}}^{\min\{t,T-\pi_i+1\}} v_{i\tau} > 0 \\ \sum_{i=1}^{n} x_{it}^p \cdot \sum_{\tau=\max\{t-\pi_i+1,1\}}^{\min\{t,T-\pi_i+1\}} v_{i\tau} & \text{otherwise;} \ \forall p \in P_t''. \end{cases}$$

The following LP is yielded for the Lagrangean dual

$$Z_{(P)/(C)} = \max_{\alpha \in \mathbb{R}^{n \times T} \wedge \beta \in \mathbb{R}^T} ZL'(\alpha, \beta) + ZL''(\alpha, \beta).$$

Note that the results of (5.70)–(5.75) are applied to ZL".

$$Z_{(P)/(C)} = \max \eta' - \sum_{t=1}^{T} \eta_t'' \qquad \text{subject to}$$

$$\eta' - \sum_{t=1}^{T} y_t^{\prime p} \cdot \beta_t - \sum_{i=1}^{n}\sum_{t=1}^{T} x_{it}^{\prime p} \cdot \alpha_{it} \leq 0 \quad \forall p \in P' \qquad \lambda_p' \qquad (5.90)$$

$$-\sum_{t=1}^{T} y_t^{\prime r} \cdot \beta_t - \sum_{i=1}^{n}\sum_{t=1}^{T} x_{it}^{\prime r} \cdot \alpha_{it} \leq 0 \quad \forall r \in R' \qquad \lambda_r'$$

$$-\eta_t'' + \beta_t + \bar{z}_t \leq 1 \qquad \forall t = 1, \ldots, T \qquad \lambda_t''^0$$

$$\sum_{i=1}^{n} x_{it}^{\prime\prime p} \cdot \alpha_{it} - \bar{z}_t \leq 0 \qquad \forall t = 1, \ldots, T; p \in P_t'' \quad \lambda_{tp}'' \qquad (5.91)$$

$$\sum_{i=1}^{n} x_{it}^{\prime\prime r} \cdot \alpha_{it} \leq 0 \qquad \forall t = 1, \ldots, T; r \in R_t'' \quad \lambda_{tr}''$$

$$\alpha_{it} \in \mathbb{R} \ \forall i = 1, \ldots, n; t = 1, \ldots, T \wedge \beta_t \in \mathbb{R} \ \forall t = 1, \ldots, T$$
$$\bar{z}_t \in \mathbb{R} \ \forall t = 1, \ldots, T \wedge \eta_t'' \geq 0 \ \forall t = 1, \ldots, T \wedge \eta' \in \mathbb{R}$$

Dualization yields the primal problem of the Lagrangean dual as

$$Z_{(P)/(C)} = \min \sum_{t=1}^{T} \lambda_t''^{0} \qquad \text{subject to}$$

$$\sum_{p \in P'} \lambda_p' = 1 \qquad\qquad\qquad\qquad\qquad\qquad\qquad\qquad\qquad \eta'$$

$$-\lambda_t''^{0} \geq -1 \qquad\qquad\qquad\qquad\qquad\qquad \forall t = 1,\ldots,T \qquad \eta_t''$$

$$-\sum_{p \in P'} x_{it}'^{p} \cdot \lambda_p' - \sum_{r \in R'} x_{it}'^{r} \cdot \lambda_r' + \sum_{p \in P_t''} x_{it}''^{p} \cdot \lambda_{tp}'' + \sum_{r \in R_t''} x_{it}''^{r} \cdot \lambda_{tr}'' = 0 \qquad \begin{array}{l} \forall i = 1,\ldots,n; \\ t = 1,\ldots,T \end{array} \quad \alpha_{it}$$

$$-\sum_{p \in P'} y_t'^{p} \cdot \lambda_p' - \sum_{r \in R'} y_t'^{r} \cdot \lambda_r' + \lambda_t''^{0} = 0 \qquad\qquad\qquad \forall t = 1,\ldots,T \qquad \beta_t$$

$$\lambda_t''^{0} - \sum_{p \in P_t''} \lambda_{tp}'' = 0 \qquad\qquad\qquad\qquad\qquad \forall t = 1,\ldots,T \qquad \bar{z}_t$$

$$\lambda_p' \geq 0 \; \forall p \in P' \wedge \lambda_r' \geq 0 \; \forall r \in R'$$

$$\lambda_t''^{0} \geq 0 \; \forall t = 1,\ldots,T \wedge \lambda_{tp}'' \geq 0 \; \forall t = 1,\ldots,T; p \in P_t'' \wedge \lambda_{tr}'' \geq 0 \; \forall t = 1,\ldots,T; r \in R_t''.$$

The interpretation of the variables is provided above for $Z_{(P)}$ and $Z_{(C)}$. In contrast to $Z_{(P)}$, no substitution of $\lambda_t''^{0}$ is done because preliminary computational studies showed that this would increase the computational time. Omit all extreme rays. The RMP is

$$Z_{(P)/(C)} = \min \sum_{t=1}^{T} \lambda_t''^{0} \qquad \text{subject to}$$

$$\sum_{p \in P'} \lambda_p' = 1 \qquad\qquad\qquad\qquad\qquad\qquad\qquad\qquad \eta'$$

$$-\lambda_t''^{0} \geq -1 \qquad\qquad\qquad\qquad \forall t = 1,\ldots,T \qquad \eta_t''$$

$$-\sum_{p \in P'} x_{it}'^{p} \cdot \lambda_p' + \sum_{p \in P_t''} x_{it}''^{p} \cdot \lambda_{tp}'' = 0 \qquad \forall i = 1,\ldots,n; t = 1,\ldots,T \qquad \alpha_{it}$$

$$-\sum_{p \in P'} y_t'^{p} \cdot \lambda_p' + \lambda_t''^{0} = 0 \qquad\qquad \forall t = 1,\ldots,T \qquad \beta_t$$

$$\lambda_t''^{0} - \sum_{p \in P_t''} \lambda_{tp}'' = 0 \qquad\qquad\qquad \forall t = 1,\ldots,T \qquad \bar{z}_t$$

$$\lambda_p' \geq 0 \; \forall p \in P' \wedge \lambda_t''^{0} \geq 0 \; \forall t = 1,\ldots,T \wedge \lambda_{tp}'' \geq 0 \; \forall t = 1,\ldots,T; p \in P_t''.$$

The reduced cost $\bar{c}'^{p}$ of an extreme point $p \in P'$ is given by (5.90). For a given $t = 1,\ldots,T$ the reduced cost $\bar{c}_t''^{p}$ of the extreme point $p \in P_t''$ is derived

5.3 Lower Bounds

from (5.91).

$$\bar{c}'^P = \min \sum_{t=1}^{T} \beta_t \cdot y'_t + \sum_{i=1}^{n}\sum_{t=1}^{T} \alpha_{it} \cdot x'_{it} - \eta' \quad \text{subject to } (P) \wedge (V) \wedge (S) \wedge (Y)$$

$$\bar{c}''^P_t = \bar{z}_t - \max\left\{ \sum_{i=1}^{n} \alpha_{it} \cdot x''_{it} \,\middle|\, \sum_{i=1}^{n} r_i \cdot x''_{it} \leq \bar{r}_t \wedge x''_{it} \in \{0,1\} \; \forall i\right\}$$

5.3.4.5 Lagrangean Dual $Z^{(Y)}_{(V)}$

The Lagrangean relaxation $ZL^{(Y)}_{(V)}$ is

$$ZL^{(Y)}_{(V)}(u) = \min \sum_{t=1}^{T} y_t + \sum_{i=1}^{n}\sum_{t=1}^{T} (x_{it} - y_t) \cdot u_{it} \quad \text{subject to } (P) \wedge (C) \wedge (S) \wedge (X).$$

Let P (R) be the index set for all extreme points for y^p_t and x^p_{it} (extreme rays for y^r_t and x^r_{it}) of $Conv(P)(C)(S)(X)$. A convexification yields

$$ZL^{(Y)}_{(V)}(u) \leq \begin{cases} -\infty & \text{if } \exists r \in R: \sum_{t=1}^{T} y^r_t + \sum_{i=1}^{n}\sum_{t=1}^{T} (x^r_{it} - y^r_t) \cdot u_{it} < 0 \\ \sum_{t=1}^{T} y^p_t + \sum_{i=1}^{n}\sum_{t=1}^{T} (x^p_{it} - y^p_t) \cdot u_{it} & \text{otherwise; } \forall p \in P \end{cases}$$

The Lagrangean dual $Z^{(Y)}_{(V)}$ is stated as

$$Z^{(Y)}_{(V)} = \max_{u \geq 0} ZL^{(Y)}_{(V)}(u)$$

$$= \max_{u \geq 0} \min_{p \in P} \sum_{t=1}^{T} y^p_t + \sum_{i=1}^{n}\sum_{t=1}^{T} (x^p_{it} - y^p_t) \cdot u_{it}.$$

A linearization as a LP yields

$$Z^{(Y)}_{(V)} = \max \eta \quad \text{subject to}$$

$$\eta + \sum_{t=1}^{T}\sum_{i=1}^{n} (y^p_t - x^p_{it}) \cdot u_{it} \leq \sum_{t=1}^{T} y^p_t \quad \forall p \in P \qquad \lambda_p \quad (5.92)$$

$$\sum_{t=1}^{T}\sum_{i=1}^{n} (y^r_t - x^r_{it}) \cdot u_{it} \leq \sum_{t=1}^{T} y^r_t \quad \forall r \in R \qquad \lambda_r$$

$$u_{it} \geq 0 \; \forall i = 1,\ldots,n; t = 1,\ldots,T \wedge \eta \in \mathbb{R}.$$

Dualization yields the primal problem of the Lagrangean dual as

$$Z^{(Y)}_{(V)} = \min \sum_{p \in P} \sum_{t=1}^{T} y_t^p \cdot \lambda_p + \sum_{r \in R} \sum_{t=1}^{T} y_t^r \cdot \lambda_r \quad \text{subject to}$$

$$\sum_{p \in P} \lambda_p = 1 \qquad \eta$$

$$\sum_{p \in P} \lambda_p \cdot \left(y_t^p - x_{it}^p\right) + \sum_{r \in R} \lambda_r \cdot \left(y_t^r - x_{it}^r\right) \geq 0 \quad \forall i = 1, \ldots, n; t = 1, \ldots, T \quad u_{it}$$

$$\lambda_p \geq 0 \; \forall p \in P \wedge \lambda_r \geq 0 \; \forall r \in R.$$

Omit all extreme rays y_t^r and x_{it}^r. The variable λ_p equals 1, if the pth maintenance plan (x^p, y^p) is realized (otherwise, 0). The RMP is

$$Z^{(Y)}_{(V)} = \min \sum_{p \in P} \sum_{t=1}^{T} y_t^p \cdot \lambda_p \qquad \text{subject to}$$

$$\sum_{p \in P} \lambda_p = 1 \qquad \eta$$

$$\sum_{p \in P} \lambda_p \cdot \left(y_t^p - x_{it}^p\right) \geq 0 \qquad \forall i = 1, \ldots, n; t = 1, \ldots, T \quad u_{it}$$

$$\lambda_p \geq 0 \qquad \forall p \in P.$$

The reduced cost $\bar{c}^p$ of the extreme point $p \in P$ is given by (5.92) as

$$\bar{c}^p = \min \sum_{t=1}^{T} \left(1 - \sum_{i=1}^{n} u_{it}\right) \cdot y_t + \sum_{i=1}^{n} \sum_{t=1}^{T} u_{it} \cdot x_{it} - \eta \quad \text{subject to (P)} \wedge \text{(C)} \wedge \text{(S)} \wedge \text{(X)}.$$

References

Ahuja, R. K., Magnanti, T. L., & Orlin, J. B. (1993). *Network flows: Theory, algorithms, and applications.* New Jersey: Prentice-Hall.

Balakrishnan, A., Magnanti, T. L., & Wong, R. T. (1989). A dual-ascent procedure for large-scale uncapacitated network design. *Operations Research, 37,* 716–740.

Barnhart, C., Hane, C. A., & Vance, P. H. (2000). Using branch-and-price-and-cut to solve origin-destination integer multicommodity flow problems. *Operations Research, 48,* 318–326.

Billionnet, A., & Costa, M.-C. (1994). Solving the uncapacited plant location problem on trees. *Discrete Applied Mathematics, 49,* 51–59.

Chvátal, V. (2002). *Linear programming.* New York: W.H. Freeman.

Cormen, T. H., Leiserson, C. E., Rivest, R. L., & Stein, C. (2001). *Introduction to algorithms* (2nd ed.). Cambridge: The MIT Press.

References

Cornuejols, G., Nemhauser, G. L., & Wolsey, L. A. (1990). The uncapacitated facility location problem. In P. B. Mirchandani & R. L. Francis (Eds.), *Discrete location theory* (pp. 119–171). New York: Wiley.

Cornuejols, G., Sridharan, R., & Thizy, J. M. (1991). A comparison of heuristics and relaxations for the capacitated plant location problem. *European Journal of Operational Research, 50*, 280–297.

Cortinhal, M. J., & Captivo, M. E. (2003). Upper and lower bounds for the single source capacitated location problem. *European Journal of Operational Research, 151*, 333–351.

Dietrich, B. L., Escudero, L. F., Garín, A., & Pérez, G. (1993). $O(n)$ procedures for identifying maximal cliques and non-dominated extensions of consecutive minimal covers and alternates. *TOP, 1*, 139–160.

Domschke, W., & Drexl, A. (2005). *Einführung in operations research* (6th ed.). Berlin: Springer.

Domschke, W., & Wagner, B. (2005). Models and methods for standardization problems. *European Journal of Operational Research, 162*, 713–726.

Garey, M. R., & Johnson, D. S. (1978). Strong NP - completeness results: Motivation, examples, and implications. *Journal of the Association for Computing Machinery, 25*, 499–508.

Garey, M. R., & Johnson, D. S. (1979). *Computers and intractability: A guide to the theory of NP-completeness.* New York: Freeman.

Gendron, B. (2002). A note on a dual-ascent approach to the fixed-charge capacitated network design problem. *European Journal of Operational Research, 138*, 671–675.

Guignard, M. (2003). Lagrangean relaxation. *TOP, 11*, 151–228.

Herrmann, J., Ioannou, G., Minis, I., & Proth, J. M. (1996). A dual ascent approach to the fixed-charge capacitated network design problem. *European Journal of Operational Research, 95*, 476–490.

Hung, M. S., & Brown, J. R. (1978). An algorithm for a class of loading problems. *Naval Research Logistics Quarterly, 25*, 289–297.

Jones, P. C., Lowe, T. J., Müller, G., Xu, N., Ye, Y., & Zydiak, J. L. (1995). Specially structured uncapacitated facility location problems. *Operations Research, 43*, 661–669.

Karmarkar, N. (1984). A new polynomial time algorithm for linear programming. *Combinatorica, 4*, 373–395.

Karp, R. M. (1972). Reducibility among combinatorial problems. In R. E. Miller & J. W. Thatcher (Eds.), *Complexity of computer computations* (pp. 85–103). New York: Plenum.

Kellerer, H., Pferschy, U., & Pisinger, D. (2004). *Knapsack problems*. Berlin: Springer.

Klose, A. (2001). *Standortplanung in distributiven systemen*. Heidelberg: Physica.

Klose, A., & Drexl, A. (2005). Facility location models for distribution system design. *European Journal of Operational Research, 162*, 4–29.

Kolen, A. (1983). Solving covering problems and the uncapacitated plant location problem on trees. *European Journal of Operational Research, 12*, 266–278.

Kolen, A., & Tamir, A. (1990). Covering problems. In P. B. Mirchandani & R. L. Francis (Eds.), *Discrete location theory* (pp. 263–304). New York: Wiley.

Koshy, T. (2009). *Catalan numbers with applications*. New York: Oxford University Press.

Krarup, J., & Pruzan, P. M. (1983). The simple plant location problem: Survey and synthesis. *European Journal of Operational Research, 12*, 36–81.

Kuschel, T., & Bock, S. (2016). *Solving the weighted capacitated planned maintenance problem and its variants*, Working paper, University of Wuppertal.

Lewis, R. T., & Parker, R. G. (1982). On a generalized bin-packing problem. *Naval Research Logistics Quarterly, 25*, 119–145.

Magnanti, T. L., & Wong, R. T. (1984). Network design and transportation planning: Models and algorithms. *Transportation Science, 18*, 1–55.

Martello, S., & Toth, P. (1990). *Knapsack problems: Algorithms and computer implementations*. New York: Wiley.

Nemhauser, G. L., & Wolsey, L. A. (1988). *Integer and combinatorial optimization*. New York: Wiley.

Ottmann, T., & Widmayer, P. (2012). *Algorithmen und datenstrukturen* (5th ed.). Heidelberg: Spektrum Akdemischer.

Rhys, J. M. W. (1970). A selection problem of shared fixed costs and network flows. *Management Science, 17*, 200–207.

Rodríguez-Martín, I., & Salazar-González, J. J. (2010). A local branching heuristic for the capacitated fixed-charge network design problem. *Computers & Operations Research, 37*, 575–581.

Sridharan, R. (1993). A Lagrangian heuristic for the capacitated plant location problem with single source constraints. *European Journal of Operational Research, 66*, 305–312.

Vavasis, S. A., & Ye, Y. (1996). A primal-dual interior point method whose running time depends only on the constraint matrix. *Mathematical Programming, 74*, 79–120.

Chapter 6
Algorithms for the Capacitated Planned Maintenance Problem

Based on the findings from Chaps. 4 and 5, specific algorithms for the CPMP are developed. Three construction heuristics are presented in Sect. 6.1. The obtained initial feasible solution is potentially improved by the following metaheuristics. Two Lagrangean heuristics are provided in Sect. 6.2 that use the Lagrangean relaxations of the capacity constraint $ZL_{(C)}$ and of the period covering constraint $ZL_{(P)}$ from Theorem 5.22. In Sect. 6.3, a general, problem independent approach that links two Lagrangean relaxations yields two Lagrangean hybrid heuristics. A tabu search heuristic is given in Sect. 6.4.

6.1 Three Construction Heuristics

Denote (x, y) as a feasible solution that is returned by one of the following construction heuristics. Section 6.1.3 presents a heuristic that iteratively applies the heuristics from Sects. 6.1.1 and 6.1.2. The information between the iterations is carried over by a so called *reference solution* $(x, y)^{ref}$ that the heuristics take in account and that may be empty or violate any constraint (P) or (C) but neither (V), (X) nor (Y). Note that all three heuristics yield an initial solution even if the reference solution is empty.

6.1.1 The First Fit Heuristic

Algorithm 6.1 is a serial heuristic[1] and maintenance activities are selected in the order of the smallest coverage. Two tasks of the same maintenance activity are planned such that the difference in periods is maximal. This corresponds to the observation that the maintenance activity $i = 1$ with the smallest coverage determines the number of maintenance slots in the UPMP (Lemma 4.17).

Commence with the first period $t = 1$ and the first maintenance activity $i = 1$. Schedule a task i in the latest possible period τ within the task horizon $t, \ldots, t + \pi_i - 1$ such that the residual capacity is larger than r_i and the number of periods to the previous task i is maximal.[2] Specifically, find the largest open period in (x, y). If no such period exists, a period from the reference solution is opened. Specifically, find the largest period that is closed in (x, y) but open in $(x, y)^{ref}$. If no such period exists, find the largest period that is closed in (x, y).[3] If no such period exists, the algorithm terminates because no feasible solution is found. Afterwards, select the next task horizon $t = \tau + 1$ until the task plan for maintenance activity i satisfies (P).

6.1 Algorithm. The First Fit Heuristic for the CPMP
  Input: An instance to the CPMP and a reference solution $(x, y)^{ref}$
  Output: A feasible solution (x, y) if one was found
  Initialization: $(x, y) \leftarrow (0, 0)$, $i \leftarrow 1$, $t \leftarrow 1$, $\tau \leftarrow 1$
  1 **for** $i \leftarrow 1$ **to** n **do**
  2     $t \leftarrow 1$
  3   **do**
  4   Find the largest open period τ in (x, y) with $\tau = t, \ldots, t + \pi_i - 1$ such the residual capacity is larger than r_i
  5   **if** *No period τ is found* **then**
  6   Find the largest closed period τ in (x, y) with $\tau = t, \ldots, t + \pi_i - 1$ that is open in $(x, y)^{ref}$ such the residual capacity is larger than r_i
  7   **if** *No period τ is found* **then**
  8   Find the largest closed period τ in (x, y) with $\tau = t, \ldots, t + \pi_i - 1$ such the residual capacity is larger than r_i
  9   **if** *A period τ is found* **then**
 10   Schedule the maintenance activity i in τ in (x, y) and set $t \leftarrow \tau + 1$
 11   **else** Terminate the algorithm because no feasible solution is found
 12   **while** *(P) is not satisfied*
 13   Right-shift and left-shift post optimization

---

[1] The term 'serial heuristic' refers to the project planning terminology (Kolisch 1996).
[2] This strategy gives the name 'First Fit Heuristic'.
[3] Implementation detail: Use a list to store all open periods.

In the *right-shift post optimization routine*, consider all open periods of (x, y) in a decreasing order of indices. For all maintenance activities $j = i, i-1, \ldots, 1$ scheduled in such a period, try to close periods by shifting each task j to the rightmost, open period and remove redundant tasks on the left and on the right. A *left-shift post optimization routine* follows that works the other way around but a task is shifted to the left. Both routines aim to close periods and to increase the capacity usage. Finally, choose the next maintenance activity $i = i + 1$. Terminate if $i > n$ and return a feasible solution (x, y) if one was found.

If the capacity constraint is not restrictive, the algorithm clearly yields an optimal solution to the UPMP (see Lemma 4.17).

6.1.2 The Overlap Heuristic

Algorithm 6.2 is a parallel heuristic[4] and constructs a feasible solution from the beginning towards the end of the planning horizon. In every iteration, a period t is selected that marks the progress of the construction towards the end of the planning horizon. For all maintenance activities that violate (P), a set of periods around the

---

6.2 Algorithm. The Overlap Heuristic for the CPMP

Input: An instance to the CPMP, a reference solution $(x, y)^{ref}$ and a priority rule
Output: A feasible solution (x, y) if one was found
Initialization: $(x, y) \leftarrow (0, 0), i \leftarrow 0, t \leftarrow 1$

1 **if** $(x, y)^{ref}$ is non-empty **then** Set t to the first open period in $(x, y)^{ref}$
2 **do**
3   **if** t is open in (x, y) **then** Set $x_{it} = 0 \; \forall i$ and $y_t = 0$
4   Label all maintenance activities, where (P) is violated for $x_{it'} = 1$ and $x_{it''}^{ref} = 1$
5   **while** *There exist labeled maintenance activities* **do**
6   **for** *All affected periods* **do**
7   Construct a temporary schedule and calculate the score for the period
8   **if** *All temporary schedules are empty but there exists a labeled maintenance activity* **then** Terminate the algorithm because no feasible solution is found
9   Select the temporary schedule with the maximal score and fix it in (x, y)
10   Update the labels and temporary schedules. Remove redundancies in (x, y)
11   **for** $i = 1$ **to** n **do**
12   Shift the left neighbor t' of the ith maintenance activity to the rightmost, open period and remove redundancies
13   Set t to the next period that is open in either in $(x, y)^{ref}$ or in (x, y)
14 **while** $t \leq T$ or (P) *is not satisfied*
15 Left-shift and right-shift post optimization

---

[4] The term 'parallel heuristic' refers to the project planning terminology (Kolisch 1996).

period t is selected such that (C) is satisfied and (P) holds for all open periods at the end of the iteration. Let (x, y) be the solution the heuristic generates.

Initialize t with the first open period in $(x, y)^{ref}$ if $(x, y)^{ref}$ is non-empty. Otherwise, set $t = 1$. At the beginning of an iteration, the opening of the period t is evaluated again. Therefore, set $x_{it} = 0$ $\forall i$ and $y_t = 0$. This step increases the flexibility of the heuristic.

Afterwards, *label* all maintenance activities that violate (P) between the left and the right neighbor (i.e., $t' + \pi_i < t''$). Note that the left neighbor t' (right neighbor t'') of a maintenance activity i is scheduled in (x, y) $((x, y)^{ref})$ and that a neighbor refers to the period t. Feasibility of (P) is partially established by opening a set of periods. This is iteratively done by the following three steps. In the first step, a temporary schedule is generated for each period by assigning labeled maintenance activities such that (C) is satisfied. The selection of the maintenance activities is determined by a priority rule. In the second step, all temporary schedules are evaluated by a score, one temporary schedule with the maximal score is selected and the variables of this temporary schedule are fixed in (x, y). The third step updates the labels and temporary schedules. In what follows, further details are presented.

In the first step, a set of periods is determined in which labeled maintenance activities could be scheduled. Specifically, define a period τ as *affected* if $\tau = t' + 1, \ldots, t' + \pi_i$ holds for at least one labeled maintenance activity i. Consider all affected periods in an increasing order of periods[5] and construct a *temporary schedule* as follows: If the residual capacity is strictly smaller than $\min\{r_i | i = 1, \ldots, n \wedge i \text{ is labeled}\}$, a temporary schedule does not need to be constructed for that period because it would be empty. Otherwise, sort all maintenance activities that can be scheduled in the affected period in a non-decreasing order of a given *priority rule*. The applied priority rules are presented below. Assign as many maintenance activities as possible to the temporary schedule without violating (C). Note that the task horizons of an assigned maintenance activity overlaps with the task horizon of its left neighbor and thus, (P) holds for the left neighbor.[6] The following two priority rules are applied: The *priority rule flexibility* $f_i = \min\{t' + \pi_i, T\} - t' - 1$ measures the width of the task horizon determined by the left neighbor. The smaller the *flexibility* f_i is, the more tasks are necessary to cover the planning horizon. The *priority rule simplicity* is given by the fraction $\frac{f_i}{r_i}$. In this priority rule, a maintenance

---

[5]Implementation detail: An efficient data structure stores only the affected periods. Let a tuple (l, r) represent an interval of periods where the periods $l, \ldots, r$ are affected by the labeled maintenance activity i. From the definition follows $a = t' + 1$ and $b = t' + \pi_i$. Assume that n maintenance activities are labeled. Sort all tuples in an increasing order of values of l in time $O(n \cdot \log n)$ (e.g., use introsort from Footnote 32 on page 58). Given an empty stack of tuples and let (a, b) be the top element of the stack. Insert the first tuple (l, r) in the stack with $(a, b) = (l, r)$. Afterwards, consider the tuples $2, \ldots, n$: If (a, b) and the tuple (l, r) have $b > l$, do not insert the tuple in the stack but update (a, b) with $(a, b) = (a, \max\{b, r\})$. Otherwise, insert (l, r) in the stack. Sorting is the most time consuming step and hence, the stack is obtained in $O(n \cdot \log n)$. Removing the tuples from the top of the stack, all affected periods are accessible in an increasing order of periods.

[6]This strategy gives the name 'Overlap Heuristic'.

6.1 Three Construction Heuristics

activity is considered as difficult to schedule if it has a low flexibility and a high maintenance time.

The second step evaluates all temporary schedules with a *score* that is defined as follows: An already open period has a higher score than a closed period. Otherwise, a period has a higher score if its temporary schedule comprises more maintenance activities with a lower flexibility (lexicographical comparison). The largest period breaks ties. Note that if a period is open and all labeled maintenance activities can be scheduled, it has the largest possible score. This can be considered already in the first step as a termination criterion for the generation of temporary schedules. Finally, select the temporary schedule with the maximal score and fix the temporary schedule in (x, y).[7]

The third step proceeds with a termination criterion. If there exists a labeled maintenance activity but all temporary schedules are empty, then terminate the algorithm because no feasible solution could be found. Otherwise, consider all maintenance activities of the fixed temporary schedule and erase their labels as well as empty all temporary schedules of the periods that are affected by them. Remove tasks in (x, y) that are redundant to (P). Execute the three steps again until there exists no labeled maintenance activity.

In a post optimization routine, consider all maintenance activities and shift the task of the left neighbor t' to the rightmost, open period and remove redundancies with respect to (P). Select the next period t that is open either in $(x, y)^{ref}$ or in (x, y).[8] Until $t \leq T$ or (P) is not satisfied, commence with the next iteration. Afterwards, apply the left-shift and the right-shift post optimization routine with all maintenance activities $n, n-1, \ldots, 1$. In Algorithm 6.2, the first and the second step are executed in parallel.

---

[7]Implementation detail: A maximum *priority queue* (Cormen et al. 2001, pp. 138–140) provides an efficient data structure. Given a data record of size n. Each element in the data record contains an ID and a key. Let a maximum heap be linearized in an array (Cormen et al. 2001, pp. 127–135). Whenever, a heap operation 'upheap' or 'downheap' is executed, a reference to the final position of the element in the heap array is stored in an additional array. Therefore, it takes $O(1)$ to access any element in the heap via the ID. The operation 'update key' has two functions. If the key of an element, which is contained in the heap, changes, the element is accessed in $O(1)$, the key is updated and an 'upheap' or respectively, 'downheap' operation is done in $O(\log n)$. If the element is not in the heap, it is inserted with $O(\log n)$. This modified heap allows an 'erase element' operation with $O(\log n)$ as follows: Access the element in $O(1)$, remove it from the heap by replacing it with the last element and perform a 'downheap' operation on the last element.

In the algorithm, the heap is emptied at the beginning of an iteration. The heap holds a temporary schedule for each affected period with the period as the ID and the score as the key. If the temporary schedule is empty, it is removed from the heap. Otherwise, it is inserted in the heap.

[8]Implementation detail: Use a list to store all open periods of $(x, y)^{ref}$ and of (x, y).

6.3 Algorithm. The Iterated Best-of-Three Heuristic for the CPMP

Input: An instance to the CPMP, a reference solution $(x, y)^{ref}$, a lower bound LB
Output: A feasible solution (x^*, y^*) if one was found
Initialization: $(x, y) \leftarrow (0, 0)$, $(x, y)^{refcopy} \leftarrow (0, 0)$, $(x, y) \leftarrow (x, y)^{ref}$

1 **repeat**
2      $(x, y)^{refcopy} \leftarrow (x, y)$
3   Apply Algorithm 6.2 with the priority rule flexibility to $(x, y)^{refcopy}$
4   Apply Algorithm 6.2 with the priority rule simplicity to $(x, y)^{refcopy}$
5   Apply Algorithm 6.1 to $(x, y)^{refcopy}$
6   Update (x^*, y^*) with the best solution of this iteration (x, y) if it is feasible
7 **until** *A termination criterion is not met*

6.1.3 The Iterated Best-of-Three Heuristic

Algorithm 6.3 combines both heuristics in an iterated improvement approach. Let LB be a given lower bound (e.g., $\lfloor \frac{T}{\pi_1} \rfloor$ from Lemma 4.17). Let UB^* be an upper bound of the best known feasible solution (x^*, y^*). In an iteration, apply the Overlap Heuristic (Algorithm 6.2) with $(x, y)^{ref}$ twice but use a different priority rule each time. Afterwards, apply the First Fit Heuristic (Algorithm 6.1). The best solution of the iteration (x, y) is used to update the best known feasible solution (x^*, y^*) and its upper bound UB^*. Commence the next iteration with $(x, y)^{ref} = (x, y)$. Repeat until either $UB^* = \lceil LB \rceil$, the best solution did not improve for ten consecutive iterations or the total number of iterations is 50.

6.2 Two Lagrangean Heuristics

Two subgradient-based Lagrangean heuristics are developed for the Lagrangean relaxations $ZL_{(C)}$ and $ZL_{(P)}$, which are introduced and discussed in Sect. 5.3. From the analysis of the Lagrangean duals in Sect. 5.3, the choice fell on these two Lagrangean relaxations for several reasons. $Z_{(C)}$ and $Z_{(P)}$ belong to the six potentially interesting Lagrangean duals from Theorem 5.22 that do not dominate each other. $ZL_{(C)}$ and $ZL_{(P)}$ guarantee the existence of a valid subgradient in the sense of Definition 3.15 because an optimal solution always exists. Furthermore, each Lagrangean relaxation covers a fundamental structural aspect of the CPMP that is the period covering constraint (P) and the capacity constraint (C).

6.2 Two Lagrangean Heuristics

Sections 6.2.1 and 6.2.2 provide lower and upper bounds to both Lagrangean relaxations. These bounds are used in Sect. 6.2.3 where a generic Lagrangean heuristic is presented. The fundamental difference of this Lagrangean heuristic is that the respective Lagrangean relaxation is solved heuristically, whereas many approaches in literature prefer an optimal solution to derive subgradients from.

6.2.1 The Lagrangean Relaxation of the Capacity Constraint

The open problem (5.51) that is

$$ZL_{(C)}(\mu) = \min \sum_{t=1}^{T} (1 - \mu_t \cdot \bar{r}_t) \cdot y_t + \sum_{i=1}^{n} \sum_{t=1}^{T} r_i \cdot \mu_t \cdot x_{it} \quad \text{subject to}$$

$$\sum_{\tau=t}^{t+\pi_i-1} x_{i\tau} \geq 1 \qquad \forall i = 1, \ldots, n; t = 1, \ldots, T - \pi_i + 1 \qquad (P)$$

$$x_{it} \leq y_t \qquad \forall i = 1, \ldots, n; t = 1, \ldots, T \qquad (V)$$

$$x_{it} \in \{0, 1\} \qquad \forall i = 1, \ldots, n; t = 1, \ldots, T \qquad (X)$$

$$y_t \in \{0, 1\} \qquad \forall t = 1, \ldots, T \qquad (Y)$$

is a specific WUPMP and the following properties are important for the algorithmic design:

- If $1 - \mu_t \cdot \bar{r}_t < 0$, a *variable fixation* is applied. Let $F = \{t = 1, \ldots, T | f_t < 0\}$. Solve the WUPMP with $f_t = 0 \ \forall t \in F$, modify the obtained solution by $y_t = 1 \ \forall t \in F$ and add a constant $\sum_{t \in F} f_t$ to the objective function value. Thus, $ZL_{(C)}$ has the WUPMP objective function.
- The single-assignment property from Lemma 4.3 holds as $ZL_{(C)}(\mu) = ZL_{(C)}^{(X)}(\mu)$.
- Observation 4.4, which states the optimal solution to the LP relaxation $ZL_{(C)}^{(X)(Y)}$ contains many binary variables, is used to construct a feasible solution to $ZL_{(C)}$.[9] The definitions are provided in Sect. 6.2.1.1.

In the following sections, five lower bounds and an upper bound to $ZL_{(C)}$ are presented. The first four lower bounds solve the dual problem to the LP relaxation $ZL_{(C)}^{(X)(Y)}$ and they deliver information that is used in Sect. 6.2.1.6 to construct a feasible solution to $ZL_{(C)}$. Initial Lagrangean multiplier μ^{init} are defined in Lemma 6.9 from Sect. 6.2.3. It can be easily verified that solving each lower bound with μ^{init} yields the optimum $ZL_{(C)}(\mu^{init})$.

---

[9] An instance with a fractional extreme point to $ZL_{(C)}^{(X)(Y)}$ is given in Example 4.3. Let $r = (1, 1)^T$, $\bar{r} = (1, 0, 1)^T$ and $\mu = (\frac{1}{2}, 1, \frac{1}{2})^T$ to obtain the WUPMP objective function coefficients.

6.2.1.1 The LP Lower Bound

The LP relaxation $ZL_{(C)}^{(X)(Y)}$ is

$$ZL_{(C)}^{(X)(Y)}(\mu) = \min \sum_{t=1}^{T}(1 - \mu_t \cdot \bar{r}_t) \cdot y_t + \sum_{i=1}^{n}\sum_{t=1}^{T} r_i \cdot \mu_t \cdot x_{it} \quad \text{subject to}$$

$$\sum_{\tau=t}^{t+\pi_i-1} x_{i\tau} \geq 1 \qquad \forall i=1,\ldots,n; t=1,\ldots,T-\pi_i+1 \qquad v_{it} \qquad \text{(P)}$$

$$-x_{it} + y_t \geq 0 \qquad \forall i=1,\ldots,n; t=1,\ldots,T \qquad u_{it} \qquad \text{(V)}$$

$$-x_{it} \geq -1 \qquad \forall i=1,\ldots,n; t=1,\ldots,T \qquad \alpha_{it}$$

$$-y_t \geq -1 \qquad \forall t=1,\ldots,T \qquad \beta_t$$

$$x_{it} \geq 0 \ \forall i=1,\ldots,n; t=1,\ldots,T \wedge y_t \geq 0 \ \forall t=1,\ldots,T.$$

The dual problem is stated as

$$\max \sum_{i=1}^{n}\sum_{t=1}^{T-\pi_i+1} v_{it} - \sum_{i=1}^{n}\sum_{t=1}^{T}\alpha_{it} - \sum_{t=1}^{T}\beta_t \quad \text{subject to}$$

$$\sum_{\tau=\max\{1,t-\pi_i+1\}}^{\min\{t,T-\pi_i+1\}} v_{i\tau} - u_{it} - \alpha_{it} \leq r_i \cdot \mu_t \qquad \forall i=1,\ldots,n; t=1,\ldots,T \qquad x_{it}$$

$$\sum_{i=1}^{n} u_{it} - \beta_t \leq 1 - \mu_t \cdot \bar{r}_t \qquad \forall t=1,\ldots,T \qquad y_t$$

$$v_{it} \geq 0 \ \forall i=1,\ldots,n; t=1,\ldots,T-\pi_i+1 \wedge u_{it} \geq 0 \ \forall i=1,\ldots,n; t=1,\ldots,T$$

$$\alpha_{it} \geq 0 \ \forall i=1,\ldots,n; t=1,\ldots,T \wedge \beta_t \geq 0 \ \forall t=1,\ldots,T$$

and simplified as follows. Given a feasible solution. Let some $(\alpha_{i,t_1},\ldots,\alpha_{i,t_2})$ be sufficiently positive to increase some $(v_{i,\tau_1},\ldots,v_{i,\tau_2})$. Since the objective function value either remains the same or decreases, set $\alpha_{it} = 0 \ \forall i; t$. Consider the variables β_t. Recall that $1 - \mu_t \cdot \bar{r}_t \in \mathbb{R} \ \forall t$ holds. Let the right-hand side be positive such that some dual variables $(\beta_{t_1},\ldots,\beta_{t_2})$ are sufficiently positive to increase some $(v_{i,\tau_1},\ldots,v_{i,\tau_2})$ via $(u_{i,t_1},\ldots,u_{i,t_2})$. Since the objective function value is either the same or decreases, set $\beta_t = 0 \ \forall t$. The other way around, if the right-hand side is strictly negative, then $u_{it} = 0 \ \forall i$ holds because of the non-negativity and

6.2 Two Lagrangean Heuristics

$\beta_t = -(1 - \mu_t \cdot \bar{r}_t)$ equals the right-hand side. Thus, the dual problem becomes

$$ZL_{(C)}^{(X)(Y)}(\mu) = \max \sum_{i=1}^{n} \sum_{t=1}^{T-\pi_i+1} v_{it} + \sum_{t=1}^{T} \min\{0, 1 - \mu_t \cdot \bar{r}_t\} \quad \text{subject to} \tag{6.1}$$

$$\sum_{\tau=\max\{1,t-\pi_i+1\}}^{\min\{t,T-\pi_i+1\}} v_{i\tau} - u_{it} \le r_i \cdot \mu_t \qquad \forall i = 1, \ldots, n; t = 1, \ldots, T \qquad x_{it} \tag{6.2}$$

$$\sum_{i=1}^{n} u_{it} \le \max\{0, 1 - \mu_t \cdot \bar{r}_t\} \qquad \forall t = 1, \ldots, T \qquad y_t \tag{6.3}$$

$$v_{it} \ge 0 \qquad \forall i = 1, \ldots, n; t = 1, \ldots, T - \pi_i + 1 \tag{6.4}$$

$$u_{it} \ge 0 \qquad \forall i = 1, \ldots, n; t = 1, \ldots, T. \tag{6.5}$$

Remark 6.1 If $ZL_{(C)}^{(X)(Y)}$ has only non-negative objective function coefficients, the simple upper bounds $x, y \le 1$ are redundant. This holds from a dualization of (6.1)–(6.5).

Lemma 6.1 *The LP (6.1)–(6.5) is solvable.*

Proof Farkas' Lemma 3.4 states that the solution space of (6.1)–(6.5) is not empty if and only if the corresponding dual problem to

$$\max \sum_{i=1}^{n} \sum_{t=1}^{T-\pi_i+1} 0 \cdot v_{it} \quad \text{subject to (6.2)–(6.5)}$$

has an optimal objective function value that is zero. The corresponding dual problem

$$\bar{Z} = \min \sum_{t=1}^{T} \max\{0, 1 - \mu_t \cdot \bar{r}_t\} \cdot y_t + \sum_{i=1}^{n} \sum_{t=1}^{T} r_i \cdot \mu_t \cdot x_{it} \quad \text{subject to}$$

$$\sum_{\tau=t}^{t+\pi_i-1} x_{i\tau} \ge 0 \qquad \forall i = 1, \ldots, n; t = 1, \ldots, T - \pi_i + 1$$

$$-x_{it} + y_t \ge 0 \qquad \forall i = 1, \ldots, n; t = 1, \ldots, T$$

$$x_{it} \ge 0 \, \forall i = 1, \ldots, n; t = 1, \ldots, T \wedge y_t \ge 0 \, \forall t = 1, \ldots, T$$

has an optimal solution $(x, y) = (0, 0)$. Thus, $\bar{Z} = 0$ and (6.1)–(6.5) is solvable.

The LP (6.1)–(6.5) is solved to optimality and yields a lower bound $ZL_{(C)}^{(X)(Y)}$ that is referred to as *LP Lower Bound*. Let d'_{it} be the slack of (6.2) and let $\bar{x}_{it}$ ($\bar{y}_t$) be the corresponding dual variables of (6.2) ((6.3)). Applying the Observation 4.4 that many variables $\bar{y}_t$ can have integer values in an optimal solution, the variables y_t with $t \in P$ and

$$P = \{t = 1, \ldots, T \mid \bar{y}_t = 1\}$$

are referred to as *potential integers*. In order to set the primal variables x_{it} in accordance with the complementary slackness criterion $d'_{it} \cdot x_{it} = 0$, define the set

$$L_i = \{t = 1, \ldots, T \mid d'_{it} = 0\} \quad \forall i = 1, \ldots, n.$$

6.2.1.2 The Dual Priority Rule Lower Bound

Algorithm 6.4 computes the *Dual Priority Rule Lower Bound*, which interprets the LP (6.1)–(6.5) as a resource allocation problem. It solves this LP heuristically and yields a lower bound to $\text{ZL}_{(C)}^{(X)(Y)}$ because of Theorem 3.3.1.

6.4 Algorithm. The Dual Priority Rule Lower Bound for $\text{ZL}_{(C)}$

Input: $\mu_t \geq 0 \; \forall t$ and an instance to the CPMP
Output: A lower bound LB to $\text{ZL}_{(C)}$ and the sets $L_i \; \forall i$, $P = \emptyset$
Initialization: $L_i \leftarrow \emptyset \; \forall i$, $P \leftarrow \emptyset$, $LB \leftarrow \sum_{t=1}^{T} \min\{0, 1 - \mu_t \cdot \bar{r}_t\}$, $d'_{it} \leftarrow r_i \cdot \mu_t \; \forall i; t$,
$d''_t \leftarrow \max\{0, 1 - \mu_t \cdot \bar{r}_t\} \; \forall t$, $\bar{d}_{it} \leftarrow 0 \; \forall i; t, \bar{I} \leftarrow \emptyset, I \leftarrow \{1, \ldots, n\}$,
$\bar{t}_i \leftarrow 0 \; \forall i, i \leftarrow 0, j \leftarrow 0, t \leftarrow 1, \tau \leftarrow 0, R \leftarrow 0, \bar{R} \leftarrow 0, \bar{v}_i \leftarrow 0 \; \forall i$

1 **do**
2   **for** $i \in I$ **do**
3         $\bar{v}_i \leftarrow \min_{\tau = t, \ldots, \min\{t + \pi_i - 1, T\}} d'_{i\tau}$
4         $d'_{i\tau} \leftarrow d'_{i\tau} - \bar{v}_i \; \forall \tau = t, \ldots, \min\{t + \pi_i - 1, T\}$
5         $LB \leftarrow LB + \bar{v}_i$
6   **do**
7         $j \leftarrow 0$
8   **for** $i \in I$ **do**
9             $\bar{t}_i \leftarrow \arg\min_{\tau = t, \ldots, \min\{t + \pi_i - 1, T\}} d'_{i\tau} + d''_\tau$
10           $\bar{v}_i \leftarrow d'_{i,\bar{t}_i} + d''_{\bar{t}_i}$
11   **if** $\bar{v}_i > 0$ **then**
12               $\bar{I} \leftarrow \bar{I} \cup \{i\}$
13               $\bar{d}_{i\tau} \leftarrow \max\{0, \bar{v}_i - d'_{i\tau}\} \; \forall \tau = t, \ldots, \min\{t + \pi_i - 1, T\}$
14   Calculate the priority rule $R \leftarrow \bar{v}_i / \sum_{\tau = t}^{\min\{t + \pi_i - 1, T\}} \bar{d}_{i\tau}$
15   **if** $R > \bar{R}$ **then** $j \leftarrow i \land \bar{R} \leftarrow R$
16   **if** $j > 0$ **then**
17             $LB \leftarrow LB + \bar{v}_j$
18             $d'_{j\tau} \leftarrow d'_{j\tau} + \bar{d}_{j\tau} - \bar{v}_j \; \forall \tau = t, \ldots, \min\{t + \pi_j - 1, T\}$
19             $d''_\tau \leftarrow d''_\tau - \bar{d}_{j\tau} \; \forall \tau = t, \ldots, \min\{t + \pi_j - 1, T\}$
20             $I \leftarrow \bar{I} \setminus \{j\} \land \bar{I} \leftarrow \emptyset \land \bar{R} \leftarrow 0$
21   **while** $j \neq 0$
22     $t \leftarrow 1 + \min_{i=1,\ldots,n} \min\{\bar{t}_i, T - \pi_i + 1\}$
23     $I \leftarrow \{i = 1, \ldots, n \mid t = \min\{\bar{t}_i + 1, T - \pi_i + 1\}\} \land \bar{I} \leftarrow \emptyset$
24 **while** $t \leq T - \pi_1 + 1$
25 Derive the set $L_i \; \forall i$

6.2 Two Lagrangean Heuristics

Initialize $LB = \sum_{t=1}^{T} \min\{0, 1 - \mu_t \cdot \bar{r}_t\}$, the slack of (6.2) $d'_{it} = r_i \cdot \mu_t \ \forall i; t$, the slack of (6.3) $d''_t = \max\{0, 1 - \mu_t \cdot \bar{r}_t\} \ \forall t$ and set $t = 1$. An iteration starts with a period t. The first step is similar to solving the dual problem to the SWUPMP LP relaxation (cf. Algorithm 4.2). The largest improvement $\bar{v}_i = \min_{\tau=t,\ldots,t+\pi_i-1} d'_{i\tau}$ feasibly sets v_{it} as high as possible. $\bar{v}_i$ is computed for all maintenance activities i to improve $LB + \bar{v}_i$ and to update $d'_{i\tau} = d'_{i\tau} - \bar{v}_i \ \forall \tau$.

In the second step, d''_t is interpreted as a common resource all maintenance activities compete for to increase v_{it} through u_{it}. Calculate the maximal improvement $\bar{v}_i$ of the lower bound from (6.8) and let $\bar{\tau}_i$ be the largest period with $\bar{v}_i = d'_{i,\bar{\tau}_i} + d''_{\bar{\tau}_i}$.

$$(6.2) \xRightarrow[v_{i\tau}=v_{i\tau}+\Delta v_{i\tau}]{u_{i\tau}=u_{i\tau}+\Delta u_{i\tau}} \sum_{\theta=\max\{1,\tau-\pi_i+1\}}^{\min\{\tau,T-\pi_i+1\}} v_{i\theta} + \Delta v_{i\tau} - u_{i\tau} - \Delta u_{i\tau} \leq r_i \cdot \lambda_\tau$$

$$\Rightarrow \Delta v_{i\tau} \leq d'_{i\tau} + \Delta u_{i\tau} \ \forall i = 1, \ldots, n; \tau = t, \ldots, t + \pi_i - 1 \quad (6.6)$$

$$(6.3) \xRightarrow{u_{i\tau}=u_{i\tau}+\Delta u_{i\tau}} \sum_{h=1}^{n} u_{h\tau} + \Delta u_{i\tau} \leq \max\{0, 1 - \mu_\tau \cdot \bar{r}_\tau\}$$

$$\Rightarrow \Delta u_{i\tau} \leq d''_\tau \ \forall i = 1, \ldots, n; \tau = t, \ldots, t + \pi_i - 1 \quad (6.7)$$

$(6.6) \wedge (6.7) \Rightarrow \Delta v_{i\tau} \leq d'_{i\tau} + d''_\tau$

$$\Rightarrow \bar{v}_i = \min_{\tau=t,\ldots,t+\pi_i-1} d'_{i\tau} + d''_\tau \ \forall i = 1, \ldots, n \quad (6.8)$$

Determine the smallest consumption $\bar{d}_{i\tau}$ regarding the slack of (6.3) that ensures feasibility of $\bar{v}_i$ is determined by (6.9).

$(6.6) \wedge \Delta u_{i\tau} \geq 0 \Rightarrow \Delta u_{i\tau} = \max\{0, \Delta v_{i\tau} - d'_{i\tau}\}$ with $\Delta u_{i\tau} = \bar{d}_{i\tau} \wedge \Delta v_{i\tau} = \bar{v}_i$

$$\Rightarrow \bar{d}_{i\tau} = \max\{0, \bar{v}_i - d'_{i\tau}\} \ \forall i = 1, \ldots, n; \tau = t, \ldots, t + \pi_i - 1 \quad (6.9)$$

Among all contributing maintenance activities $\bar{v}_i > 0$, the maintenance activity j is chosen that has the largest relative improvement $\bar{v}_i / \sum_{\tau=t}^{\min\{t+\pi_i-1,T\}} \bar{d}_{i\tau}$ of the lower bound in terms of consumption $\bar{d}_{i\tau}$ (priority rule). The tie breaker is the smallest coverage. The lower bound LB is improved by $\bar{v}_j$ and an update of the slack is done for all $\tau = t, \ldots, t + \pi_j - 1$ with

$d'_{j\tau} = d'_{j\tau} + \bar{d}_{j\tau} - \bar{v}_j$ because of (6.9)

$d''_\tau = d''_\tau - \bar{d}_{j\tau}$ from (6.7) with $\Delta u_{j\tau} = \bar{d}_{j\tau}$.

Repeat the second step with maintenance activities that have $\bar{v}_i > 0$ until there exists no maintenance activity j. The periods $t, \ldots, \min_{i=1,\ldots,n} \bar{t}_i$ yield no improvement because $d'_{i,\bar{t}_i} + d''_{\bar{t}_i} = 0 \; \forall i$. Consequently, a new iteration commences at $t = 1 + \min_{i=1,\ldots,n} \bar{t}_i$ and only with maintenance activities that have $t = 1 + \bar{t}_i$. The algorithm terminates when all planning horizons were examined. It returns a lower bound LB, $L_i \; \forall i$ and $P = \emptyset$.

6.2.1.3 The Combined Lower Bound

The *Combined Lower Bound* alternates between the LP Lower Bound and the Priority Rule Lower Bound. In the first three iterations of the Lagrangean heuristic presented in Sect. 6.2.3, a lower bound is calculated from the LP Lower Bound. Afterwards, the Dual Priority Rule Lower Bound is calculated for three iterations before using the LP Lower Bound again. This alternation proceeds until the Lagrangean heuristic terminates.

The Dual Priority Rule Lower Bound is applied because preliminary computational studies showed that it has a better performance than the Primal-dual Lower Bound or the Shortest Path Lower Bound presented below.

6.2.1.4 The Primal-Dual Lower Bound

The *Primal-dual Lower Bound* from algorithm (6.5) is based on a modified Primal-dual Simplex algorithm and yields a lower bound to $ZL_{(C)}^{(X)(Y)}$. Assume that $ZL_{(C)}$ has only non-negative objective function coefficients. Denote $(\bar{v}^{iter}, \bar{u}^{iter})$ as a feasible solution to the corresponding dual problem of $ZL_{(C)}^{(X)(Y)}$ (6.1)–(6.5) at the beginning of an iteration $iter = 0, 1, 2, \ldots$. Using the slack of (6.2) d'_{it}, the slack of (6.3) d''_t and $\bar{v}_{it}^{iter}$, an iteration starts with three sets that contain potential primal basic variables.

$$J^x = \{(i, t) \in (n \times T) \mid d'_{it} = 0\} \quad (6.10)$$

$$J^y = \{t \in T \mid d''_t = 0\} \quad (6.11)$$

$$J^z = \{(i, t) \in (n \times T) \mid \bar{v}_{it}^{iter} = 0\}$$

Define the corresponding constraint coefficients $a_{it}^x = 1$ if $(i, t) \in J^x$ for all i, t (otherwise, 0), $a_t^y = 1$ if $t \in J^y$ for all t (otherwise, 0) and $a_{it}^z = 1$ if $(i, t) \in J^z$ for all i, t (otherwise, 0). In contrast to the Primal-dual Simplex algorithm, the constraint associated with (V) in the RP is relaxed to an inequality (6.13). Consequently, a lower bound to $ZL_{(C)}^{(X)(Y)}$ is obtained from $(\bar{v}^{iter}, \bar{u}^{iter})$ after termination. Furthermore, the resulting RDP decomposes and a feasible solution to the RDP is calculated with

6.2 Two Lagrangean Heuristics

a fast procedure. This is shown in what follows. The *Relaxed RP* is formulated as

$$\xi = \min \sum_{i=1}^{n} \sum_{t=1}^{T-\pi_i+1} \alpha_{it} + \sum_{i=1}^{n} \sum_{t=1}^{T} \beta_{it} \quad \text{subject to}$$

$$\alpha_{it} + \sum_{\tau=t}^{t+\pi_i-1} a_{i\tau}^x \cdot x_{i\tau} - a_{it}^z \cdot z_{it} = 1 \quad \forall i=1,\ldots,n; t=1,\ldots,T-\pi_i+1 \quad v_{it} \quad (6.12)$$

$$\beta_{it} - a_{it}^x \cdot x_{it} + a_t^y \cdot y_t \geq 0 \quad \forall i=1,\ldots,n; t=1,\ldots,T \quad u_{it} \quad (6.13)$$

$$\alpha_{it} \geq 0 \; \forall i=1,\ldots,n; t=1,\ldots,T-\pi_i+1 \land \beta_{it} \geq 0 \; \forall i=1,\ldots,n; t=1,\ldots,T$$

$$x_{it} \geq 0 \; \forall (i,t) \in J^x \land y_t \geq 0 \; \forall t \in J^y \land z_{it} \geq 0 \; \forall (i,t) \in J^z.$$

The corresponding dual problem that is referred to as *Restricted RDP* becomes[10]

$$\xi = \max \sum_{i=1}^{n} \sum_{t=1}^{T-\pi_i+1} v_{it} \quad \text{subject to} \quad (6.14)$$

$$v_{it} \leq 1 \quad \forall i=1,\ldots,n; t=1,\ldots,T-\pi_i+1 \quad \alpha_{it} \quad (6.15)$$

$$u_{it} \leq 1 \quad \forall i=1,\ldots,n; t=1,\ldots,T \quad \beta_{it} \quad (6.16)$$

$$\sum_{\tau=\max\{1,t-\pi_i+1\}}^{\min\{t,T-\pi_i+1\}} v_{i\tau} - u_{it} \leq 0 \quad \forall (i,t) \in J^x \quad x_{it} \quad (6.17)$$

$$\sum_{i=1}^{n} u_{it} \leq 0 \quad \forall t \in J^y \quad y_t \quad (6.18)$$

$$-v_{it} \leq 0 \quad \forall (i,t) \in J^z \quad z_{it} \quad (6.19)$$

$$v_{it} \in \mathbb{R} \quad \forall i=1,\ldots,n; t=1,\ldots,T-\pi_i+1 \quad (6.20)$$

$$u_{it} \geq 0 \quad \forall i=1,\ldots,n; t=1,\ldots,T. \quad (6.21)$$

A simplification yields the LP (6.23)–(6.27). In (6.17), at most π_i variables of v are summed up. Because of (6.15), the largest possible value of this sum is π_i.

---

[10]The Relaxed RP has a larger solution space than the RP because (6.13) is not an equality. Therefore, the corresponding dual problem has a smaller solution space than the RDP because the corresponding dual variable u is non-negative (6.21) and not a variable with real values.

Therefore,

$$\sum_{\tau=\max\{1,t-\pi_i+1\}}^{\min\{t,T-\pi_i+1\}} v_{i\tau} - u_{it} \leq \pi_i \quad \forall i=1,\ldots,n; t=1,\ldots,T \quad (6.22)$$

is a valid inequality. A combination of (6.17) and (6.22) yields (6.24). From (6.21) follows that (6.18) is identical to (6.25). Since the constraint coefficients as well as the objective function coefficients of v are non-negative and the objective function (6.14) maximizes, (6.15) as well as (6.20) become (6.26) and (6.19) is redundant. The Restricted RDP simplifies to

$$\xi = \max \sum_{i=1}^{n} \sum_{t=1}^{T-\pi_i+1} v_{it} \qquad \text{subject to} \qquad (6.23)$$

$$\sum_{\tau=\max\{1,t-\pi_i+1\}}^{\min\{t,T-\pi_i+1\}} v_{i\tau} - u_{it} \leq \begin{cases} 0 & \text{if } (i,t) \in J^x \\ \pi_i & \text{otherwise} \end{cases} \quad \forall i=1,\ldots,n; t=1,\ldots,T \quad (6.24)$$

$$u_{it} = 0 \qquad \forall i=1,\ldots,n; t \in J^y \quad (6.25)$$

$$0 \leq v_{it} \leq 1 \qquad \begin{array}{l} \forall i=1,\ldots,n; \\ t=1,\ldots,T-\pi_i+1 \end{array} \quad (6.26)$$

$$0 \leq u_{it} \leq 1 \qquad \forall i=1,\ldots,n; t=1,\ldots,T. \quad (6.27)$$

The LP (6.23)–(6.27) decomposes into n independent subproblems.[11] Since the objective function coefficients of u are zero, the constraints (6.24), (6.25) and (6.27) yield

$$\sum_{\tau=\max\{1,t-\pi_i+1\}}^{\min\{t,T-\pi_i+1\}} v_{i\tau} \leq \begin{cases} 0 & \text{if } (i,t) \in J^x \wedge t \in J^y \\ 1 & \text{if } (i,t) \in J^x \wedge t \notin J^y \\ \pi_i & \text{otherwise} \end{cases} \quad \forall t=1,\ldots,T \quad (6.28)$$

for all $i=1,\ldots,n$. Let (v^*, u^*) be an optimal solution to the Restricted RDP. Having solved (6.29), u^* is obtained from (6.30). The decomposed Restricted RDP

---

[11] If the RP instead of the Relaxed RP was formulated, constraint (6.13) would be an equality. But then $u \in \mathbb{R}$ would follow instead of (6.21) and the decomposition would not be possible.

6.2 Two Lagrangean Heuristics

simplifies to

$$\xi = \sum_{i=1}^{n} \max \left\{ \sum_{t=1}^{T-\pi_i+1} v_{it} \middle| (6.28) \wedge 0 \leq v_{it} \leq 1 \ \forall t = 1,\ldots,T-\pi_i+1 \right\} \quad (6.29)$$

$$u_{it}^* = \begin{cases} \sum_{\tau=\max\{1,t-\pi_i+1\}}^{\min\{t,T-\pi_i+1\}} v_{i\tau}^* & \text{if } (i,t) \in J^x \wedge t \notin J^y \quad \forall i=1,\ldots,n; \\ 0 & \text{otherwise} \end{cases} \quad t = 1,\ldots,T. \quad (6.30)$$

Lemma 6.2 $v_{it} \leq 1 \ \forall i;t$ is redundant in (6.29) if $\bar{u}_{it}^0 = 0 \ \forall i;t$ and $\bar{v}^0$ is calculated by using n times Algorithm 4.2 with $c_{it} = r_i \cdot \mu_t \ \forall i;t$.

Proof Recall that the Primal-dual Simplex algorithm requires an initial feasible dual solution $(\bar{v}^0, \bar{u}^0)$. Setting $\bar{u}_{it}^0 = 0 \ \forall i;t$ simplifies the LP (6.1)–(6.5) and $(\bar{v}^0, \bar{u}^0)$ is feasible. Because of Remark 4.7, there exists at least one $d'_{it} = 0$ every π_i periods for all i and the index (i,t) enters J^x. There exists at least one upper bound that is either 0 or 1 in (6.28) since (6.28) has the stairway structure (Remark 4.1). Hence, $\sum_{\tau=\max\{1,t-\pi_i+1\}}^{\min\{t,T-\pi_i+1\}} v_{i\tau} \leq 1 \ \forall i;t$ and $v \leq 1$ is satisfied in the first iteration. Lemma 6.5, which is presented later in this section, states that once d'_{it} with $(i,t) \in J^x$ becomes zero in some iteration, it remains zero until termination. Hence, $v \leq 1$ holds for all iterations.

The Primal-dual Lower Bound uses Lemma 6.2 and hence, an optimal solution (v^*, u^*) to the Restricted RDP (6.14)–(6.21) is obtained from (6.30) and

$$\xi = \sum_{i=1}^{n} \max \left\{ \sum_{t=1}^{T-\pi_i+1} v_{it} \middle| (6.28) \wedge v_{it} \geq 0 \ \forall t = 1,\ldots,T-\pi_i+1 \right\}. \quad (6.31)$$

The ith subproblem of (6.31) has the structure of the dual problem of the SWUPMP LP relaxation (4.42)–(4.44). Hence, it is solved to optimality with Algorithm 4.2 that requires

$$c_{it} = \begin{cases} 0 & \text{if } (i,t) \in J^x \wedge t \in J^y \\ 1 & \text{if } (i,t) \in J^x \wedge t \notin J^y \quad \forall t = 1,\ldots,T \\ \pi_i & \text{otherwise} \end{cases} \quad (6.32)$$

and returns the optimal slack of (6.28) to efficiently determine u^* from (6.30). Note that $v_{it}^* \in \{0,1\} \ \forall i;t$ and $u_{it}^* \in \{0,1\} \ \forall i;t$ holds because of Corollary 4.2 for (6.31).

Lemma 6.3 An optimal solution to the Restricted RDP where u^* is obtained from (6.30) has an optimal slack of constraint (6.17) that is zero.

Proof Let $d_{it}^{(6.17)}$ be the slack of (6.17) that is non-negative. Given $(i,t) \in J^x$.

$$d_{it}^{(6.17)} = -\sum_{\tau=\max\{1, t-\pi_i+1\}}^{\min\{t, T-\pi_i+1\}} v_{i\tau}^* + u_{it}^*$$

$$= -\sum_{\tau=\max\{1, t-\pi_i+1\}}^{\min\{t, T-\pi_i+1\}} v_{i\tau}^* + \begin{cases} \sum_{\tau=\max\{1, t-\pi_i+1\}}^{\min\{t, T-\pi_i+1\}} v_{i\tau}^* & \text{if } t \notin J^y \\ 0 & \text{otherwise} \end{cases} \quad \text{from (6.30)}$$

$$= 0$$

Having solved the Restricted RDP, either $\xi = 0$ or $\xi > 0$ holds. Consider $\xi = 0$. At this stage, there exists a feasible solution to the $ZL_{(C)}$ LP relaxation, the dual solution $(\bar{v}^{iter}, \bar{u}^{iter})$ is feasible and six from seven complementary slackness criteria between the Relaxed RP and the Restricted RDP are satisfied. Denote $d^{(K)}$ as the optimal slack of a constraint (K). Let (α, β, x, y, z) be a corresponding optimal solution of the Relaxed RP to (v^*, u^*). In particular, the complementary slackness criteria

$d_{it}^{(6.15)} \cdot \alpha_{it} = 0 \quad \forall i; t \qquad \xi = 0 \Rightarrow \alpha_{it} = 0 \; \forall i; t$

$d_{it}^{(6.16)} \cdot \beta_{it} = 0 \quad \forall i; t \qquad \xi = 0 \Rightarrow \beta_{it} = 0 \; \forall i; t$

$d_{it}^{(6.17)} \cdot x_{it} = 0 \quad \forall (i,t) \in J^x \qquad \text{because of Lemma 6.3}$

$d_t^{(6.18)} \cdot y_t = 0 \quad \forall t \in J^y \qquad \text{the slack of (6.18) is zero because of (6.21)}$

$d_{it}^{(6.19)} \cdot z_{it} = 0 \quad \forall (i,t) \in J^z \qquad \text{(6.19) is redundant}$

$d_{it}^{(6.12)} \cdot v_{it}^* = 0 \quad \forall i; t \qquad \text{(6.12) is an equality}$

are satisfied despite of $d_{it}^{(6.13)} \cdot u_{it}^* \geq 0 \; \forall i; t$ since (6.13) is an inequality. Altogether, the complementary slackness Theorem 3.3.3 is almost satisfied and $(\bar{v}^{iter}, \bar{u}^{iter})$ yields a lower bound to $ZL_{(C)}^{(X)(Y)}$. Terminate the algorithm, return *LB* and the sets $L_i \; \forall i$ as well as $P = \emptyset$.

Consider $\xi > 0$. The optimal solution (v^*, u^*) is used in the update (6.33) to obtain a feasible solution $(\bar{v}^{iter+1}, \bar{u}^{iter+1})$ of the next iteration $iter + 1$.

$$\begin{pmatrix} \bar{v}^{iter+1} \\ \bar{u}^{iter+1} \end{pmatrix} = \begin{pmatrix} \bar{v}^{iter} \\ \bar{u}^{iter} \end{pmatrix} + \lambda \cdot \begin{pmatrix} v^* \\ u^* \end{pmatrix} \quad \forall iter \in \{0, 1, 2, \ldots\} \qquad (6.33)$$

The scalar λ is chosen such that feasibility of the novel dual solution is guaranteed. Hence, λ is set to the lowest upper bound of the update formulas (6.35) and (6.36).

6.2 Two Lagrangean Heuristics

The simplification (6.35) is obtained as follows: Lemma 6.2 uses the argument that $\sum_{\tau=\max\{1,t-\pi_i+1\}}^{\min\{t,T-\pi_i+1\}} v_{i\tau} \leq 1 \ \forall i; t$. In (6.34), the strictly positive denominator yields $u_{it}^* < \sum_{\tau=\max\{1,t-\pi_i+1\}}^{\min\{t,T-\pi_i+1\}} v_{i\tau}^* \leq 1$ for a given i and t. Because of $v_{it}^* \in \{0,1\} \ \forall i; t$ there exists a $\theta = t - \pi_i + 1, \ldots, t$ with $v_{i\theta}^* = 1 \wedge u_{it}^* = 0$ and (6.35) is obtained.

$$(6.2) \xRightarrow{\substack{v_{i\tau}=\bar{v}_{i\tau}^{iter}+\lambda\cdot v_{i\tau}^* \\ u_{it}=\bar{u}_{it}^{iter}+\lambda\cdot u_{it}^*}} \lambda \leq \frac{r_i \cdot \mu_t - \sum_{\tau=\max\{1,t-\pi_i+1\}}^{\min\{t,T-\pi_i+1\}} \bar{v}_{i\tau}^{iter} + \bar{u}_{it}^{iter}}{\sum_{\tau=\max\{1,t-\pi_i+1\}}^{\min\{t,T-\pi_i+1\}} v_{i\tau}^* - u_{it}^*} \quad (6.34)$$

$$= d_{it}' \ \forall (i,t) \notin J^x \quad (6.35)$$

$$(6.3) \xRightarrow{u_{it}=\bar{u}_{it}^{iter}+\lambda\cdot u_{it}^*} \lambda \leq \frac{\max\{0, 1-\mu_t \cdot \bar{r}_t\} - \sum_{i=1}^{n} \bar{u}_{it}^{iter}}{\sum_{i=1}^{n} u_{it}^*} \quad \text{with (6.30)}$$

$$= \frac{d_t''}{\sum_{(i,t) \in J^x \wedge t \notin J^y} u_{it}^*} \ \forall t \notin J^y \quad (6.36)$$

Update the lower bound $LB + \lambda \cdot \xi$ and commence a new iteration. Lemma 6.4 also states that the maximal improvement of the lower bound LB is reached after termination.

Lemma 6.4 *If $\xi > 0$, every iteration increases LB by $\lambda \cdot \xi > 0$ and Algorithm 6.5 terminates after a finite number of iterations.*

Proof Consider iteration $iter \in \{0, 1, 2, \ldots\}$ and the obtained lower bound.

$$\sum_{i=1}^{n} \sum_{t=1}^{T-\pi_i+1} v_{it}^{iter+1} = \sum_{i=1}^{n} \sum_{t=1}^{T-\pi_i+1} v_{it}^{iter} + \lambda \cdot \sum_{i=1}^{n} \sum_{t=1}^{T-\pi_i+1} v_{it}^* \quad \text{because of (6.33)}$$

$$= \sum_{i=1}^{n} \sum_{t=1}^{T-\pi_i+1} v_{it}^{iter} + \lambda \cdot \xi \quad \text{because of (6.14)}$$

$$> \sum_{i=1}^{n} \sum_{t=1}^{T-\pi_i+1} v_{it}^{iter} \quad \xi > 0 \wedge \lambda > 0 \text{ (see below)}$$

(6.35) and (6.36) only consider dual constraints for which the corresponding primal variable is *not* a potential primal basic variable. Hence, $d_{it}' > 0 \ \forall (i,t) \notin J^x$ and $d_t'' > 0 \ \forall t \notin J^y$ holds per definition (6.10) and (6.11). There exists at least one positive denominator in (6.36) because the LP (6.1)–(6.5) is solvable (Lemma 6.1). Thus, $\lambda > 0$. LB has a finite value because the Relaxed RP is solvable (Table 3.2). Therefore, the best value for LB is reached after a finite number of iterations since $\lambda \cdot \xi > 0$ (cf. geometric improvement approach in Ahuja et al. (1993, p. 67)).

The Algorithm 6.5 runs in polynomial time. Setting λ to the smallest value of (6.35) and (6.36), one constraint from either (6.2) or (6.3) has a zero slack after

6.5 Algorithm. The Primal-dual Lower Bound for $ZL_{(C)}$

Input: $\mu_t \geq 0 \ \forall t$ and an instance to the CPMP
Output: A lower bound LB to $ZL_{(C)}$ and the sets $L_i \ \forall i, P = \emptyset$
Initialization: $L_i \leftarrow \emptyset \ \forall i, P \leftarrow \emptyset, LB \leftarrow \sum_{t=1}^{T} \min\{0, 1 - \mu_t \cdot \bar{r}_t\}, c_{it} \leftarrow 0 \ \forall i; t,$
$\bar{v}_{it}^{iter} \leftarrow 0 \ \forall i; t; iter, \bar{u}_{it}^{iter} \leftarrow 0 \ \forall i; t; iter, v_{it}^{*} \leftarrow 0 \ \forall i; t, u_{it}^{*} \leftarrow 0 \ \forall i; t,$
$J^x \leftarrow \emptyset, J^y \leftarrow \emptyset, iter \leftarrow 0, i \leftarrow 0, t \leftarrow 0, \xi \leftarrow 0, \lambda \leftarrow 0$

1 **for** $i = 1$ **to** n **do**
2 $\quad$ Use Algorithm 4.2 with $c_{it} \leftarrow r_i \cdot \mu_t \ \forall t = 1, \ldots, T$ to calculate $\bar{v}_{it}^{0} \ \forall t$ and to increase LB by the obtained SWUPMP objective function value

3 **for** $iter \in \{1, 2, 3, \ldots\}$ **do**
4 $\quad$ Determine J^x, J^y from (6.10) and (6.11) with $(\bar{v}^{iter}, \bar{u}^{iter})$
5 $\quad$ **for** $i = 1$ **to** n **do**
6 $\quad\quad$ Set $c_{it} \ \forall t$ with (6.32)
7 $\quad\quad$ Solve the ith subproblem of (6.31) to optimality with Algorithm 4.2 and obtain $v_{it}^{*} \ \forall t$ and the optimal slack of (6.28)
8 $\quad\quad$ Derive $u_{it}^{*} \ \forall t$ from (6.30)
9 $\quad$ **if** $\xi = 0$ **then** Derive $L_i \ \forall i$ and terminate the algorithm
10 $\quad$ Set λ to the lowest upper bound of (6.35) and (6.36)
11 $\quad$ Update the dual solution with (6.33)
12 $\quad$ $LB \leftarrow LB + \lambda \cdot \xi$

---

the update. Therefore, a novel potential basic variable is introduced in either J^x or J^y and it remains in the respective set until termination (Lemma 6.5). In the worst case, all possible primal basic variables are examined. Polynomial solvability follows since (6.1)–(6.5) is bounded (Lemma 6.1).[12]

Lemma 6.5 *If $d'_{it} = 0 \ \forall (i, t) \in J^x$ or $d''_t = 0 \ \forall t \in J^y$ holds in an iteration, the value remains zero until termination.*

Proof Denote d'^{iter}_{it} as d'_{it} at the beginning of an iteration $iter \in \{0, 1, 2, \ldots\}$. Let $d'^{0}_{it} = 0$ with $(i, t) \in J^x$ and the induction hypothesis holds for $iter = 0$. Let $iter = iter + 1$.

$$d'^{iter+1}_{it} = r_i \cdot \mu_t - \sum_{\tau = \max\{1, t - \pi_i + 1\}}^{\min\{t, T - \pi_i + 1\}} \bar{v}^{iter+1}_{i\tau} + \bar{u}^{iter+1}_{it}$$

$$= r_i \cdot \mu_t - \sum_{\tau = \max\{1, t - \pi_i + 1\}}^{\min\{t, T - \pi_i + 1\}} (\bar{v}^{iter}_{i\tau} + \lambda \cdot v^{*}_{i\tau}) + (\bar{u}^{iter}_{it} + \lambda \cdot u^{*}_{it}) \quad \text{from (6.33)}$$

---

[12]This behavior is observed for other combinatorial optimization problems that are in $\mathcal{P}$ and polynomially solvable with the Primal-dual Simplex algorithm such as the SPP (Papadimitriou and Steiglitz 1998, pp. 109–113; Chvátal 2002, pp. 390–400). However, the Maximum Flow Problem is in $\mathcal{P}$ but the Ford-Fulkerson algorithm that is based on a Primal-dual Simplex algorithm has a pseudo-polynomial worst case running time (Ahuja et al. 1993, p. 186).

$$= d'^{iter}_{it} + \lambda \cdot d^{(6.17)}_{it} \quad d'^{iter}_{it} = 0 \text{ (induction start)}, d^{(6.17)}_{it} = 0 \text{ (Lemma 6.3)}$$
$$= 0$$

Observe that $\bar{u}^0_{it} \geq 0 \; \forall i; t$. After the update of $\bar{u}^{iter}_{it}$ with (6.33), $\bar{u}^{iter+1}_{it} \geq 0 \; \forall i; t$ holds because $\lambda > 0$ from Lemma 6.4. Thus, d''_t decreases over the iterations.

6.2.1.5 The Shortest Path Lower Bound

Add (P^y) to $ZL_{(C)}$, neglect (V) completely and obtain the *Shortest Path Lower Bound* as

$$LB = \min \sum_{t=1}^{T} (1 - \mu_t \cdot \bar{r}_t) \cdot y_t + \sum_{i=1}^{n} r_i \cdot \min \sum_{t=1}^{T} \mu_t \cdot x_{it} \quad \text{subject to } (P^y) \wedge (P) \wedge (X) \wedge (Y).$$

The problem decomposes into $n+1$ subproblems. A reformulation from Remark 4.2 yields n SPPs for x with arc weights $c_{it} = \mu_t \; \forall i; t$ and one SPP for y in the graph $G_1(V, E_1)$ with non-negative arc weights $c_{1t} = \max\{1 - \mu_t \cdot \bar{r}_t, 0\} \; \forall t$. Add $\sum_{t=1}^{T} \min\{1 - \mu_t \cdot \bar{r}_t, 0\}$ to the total length of all shortest paths. Return the lower bound LB, $P = \emptyset$ and $L_i = \{1, \ldots, T\} \; \forall i$.

Lemma 6.6 $LB \leq ZL^{(X)(Y)}_{(C)}$

Proof $H(P)(V)(S) = H(P)(P^y)(V)(S) \subseteq H(P)(P^y)(S) = Conv(P)(P^y)(X)(Y)$ holds since $H(P)(S^x) = Conv(P)(X)$ and $H(P^y)(S^y) = Conv(P^y)(Y)$ hold by Theorem 4.1.

6.2.1.6 The Upper Bound Heuristic

A dual solution to the LP relaxation might provide valuable information about the structure of an optimal integer solution. This is particularly interesting for $ZL_{(C)}$ because preliminary computational studies revealed that the LP Lower Bound is close to the optimum of $ZL_{(C)}$ (see Sect. 7.3.1.1) and because of Observation 4.4. The *Upper Bound Heuristic* from Algorithm 6.6 identifies 'good' primal variables with two strategies. Potential integers from P are preferably set and two of the four complementary slackness criteria are taken into account via $L_i \; \forall i$ (Lemma 6.7). Considering the complementary slackness is successful in dual-ascent procedures for related problems such as the UFLP (Erlenkotter 1978), the UNDP (Balakrishnan et al. 1989) and the capacitated variants, which are the CFLP (Guignard and Spielberg 1979) and the FCNDP (Herrmann et al. 1996; Gendron 2002).

6.6 Algorithm. The Upper Bound Heuristic for $ZL_{(C)}$
Input: $\mu_t \geq 0 \ \forall t, L_i \neq \emptyset \ \forall i, P$ and an instance to the CPMP
Output: A feasible solution (x, y) to $ZL_{(C)}$ and if necessary its objective function value UBL
Initialization: $(x, y) \leftarrow (0, 0), I \leftarrow \{1, \ldots, n\}, \bar{I} \leftarrow \{1, \ldots, n\}, Path_i \leftarrow \emptyset \ \forall i,$
$c_{it} \leftarrow 0 \ \forall i; t = 1, \ldots, T+2, SP_{it} \leftarrow 0 \ \forall t = 1, \ldots, T+3, i \leftarrow 0, j \leftarrow 0,$
$t \leftarrow 0, UBL \leftarrow 0$

1   $UBL \leftarrow \sum_{t=1}^{T} \min\{0, 1 - \mu_t \cdot \bar{r}_t\}$ // Calculate iff UBL is an output
2   $y_t \leftarrow 1 \ \forall t = 1, \ldots, T : 1 - \mu_t \cdot \bar{r}_t < 0$
3   Set arc weights $c_{it} \ \forall i; t$ with (6.38)
4   **do**
5   **for** $i \in \bar{I}$ **do**
6   Solve the ith SPP to optimality with Algorithm 4.1 and obtain the shortest path $Path_i$ as well its length $SP_{i,T+1}$
7      $UBL \leftarrow UBL + SP_{j,T+1} + \sum_{t \in \{\tau \in Path_j | y_\tau = 0 \wedge \tau \in P\}} 1 - \mu_t \cdot \bar{r}_t$
                                                                                     // Iff UBL is an output
8      $j \leftarrow \arg\min_{i \in I} SP_{i,T+1}$
9      $I \leftarrow I \setminus \{j\} \wedge \bar{I} \leftarrow \emptyset \wedge x_{jt} \leftarrow 1 \ \forall t \in Path_j \wedge y_t \leftarrow 1 \ \forall t \in Path_j$
10   **for** $i \in I$ **do**
11   **if** $L_i \cap Path_j \neq \emptyset$ **then** $\bar{I} \leftarrow \bar{I} \cup \{i\} \wedge c_{it} \leftarrow r_i \cdot \mu_t \ \forall t \in Path_j$
12 **while** $I \neq \emptyset$

Lemma 6.7 *If d'_{it} is obtained from one of the lower bounds presented in the Sects. 6.2.1.1–6.2.1.5, a feasible solution to $ZL_{(C)}$ exists that satisfies the complementary slackness criteria*[13]

$$d'_{it} \cdot x_{it} = 0 \quad \forall i = 1, \ldots, n; t = 1, \ldots, T \quad (6.37a)$$

$$d''_t \cdot y_t = 0 \quad \forall t = 1, \ldots, T. \quad (6.37b)$$

Proof Replace (S^y) with $y_t \in \mathbb{R}$ in the LP relaxation because (S^y) is implied by (V) and $x_{it} \geq 0 \ \forall i; t$. (6.3) becomes an equality with $d''_t = 0 \ \forall t$ and (6.37b) always holds. Let L_i be sorted according to increasing values. Assume there exists an i that has two successive periods $\tau, \theta \in L_i$ with $\tau + \pi_i < \theta$. LB could be feasibly improved by $\min_{t=\tau+1,\ldots,\tau+\pi_i} d'_{it} > 0$. Obviously, this case never occurs in the presented lower bounds. Thus, $\tau + \pi_i \geq \theta \ \forall i$, which ensures feasibility of (P). The solution (x, y) defined as $x_{it} = 1 \wedge y_t = 1$ if $d'_{it} = 0$ (otherwise, 0) $\forall i; t$ is feasible to $ZL_{(C)}$ and (6.37a) holds.

The heuristic uses the fact that $ZL_{(C)}$ decomposes into n SWUPMPs if y is fixed. Given the sets P and $L_i \ \forall i$. The initialization sets $y_t = 1 \ \forall t$ if $1 - \mu_t \cdot \bar{r}_t < 0$ and the

---

[13]Erratum: Only (6.37a) holds. However, (6.37b) is not considered as insightful because of the reason mentioned in the proof.

6.2 Two Lagrangean Heuristics

arc weights become

$$c_{it} = \begin{cases} r_i \cdot \mu_t & \text{if } 1 - \mu_t \cdot \bar{r}_t < 0 \text{ or } t \in P \\ \infty & \text{if } t \notin L_i \\ 1 - \mu_t \cdot \bar{r}_t + r_i \cdot \mu_t & \text{otherwise} \end{cases} \quad \begin{array}{l} \forall i = 1, \ldots, n; \\ t = 1, \ldots, T. \end{array} \quad (6.38)$$

In the first case, the fixed costs are zero if the period is already open or integer in the LP Lower Bound (i.e., a potential integer). Infinite arc weights allow only periods in L_i, which satisfies (6.37a) from Lemma 6.7. Note that (6.37b) is always satisfied (proof of Lemma 6.7). In the last case, the regular arc weight is set. Set $I = \{1, \ldots, n\}$ and $\bar{I} = \{1, \ldots, n\}$. In each iteration, a SWUPMP is solved as a SPP for every maintenance activity $i \in \bar{I}$ with Algorithm 4.1.[14] $Path_i$ denotes the set of nodes of the shortest path. Let j be the maintenance activity with the shortest path among all maintenance activities $i \in I$ (priority rule).[15] Set $x_{jt} = 1 \wedge y_t = 1$ if $t \in Path_j$ (otherwise, 0) and remove j from I. The newly opened periods affect only maintenance activities i with $L_i \cap Path_j \neq \emptyset$ for which arc weights are updated with $c_{it} = r_i \cdot \mu_t \; \forall t \in Path_j$. In the next iteration, a shortest path is only recomputed for these maintenance activities $i \in \bar{I}$. Note that a calculation of UBL in lines 1 and 7 is not necessary for a Lagrangean heuristic but required for the computational results of Sect. 7.3.1. The algorithm terminates when n paths are fixed and returns a feasible solution to $ZL_{(C)}$.

If the LP Lower Bound is used, Algorithm 6.6 is repeatedly executed. P and $L_i \; \forall i$ provide three alternatives to consider the dual information[16]:

- Consider the complementary slackness (6.37a) and the potential integers
- Ignore the complementary slackness (6.37a) but consider the potential integers
- Consider the complementary slackness (6.37a) but ignore the potential integers

Since no dominance between the three alternatives exists, three feasible solutions are computed and the best feasible solution is chosen. Note that ignoring the complementary slackness requires $L_i = \{1, \ldots, T\} \; \forall i$.

---

[14] Implementation detail: Algorithm 4.1 examines *all* periods $1, \ldots, T$ in a breadth-first search. An improvement is to exclude the periods $t \notin L_i$ since they do not improve the shortest path due to infinite arc lengths: Order the set L_i with respect to increasing values of periods when it is obtained from d'_{it}. Let p be an index in L_i such that $L_i(p)$ returns the period at the pth position in L_i. A simple modification now traverses only the periods $L_i(t), \ldots, L_i(t) + \pi_i$ where t is an index instead of all periods $t, \ldots, t + \pi_i$.

[15] Implementation detail: An efficient implementation uses a minimum priority queue. In every iteration, the heap holds one shortest path for each maintenance activity $i \in J$ with i as the ID and the length of the shortest path as the key. Recomputed shortest paths for the maintenance activities $i \in \bar{I}$ are inserted with the heap operation 'update key'.

[16] The fourth alternative ignores the complementary slackness (6.37a) as well as potential integers but it does not yield satisfying results. This is discussed in Sect. 7.3.1.1.

6.2.2 The Lagrangean Relaxation of the Period Covering Constraint

Lemma 5.23 shows that

$$ZL_{(P)}(\upsilon) = \sum_{t=1}^{T} \min\{0, 1 - z_t\} + \sum_{i=1}^{n} \sum_{t=1}^{T-\pi_i+1} \upsilon_{it} \qquad (5.44)$$

$$\forall t : z_t = \max\left\{\sum_{i=1}^{n} x_{it} \cdot \sum_{\tau=\max\{1,t-\pi_i+1\}}^{\min\{t,T-\pi_i+1\}} \upsilon_{i\tau} \,\middle|\, \sum_{i=1}^{n} r_i \cdot x_{it} \leq \bar{r}_t \wedge x_{it} \in \{0,1\} \;\forall i\right\}$$

is binary $\mathcal{NP}$-hard, solvable in time $O(n \cdot T + n \cdot \bar{r}^{sum})$ and every subproblem z_t is a KP. Let LB_t^{KP} (UB_t^{KP}) be a lower (an upper) bound to z_t. Note that z_t contributes to $ZL_{(P)}(\upsilon)$ if $z_t > 1$ holds. Therefore, if $UB_t^{KP} \leq 1$ holds, z_t is pruned.

The *Knapsack Heuristic* in Algorithm 6.7 provides a feasible solution to $ZL_{(P)}$ and a lower bound. All periods t are considered one after another and z_t is solved heuristically. Define $I_t = \{i = 1, \ldots, n | r_i \leq \bar{r}_t\} \;\forall t$, $r_t^{min} = \min_{i \in I_t} r_i \;\forall t$ and $r_t^{sum} = \sum_{i \in I_t} r_i \;\forall t$.

6.7 Algorithm. The Knapsack Heuristic for $ZL_{(P)}$

Input: $\upsilon_{it} \geq 0 \;\forall i; t$ and an instance to the CPMP
Output: A feasible solution (x, y) as well as a lower bound LB to $ZL_{(P)}$ and if necessary its objective function value UBL
Initialization: $(x, y) \leftarrow (0, 0)$, $LB \leftarrow \sum_{i=1}^{n} \sum_{t=1}^{T-\pi_i+1} \upsilon_{it}$, $x_{it}^{KP} \leftarrow 0 \;\forall i; t$, $UB_t^{KP} \leftarrow 0 \;\forall t$,
$LB_t^{KP} \leftarrow 0 \;\forall t$, $I_t \leftarrow \{i = 1, \ldots, n | r_i \leq \bar{r}_t\} \;\forall t$, $r_t^{sum} \leftarrow \sum_{i \in I_t} r_i \;\forall t$,
$r_t^{min} \leftarrow \min_{i \in I_t} r_i \;\forall t$, $t \leftarrow 0$, $i \leftarrow 0$, $UBL \leftarrow 0$

1    $UBL \leftarrow LB$ // Calculate iff UBL is an output
2   **for** $t = 1$ **to** T **do**
3   **if** $r_t^{min} \leq \bar{r}_t$ **then**
4   **if** $r_t^{sum} \leq \bar{r}_t$ **then**
5                $UB_t^{KP} \leftarrow \sum_{i \in I_t} \sum_{\tau=\max\{1,t-\pi_i+1\}}^{\min\{t,T-\pi_i+1\}} \upsilon_{i\tau}$
6   **if** $UB_t^{KP} > 1$ **then**
7                    $LB \leftarrow LB + 1 - UB_t^{KP} \wedge y_t \leftarrow 1 \wedge x_{it} \leftarrow 1 \;\forall i \in I_t$
8                    $UBL \leftarrow UBL + 1 - UB_t^{KP}$ // Iff UBL is an output
9   **else**
10   Calculate UB_t^{KP} for z_t with (3.58)
11   **if** $UB_t^{KP} > 1$ **then**
12   Calculate LB_t^{KP} for z_t with (3.59) and obtain the best feasible solution $(x_{1t}^{KP}, \ldots, x_{nt}^{KP})$ to the tth KP
13                    $LB \leftarrow LB + 1 - UB_t^{KP} \wedge y_t \leftarrow 1 \wedge x_{it} \leftarrow x_{it}^{KP} \;\forall i$
14                    $UBL \leftarrow UBL + \min\{0, 1 - LB_t^{KP}\}$ // Iff UBL is an output

6.2 Two Lagrangean Heuristics

Initialize with $t = 1$. In an iteration, set $UB_t^{KP} = 0$, $LB_t^{KP} = 0$, $y_t = 0$ and $x_{it} = 0$ $\forall i$. If it is not possible to schedule at least one maintenance activity since $r_t^{min} > \bar{r}_t$ holds, proceed with the next period $t = t+1$. Otherwise, if all maintenance activities can be scheduled because of $r_t^{sum} \leq \bar{r}_t$ and if $UB_t^{KP} = \sum_{i \in I_t} \sum_{\tau=\max\{1,t-\pi_i+1\}}^{\min\{t,T-\pi_i+1\}} v_{i\tau} > 1$ holds, set $y_t = 1$, set $x_{it} = 1$ if $i \in I_t$ (otherwise, 0) for all i and proceed with the next period $t = t + 1$. Otherwise, if $r_t^{sum} > \bar{r}_t$, the tth KP is not trivially solvable and the following applies: Calculate UB_t^{KP} as the upper bound (3.58) and obtain the critical item.[17] If $UB_t^{KP} \leq 1$ holds, proceed with the next period $t = t+1$. Otherwise, determine the best feasible KP solution $(x_{1t}^{KP}, \ldots, x_{nt}^{KP})$ with LB_t^{KP} from (3.59), set $x_{it} = x_{it}^{KP}$ $\forall i$ as well as $y_t = 1$. Proceed with the next period $t = t+1$. The algorithm terminates if $t > T$, returns a feasible solution to $ZL_{(P)}$ and the lower bound

$$LB = \sum_{t=1}^{T} \min\{0, 1 - UB_t^{KP}\} + \sum_{i=1}^{n} \sum_{t=1}^{T-\pi_i+1} v_{it}.$$

The upper bound UBL that is calculated from the lines 1, 8 and 14 is not necessary for a Lagrangean heuristic but used in the computational results of Sect. 7.3.1.[18]

6.2.3 The Lagrangean Heuristic for Both Relaxations

Subgradient optimization commences from given Lagrangean multiplier $\mu \in \mathbb{R}^m$ and seeks the next multiplier $\mu^{iter} \in \mathbb{R}^m$ such that the Lagrangean lower bound is maximized. Most approaches in literature derive a valid subgradient from an optimal solution to the Lagrangean relaxation (Lemma 3.16) because the subdifferential can be obtained from all optimal solutions (Klose 2001, pp. 174–176). In the presented approach, the Lagrangean relaxations are solved heuristically because of the computational complexity (cf. Sect. 5.3.2). $ZL_{(C)}$ is assumed to be $\mathcal{NP}$-hard and $ZL_{(P)}$ is binary $\mathcal{NP}$-hard (Lemma 5.23). Note that the computational time, which is required to optimally solve an instance, can significantly increase with the instance size and this would lead to only few iterations in a Lagrangean heuristic.

Using the definitions from Sect. 3.3.2, let a concave function $f : \mathbb{R}_+^m \mapsto \mathbb{R}$ be non-differentiable at the position μ^{iter} and let UB be an upper bound with $UB \geq f(\mu)$.

Definition 6.8 If $f(\mu)$ is concave, $\hat{g} \in \mathbb{R}^m$ is a *pseudo-subgradient* of $f(\mu^{iter})$ at the position μ^{iter} if $f(\mu) \leq UB + \hat{g}^T \cdot (\mu - \mu^{iter})$.

---

[17] Implementation detail: The KP objective function coefficients $c_{it} = \sum_{\tau=\max\{1,t-\pi_i+1\}}^{\min\{t,T-\pi_i+1\}} v_{i\tau}$ $\forall i; t$ are efficiently calculated as follows. Set $c_{i0} = 0$, $v_{i0} = 0$ and $v_{i,T-\pi_i+2} = 0$. Calculate $c_{it} = c_{i,t-1} - v_{i,\max\{0,t-\pi_i\}} + v_{i,\min\{t,T-\pi_i+2\}}$ $\forall i; t$.

[18] $UBL = \sum_{t=1}^{T} \min\{0, 1 - LB_t^{KP}\} + \sum_{i=1}^{n} \sum_{t=1}^{T-\pi_i+1} v_{it}$.

Strictly speaking $\hat{g}$ is not a subgradient in the sense of Definition 3.15. Let a heuristic construct a feasible solution x to $ZL_{(I)}$ with an objective function value UB and let LB be a lower bound to $ZL_{(I)}\left(\mu^{iter}\right)$ where constraint (I) is Lagrangean relaxed. The best known upper bound is denoted as UB^*. Using an analogous proof of Lemma 3.16, a valid pseudo-subgradient $\hat{g}$ is derived from a feasible solution to $ZL_{(I)}$ as the negative slack of constraint (I). The Lagrangean multiplier of the next iteration are obtained from a steepest ascent approach (3.29) that yields (6.39) for $ZL_{(C)}(\mu)$ and (6.40) for $ZL_{(P)}(\nu)$. Both formulas are slight modifications of the Held-Karp step size (3.34).

$$\mu_t = \max\left\{0, \mu_t + \text{STEPSIZE} \cdot \hat{g}_t \cdot \frac{UB^* - LB}{\sum_{\tau=1}^{T} \hat{g}_\tau^2}\right\} \text{ with } \hat{g}_t = \sum_{i=1}^{n} r_i \cdot x_{it} - \bar{r}_t \cdot y_t \; \forall t \tag{6.39}$$

$$\nu_{it} = \max\left\{0, \nu_{it} + \text{STEPSIZE} \cdot \hat{g}_{it} \cdot \frac{UB^* - LB}{\sum_{\tau=1}^{T} \hat{g}_{i\tau}^2}\right\} \text{ with } \hat{g}_{it} = 1 - \sum_{\tau=t}^{t+\pi_i-1} x_{i\tau} \; \forall i; t \tag{6.40}$$

Preliminary computational studies showed that a pseudo-subgradient is not as 'good' as a subgradient obtained from an optimal solution (see Sects. 7.3.1.2 and 7.3.1.3). Therefore, the *dynamic step size strategy* adjusts STEPSIZE according to the obtained Lagrangean bound and the approach from Caprara et al. (1999) is modified to yield better results for the developed Lagrangean heuristics. Compared to the constant size strategy, the convergence is slower but more controlled. Denote $\overline{LB}$ ($\underline{LB}$) as the largest lower bound (smallest lower bound) to $ZL_{(I)}(\mu)$ for a given μ from five preceding iterations. Let $\overline{iter}$ ($\underline{iter}$) be the iteration where $\overline{LB}$ ($\underline{LB}$) is obtained. The step size is updated every five iterations of the Lagrangean heuristic by[19]

$$\text{STEPSIZE} = \begin{cases} 0.5 \cdot \text{STEPSIZE} & \text{if } \overline{LB} - \underline{LB} > 0.01 \cdot \underline{LB} \wedge \overline{iter} < \underline{iter} \\ 1.5 \cdot \text{STEPSIZE} & \text{if } \overline{LB} - \underline{LB} < 0.005 \cdot \underline{LB} \wedge \overline{iter} > \underline{iter} \\ \text{STEPSIZE} & \text{otherwise.} \end{cases} \tag{6.41}$$

The first case of (6.41) is used to detect the tendency that the gap is becoming larger but the lower bound is not increasing. Hence, the step size is reduced. The second case is relevant because of the following observation. If the lower bound

---

[19] Implementation detail: (6.41) requires to find $\overline{LB}$ and $\underline{LB}$. An efficient implementation uses a *priority queue of a fixed size*. This variant is a priority queue, which is modified to hold a restricted, maximal number of elements. The array, which is used to access the elements in the heap, is cyclically traversed such that the priority queue has a fixed size. Thus, the ID of an element is already used for the data structure. Every element holds the value of the lower bound as a key and the iteration in which the lower bound was obtained.

does not improve anymore, then either the lower bound is already converging to the best possible value or the step size is so small that the influence of $\hat{g}$ is insignificant. Hence, the dynamic step size strategy increases the step size if the lower bound is increasing. Note that this case is not detected by the constant step size strategy. The step size is further modified because it can become arbitrarily small or large. If STEPSIZE < 0.001 or STEPSIZE > 1, reset STEPSIZE to the initial value STEPSIZE^{init} and record $\overline{LB}$ as well as $\underline{LB}$ anew.

6.2.3.1 Initial Lagrangean Multiplier

Lemma 6.9 defines initial Lagrangean multiplier μ^{init}, v^{init} and the *initial Lagrangean lower bound* ZL^{init} (6.42) that is a WUPMP with $f_t = 0$ $\forall t$.[20] For a comparison of the Lagrangean heuristics for ZL$_{(C)}$ and ZL$_{(P)}$ it is particularly interesting that ZL$_{(C)}\left(\mu^{init}\right) =$ ZL$_{(P)}\left(v^{init}\right)$ holds.

$$ZL^{init} = \min \sum_{t=1}^{T} \sum_{i=1}^{n} \frac{r_i}{\bar{r}_t} \cdot x_{it} \quad \text{subject to (P)} \wedge \text{(X)} \tag{6.42}$$

Lemma 6.9 The initial Lagrangean multiplier

$$\mu_t^{init} = \frac{1}{\bar{r}_t} \qquad \forall t = 1, \ldots, T$$

$$v_{it}^{init} = \min_{\tau = t, \ldots, t + \pi_i - 1} \frac{r_i}{\bar{r}_\tau} - \sum_{\theta = \max\{1, \tau - \pi_i + 1\}}^{t-1} v_{i\theta}^{init} \quad \forall i = 1, \ldots, n; t = 1, \ldots, T - \pi_i + 1$$

yield ZL$^{init} =$ ZL$_{(C)}\left(\mu^{init}\right) =$ ZL$_{(P)}\left(v^{init}\right) = \sum_{i=1}^{n} \sum_{t=1}^{T-\pi_i+1} v_{it}^{init} \leq Z^{(X)(Y)}$. The solution $(x, y) = (0, 0)$ is optimal to ZL$_{(P)}\left(v^{init}\right)$.

Proof (6.42) has $Conv(P)(X) = H(P)(S^x)$ and coincides with its LP relaxation because of Theorem 4.1 with $a_{it} = 1$ $\forall i; t$ and $b_{it} = 1$ $\forall i; t$. Two conclusions follow. First, ZL$^{init} \leq Z^{(X)(Y)}$ because $H(P)(S^x) \subseteq H(P)(V)(S)$. Second, the optimal objective function value of the corresponding dual problem of the LP relaxation of (6.42) is ZL^{init}. This fact is used to show that v^{init} solves the dual problem to the LP relaxation of (6.42) to optimality. The *i*th dual problem of the LP relaxation of

---

[20]ZL^{init} (6.42) is directly obtained from ZL$_{(C)}$ with μ^{init} (Lemma 6.9). Alternatively, consider the CPMP LP relaxation of (Y) and completely relax (V) as well as (S^y). Since the objective function coefficients of y are non-negative and y is continuous, there exists an optimal solution that satisfies (C) an equality. A substitution with $y_t = \sum_{i=1}^{n} \frac{r_i}{\bar{r}_t} \cdot x_{it}$ $\forall t$ yields (6.42).

ZL^{init} is

$$Z_i = \max \sum_{t=1}^{T-\pi_i+1} v_{it} \qquad \text{subject to}$$

$$\sum_{\tau=\max\{1,t-\pi_i+1\}}^{\min\{t,T-\pi_i+1\}} v_{i\tau} \leq \frac{r_i}{\bar{r}_t} \qquad \forall t = 1, \ldots, T \qquad (6.43)$$

$$v_{it} \geq 0 \qquad \forall t = 1, \ldots, T-\pi_i+1$$

and Lemma 4.15 shows that $(v_{i1}^{init}, \ldots, v_{i,T-\pi_i+1}^{init})$ is an optimal solution to Z_i. Hence, v^{init} solves (6.42) to optimality with $ZL^{init} = \sum_{i=1}^{n} Z_i$.

Consider $ZL_{(P)}(v^{init})$ in (5.44). Since v^{init} is optimal, the objective function coefficients of x_{it} in z_t satisfy (6.43) for all t and z_t is bounded by the SSP

$$z_t(v^{init}) \leq \max\left\{\sum_{i=1}^{n} \frac{r_i}{\bar{r}_t} \cdot x_{it} \middle| \sum_{i=1}^{n} \frac{r_i}{\bar{r}_t} \cdot x_{it} \leq 1 \wedge x_{it} \in \{0,1\} \; \forall i\right\} \leq 1. \qquad (6.44)$$

Using (5.44) and (6.44), $(x, y) = (0, 0)$ is optimal to $ZL_{(P)}$. Hence, $ZL_{(P)}(v^{init}) = \sum_{i=1}^{n} \sum_{t=1}^{T-\pi_i+1} v_{it}^{init} = \sum_{i=1}^{n} Z_i = ZL^{init}$ holds. $ZL^{init} = ZL_{(C)}(\mu^{init})$ because of (5.51).

The capacity $\bar{r}_t$ appears in the denominator of (6.42) and (6.43). Hence, $\bar{r}_t \leq r^{sum}$ $\forall t$ from Assumption (2.3) tightens ZL^{init} to some extent. In order to have defined ratios for $\frac{1}{\bar{r}_t}$ and $\frac{r_i}{\bar{r}_t}$, set $\bar{r}_t = \max\{\epsilon, \bar{r}_t\}$ where ϵ is a very small, strictly positive value such that $\bar{r}_t = \epsilon$ indicates that no maintenance activity can be scheduled in period t. The initial Lagrangean lower bound ZL^{init} yields a lower bound to the CPMP, which gets arbitrarily bad. This is demonstrated with a series of CPMP instances.

Example 6.1 Given a constant $K \in \mathbb{N}$. Let $n = 2$, $T = 2 \cdot K$, $\pi = (2, 2 \cdot K)^T$, $r = (1, 2 \cdot K)^T$ and $\bar{r}_t = 2 \cdot K$ $\forall t$. An instance satisfies Assumption 2.4. A decomposition into two SPPs with $\frac{T}{2}$ tasks for $i = 1$ and one task for $i = 2$ follows from Remark 4.2.[21] Hence, $ZL^{init} = \frac{r_1}{\bar{r}_1} \cdot \frac{T}{2} + \frac{r_2}{\bar{r}_1} = \frac{1}{2 \cdot K} \cdot K + 1 = \frac{3}{2}$. Since both maintenance activities cannot be scheduled in one period, $Z = \frac{T}{2} + 1 = K + 1$ holds. The worst case performance ratio is arbitrarily close to 0 because $\lim_{K \to \infty} \frac{ZL^{init}}{Z} = \lim_{K \to \infty} \frac{3}{2} \cdot \frac{1}{K+1} = 0$.

Remark 6.2 $ZL_{(C)}$ with μ_t^{init} becomes a WUPMP with $f_t = 0$ $\forall t$ that is solvable by Lemma 4.11 with $c_{it} = \frac{1}{\bar{r}_t}$ $\forall i; t$. Applying Algorithm 4.1 n-times calculates $ZL_{(C)}(\mu^{init})$. v^{init} is obtained from the dual problem of the LP relaxation of (6.42).

---

[21] This can be concluded from a proof similar to Lemma 4.13.

6.2 Two Lagrangean Heuristics
191

Set $c_{it} = \frac{r_i}{\bar{r}_t}$ $\forall i; t$ and use Algorithm 4.2 n-times to calculate v^{init}. $ZL_{(P)}(v^{init})$ is computed as presented in Sect. 6.2.2.

6.2.3.2 A Lagrangean Heuristic

The *Lagrangean Heuristic* from Algorithm 6.8 is based on the Lagrangean subgradient heuristic of Algorithm 3.2 and used to separately approximate both Lagrangean duals $Z_{(C)}$ and $Z_{(P)}$. However, the algorithm is presented such that the Lagrangean relaxation can be heuristically or optimally solved.

Remark 6.3 The following modifications apply if a Lagrangean relaxation is solved to optimality. For $ZL_{(C)}$ apply the aforementioned variable fixation to yield non-negative objective function coefficients and solve

$$ZL_{(C)}(\mu) = \min \sum_{t=1}^{T}(1 - \mu_t \cdot \bar{r}_t) \cdot y_t + \sum_{i=1}^{n}\sum_{t=1}^{T} r_i \cdot \mu_t \cdot x_{it} \text{ subject to } (P) \wedge (V) \wedge (S) \wedge (Y).$$

The obtained optimal solution is modified such that $y_t = 0$ and $x_{it} = 0$ $\forall i$ holds for any period t with $1 - \mu_t \cdot \bar{r}_t = 0$. This prevents redundant information in the subgradient.

In $ZL_{(P)}$, each KP from (5.44) is solved with the Combo Algorithm 3.7, which is initialized with the lower and upper bounds as well as pruned with $UB_t^{KP} \leq 1$ as presented in Sect. 6.2.2.

Initial step size parameter are STEPSIZE^{init} = 0.5 for $Z_{(C)}$ and STEPSIZE^{init} = 0.05 for $Z_{(P)}$. Set STEPSIZE = STEPSIZE^{init}. Let (x^*, y^*) be the best known feasible solution to the CPMP with an objective function value UB^*. Let LB^* be the best Lagrangean lower bound and ensure $LB^* \geq \lfloor \frac{T}{\pi_1} \rfloor$ (Lemma 4.17). Given initial Lagrangean multiplier (cf. Lemma 6.9). If the First Fit Heuristic (Algorithm 6.1) finds an initial feasible solution, then update (x^*, y^*) and UB^*. Otherwise, set $UB^* = T + 1$.

At the beginning of an iteration, solve the Lagrangean relaxation and obtain LB as well as a feasible solution (x, y). Update LB^* with LB. Calculate the pseudo-subgradient $\hat{g}$ from either (6.39) or (6.40), respectively. If (x, y) has less open periods than (x^*, y^*) and the solution is feasible to the CPMP because of $\hat{g} \leq 0$,[22] update (x^*, y^*) and UB^*. Repair (x, y) to make it feasible to the CPMP with the Iterated Best-of-three Heuristic (Algorithm 6.3) and update (x^*, y^*) as well as UB^*. If $UB^* = \lceil LB^* \rceil$, then terminate because (x^*, y^*) is optimal (Remark 2.1). Update the step size with (6.41) and obtain the next Lagrangean multiplier from either (6.39) or (6.40), respectively. Repeat until a certain time limit or a maximal number of

---

[22] Implementation detail: It is efficient to verify $\sum_{t=1}^{T} y_t < UB^*$ before $\hat{g} \leq 0$. Whereas the latter is executed in time $O(T)$ for $ZL_{(C)}$, it takes the time $O(n \cdot T)$ for $ZL_{(P)}$.

iterations is reached, then terminate. The algorithm returns LB^* and (x^*, y^*) if a feasible solution was found.

In the first iteration, optimal solutions to both initial Lagrangean relaxations are obtained even if the Lagrangean relaxation is heuristically solved (Lemma 6.9). In the first iteration, $ZL_{(C)}(\mu^{init})$ is computed as presented in Remark 6.2. Afterwards, the Combined Lower Bound (Sect. 6.2.1.3) is applied. For $ZL_{(P)}(\upsilon^{init})$, the solution is empty $(x, y) = (0, 0)$ as a consequence of Lemma 6.9 and of the pruning criterion $UB_t \leq 1$ in the Knapsack Heuristic (Sect. 6.2.2). However, the update from (6.40) yields $\upsilon \neq \upsilon^{init}$ because $\hat{g} \neq 0$ and $(x, y) \neq (0, 0)$ usually holds in the second iteration. After the first iteration, an *initial lower bound* $LB^{init} = \max \{ZL^{init}, Z^{(C)}\}$ is obtained, which is characterized by Lemma 6.10. This lower bound is identical for both Lagrangean relaxations and valid to $Z_{(C)}$ and $Z_{(P)}$ because of Lemma 4.17 with Geoffrion's Theorem 3.11 and Lemma 6.9.

Lemma 6.10 No dominance exists between ZL^{init} and $Z^{(C)}$.

Proof Consider ZL^{init} from Lemma 6.9 and $Z^{(C)}$ from Lemma 4.17.

'$ZL^{init} > Z^{(C)}$': Assume $\pi_i = \pi_1 \, \forall i$, $r_i = r_1 \, \forall i$, $\bar{r}_t = \bar{r}_1 \, \forall t$ and $\bar{r}_1 < n \cdot r_i$. A decomposition of (6.42) with Remark 4.2 yields n SPPs with $\lfloor \frac{T}{\pi_1} \rfloor$ tasks for maintenance activity i.[23] It follows

$$ZL^{init} = \frac{n \cdot r_1}{\bar{r}_1} \cdot \left\lfloor \frac{T}{\pi_1} \right\rfloor > \left\lfloor \frac{T}{\pi_1} \right\rfloor = Z^{(C)}.$$

'$ZL^{init} < Z^{(C)}$': Assume $\pi_i = \pi_1 \, \forall i = 1, \ldots, n-1$, $\pi_n = 2 \cdot \pi_1$, $r_i = r_1 \, \forall i$, $\bar{r}_t = \bar{r}_1 \, \forall t$ and $\bar{r}_1 = n \cdot r_i$. The decomposition of (6.42) yields n SPPs with $\lfloor \frac{T}{\pi_1} \rfloor$ tasks for maintenance activity $i = 1, \ldots, n-1$ and $\lfloor \frac{T}{2 \cdot \pi_1} \rfloor$ tasks for $i = n$.[24] It follows

$$ZL^{init} = \frac{r_1}{\bar{r}_1} \cdot \left((n-1) \cdot \left\lfloor \frac{T}{\pi_1} \right\rfloor + \left\lfloor \frac{T}{2 \cdot \pi_1} \right\rfloor \right) < \frac{n \cdot r_1}{\bar{r}_1} \cdot \left\lfloor \frac{T}{\pi_1} \right\rfloor = \left\lfloor \frac{T}{\pi_1} \right\rfloor = Z^{(C)}.$$

The Example 6.2 as well as Theorem 5.22 show that both Lagrangean relaxations do not have the integrality property (see Corollary 3.12). Thus, the subgradient algorithm might yield a lower bound superior to the one provided by the CPMP LP relaxation $Z^{(X)(Y)}$.

Example 6.2 This instance shows $Z_{(C)}, Z_{(P)} > Z^{(X)(Y)}$. Let $n = 2$, $T = 3$, $\pi = (2, 3)^T$, $r = (4, 5)^T$ and $\bar{r} = (9, 5, 9)^T$. The optimal integer solution has $Z = 2$ with $x_{12} = x_{23} = 1$ and $y_2 = y_3 = 1$. The LP relaxation has $Z^{(X)(Y)} = 1\frac{4}{9}$ with $x = \begin{pmatrix} \frac{4}{9} & \frac{5}{9} & \frac{4}{9} \\ \frac{4}{9} & \frac{1}{9} & \frac{4}{9} \end{pmatrix}$ with $y = \left(\frac{4}{9}, \frac{5}{9}, \frac{4}{9}\right)^T$. $Z_{(C)}$ and $Z_{(P)}$ are solved via column

---

[23] See Footnote 21 on page 190.
[24] See Footnote 21 on page 190.

6.2 Two Lagrangean Heuristics

6.8 Algorithm. The Lagrangean Heuristic to the CPMP for $Z_{(C)}$ and $Z_{(P)}$

Input: An instance to the CPMP, an initial feasible solution to the CPMP if one was found, initial Lagrangean multiplier and an initial lower bound LB^*

Output: The best known lower bound LB^* and the best known feasible solution (x^*, y^*) with UB^* if one was found

Initialization: $(x, y) \leftarrow (0, 0)$, $(x^*, y^*) \leftarrow (0, 0)$, $LB \leftarrow 0$, $UB^* \leftarrow T + 1$, $\hat{g} \leftarrow 0$

1   **if** *An initial feasible solution to the CPMP exists* **then** Update (x^*, y^*) and UB^*
2   **repeat**
3   Solve the Lagrangean relaxation and obtain a Lagrangean lower bound LB as well as a feasible solution (x, y)
4       $LB^* \leftarrow \max\{LB, LB^*\}$
5   Calculate the pseudo-subgradient $\hat{g}$ (6.39), (6.40)
6   **if** $\sum_{t=1}^{T} y_t < UB^* \wedge \hat{g} \leq 0$ **then**
7   Update the best known feasible solution (x^*, y^*) and UB^*
8   Repair (x, y) with the Iterated Best-of-three Heuristic (Algorithm 6.3)
9   Update the best known feasible solution (x^*, y^*) and UB^*
10   **if** $UB^* = \lceil LB^* \rceil$ **then** Terminate because (x^*, y^*) is optimal
11   Update the step size (6.41) and the Lagrangean multiplier (6.39), (6.40)
12 **until** *A time limit is reached*

---

generation as presented in Sect. 5.3.4. This yields $Z_{(C)} = 1\frac{4}{5}$ and two maintenance plans for (5.82)–(5.85), which are

$$\lambda_1 = 0.2 \wedge x^1 = \begin{pmatrix} 0 & 1 & 0 \\ 0 & 1 & 0 \end{pmatrix} \wedge y^1 = \begin{pmatrix} 0 \\ 1 \\ 0 \end{pmatrix} \text{ and } \lambda_2 = 0.8 \wedge x^2 = \begin{pmatrix} 0 & 1 & 1 \\ 0 & 0 & 1 \end{pmatrix} \wedge y^2 = \begin{pmatrix} 0 \\ 1 \\ 1 \end{pmatrix}.$$

$Z_{(P)} = 1\frac{1}{2}$ and one plan for each period are obtained for (5.77)–(5.80). These are

$$\lambda_{11} = 0.5 \wedge x^1 = \begin{pmatrix} 1 \\ 1 \end{pmatrix}, \lambda_{21} = 0.5 \wedge x^2 = \begin{pmatrix} 1 \\ 0 \end{pmatrix} \text{ and } \lambda_{31} = 0.5 \wedge x^3 = \begin{pmatrix} 1 \\ 1 \end{pmatrix}.$$

In what follows, the proposed solution techniques are applied in Algorithm 6.8 to solve one instance. Figure 6.1 shows the evolution of the upper and lower bounds over the iterations. Due to Lemma 6.9, all algorithms provide the same initial lower bound LB^{init} but this value is not shown in Fig. 6.1. All algorithms find the optimal solution of the instance, although only Ke proves optimality because $\lceil LB_{(P)} \rceil = UB^*$ holds. In the example, Ke yields a better lower bound than Ue. Furthermore, the lower bounds and consequently, the upper bounds to the Lagrangean relaxations show the zig-zagging evolution of the lower bound that is well-known from subgradient algorithms. This zig-zagging is more distinct for $ZL_{(P)}$ than for $ZL_{(C)}$ because $LB_{(C)}$ is already close to $Z_{(C)}$ after the first iteration. Uh with the LP Lower Bound provides an approximation of $Z_{(C)}$ that is comparable to Ue. $LB_{(C)}$ from Uh with the Dual Priority Rule Lower Bound worsens after the first iterations. Interestingly, $UB_{(C)}$ follows this tendency and the gap between both bounds increase over the iterations. Note that dual information is used to derive $UB_{(C)}$. The alternation between two lower bounds in the Combined Lower Bound is

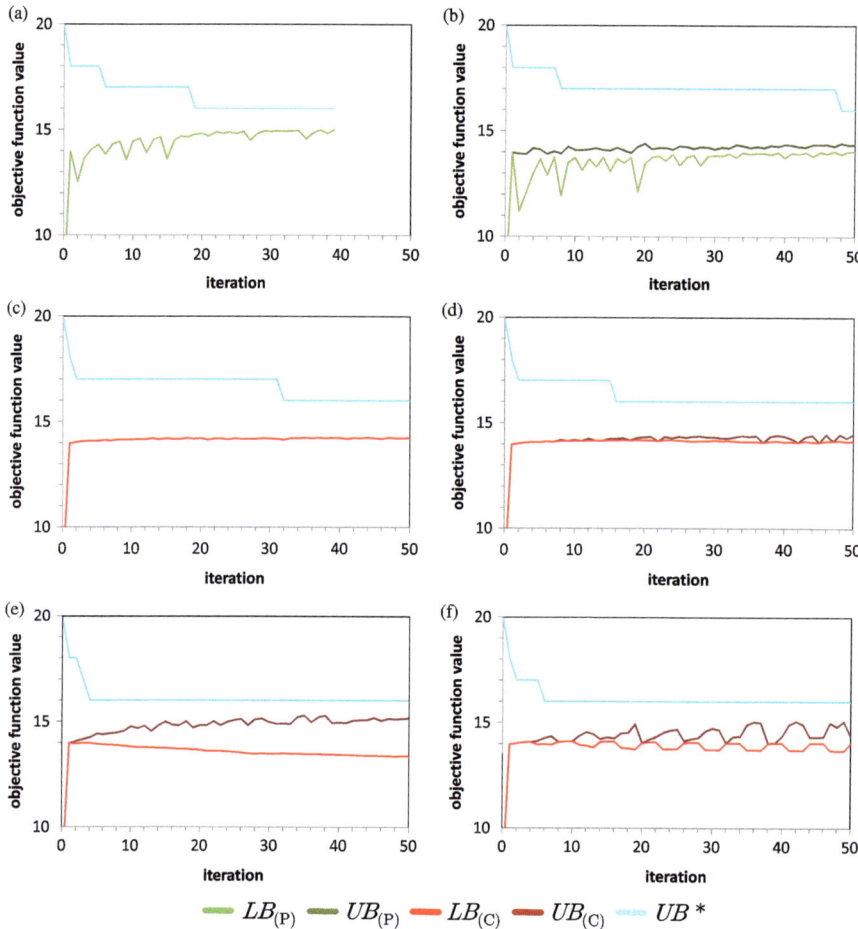

Fig. 6.1 Algorithm 6.8 approximates $Z_{(P)}$ in (**a**)–(**b**) and $Z_{(C)}$ in (**c**)–(**f**) of one instance. The upper and lower bound of $ZL_{(P)}$ and $ZL_{(C)}$ are denoted as $LB_{(P)}$, $UB_{(P)}$ and $LB_{(C)}$, $UB_{(C)}$, respectively. UB^* is the upper bound of the best feasible solution to the CPMP. (**a**) In Ke, $ZL_{(P)}$ is solved to optimality. (**b**) In Kh, $ZL_{(P)}$ is solved heuristically. (**c**) In Ue, $ZL_{(C)}$ is solved to optimality. (**d**)–(**f**) In Uh, $ZL_{(C)}$ is heuristically solved with different lower bounds, namely the LP Lower Bound in (**d**), the Dual Priority Rule Lower Bound in (**e**) and the Combined Lower Bound in (**f**)

seen in Fig. 6.1f where the LP Lower Bound provides better bounds than the Dual Priority Rule Lower Bound. Compared to subgradient optimization in Ke and Ue, pseudo-subgradient optimization applied in Kh and in Uh with the LP Lower Bound or the Combined Lower Bound approximates the Lagrangean dual well.

6.3 A More Sophisticated Lagrangean Hybrid Heuristic

Section 6.3.1 presents a Lagrangean hybrid approach that exploits the duality between Benders' decomposition and Lagrangean relaxation. Two Lagrangean hybrid heuristics are developed in Sect. 6.3.2 and the required auxiliary LPs are provided in Sect. 6.3.3.

6.3.1 A General Approach to Link Two Lagrangean Relaxations

Given an IP Z with n variables $x \in \mathbb{N}^n$ and two Lagrangean relaxations $ZL_{(I)}$ and $ZL_{(II)}$. Let $c \in \mathbb{R}^n$, $A \in \mathbb{R}^{m \times n}$ and $b \in \mathbb{R}^m$ in constraint (I), $B \in \mathbb{R}^{k \times n}$ and $d \in \mathbb{R}^k$ in constraint (II). The variable domain (III) is integer. The Lagrangean multiplier are $\mu \in \mathbb{R}^m_+$ and $\upsilon \in \mathbb{R}^k_+$.

$$Z = \min\left\{ c^T \cdot x \,\middle|\, A \cdot x \geq b \text{ (I)} \wedge B \cdot x \geq d \text{ (II)} \wedge x \in \mathbb{N}^n \text{ (III)} \right\}$$

$$ZL_{(I)}(\mu) = \min\left\{ c^T \cdot x + (b - A \cdot x)^T \cdot \mu \,\middle|\, B \cdot x \geq d \text{ (II)} \wedge x \in \mathbb{N}^n \text{ (III)} \right\}$$

$$ZL_{(II)}(\upsilon) = \min\left\{ c^T \cdot x + (d - B \cdot x)^T \cdot \upsilon \,\middle|\, A \cdot x \geq b \text{ (I)} \wedge x \in \mathbb{N}^n \text{ (III)} \right\}$$

Let the respective Lagrangean duals be approximated by a Lagrangean heuristic. Assume the Lagrangean heuristic with $ZL_{(I)}$ can be terminated because the lower bound converges, does not improve or a repair heuristic does not find improving feasible solutions. As shown in Fig. 6.2, an optimal solution of the *auxiliary LP* $DL_{(I)}$ provides variables υ that are used to yield novel initial Lagrangean multiplier for $ZL_{(II)}$. Improving the estimation, already obtained information is used in $DL_{(I)}$. Benders' cuts comprise Lagrangean multiplier of $ZL_{(I)}$ and cutting planes contain feasible solutions to $ZL_{(II)}$ and to Z. Afterwards, $ZL_{(II)}$ is solved by a Lagrangean heuristic.

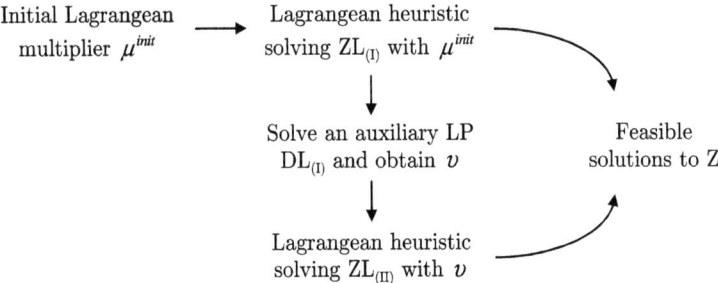

Fig. 6.2 Linking two Lagrangean relaxations

The Lagrangean multiplier for $ZL_{(II)}$ are estimated as the opportunity costs υ of the resource d in a situation where opportunity costs μ of the resource b are given. $DL_{(I)}$ is obtained in two steps. First, consider $ZL_{(II)}$. Let $\bar{P}$ ($\bar{R}$) be the index set of all extreme points x^p (extreme rays x^r) to $ZL_{(II)}$. Because of (3.12), the convexified Lagrangean dual becomes

$$Z_{(II)} = \max_{\upsilon \geq 0} \min_{p \in \bar{P}} \left\{ c^T \cdot x^p + (d - B \cdot x^p)^T \cdot \upsilon \,\middle|\, (c - B^T \cdot \upsilon)^T \cdot x^r \geq 0 \,\forall r \in \bar{R} \right\}. \tag{6.45}$$

Let $F_{(II)} \subseteq \bar{P}$ be the index set of *some* solutions $x^p \in Conv(I)(III)$ to $ZL_{(II)}$ that satisfy $Z_{(II)} \leq c^T \cdot x^p$. In practice, $Z_{(II)}$ is not known in advance. Hence, $F_{(II)}$ can contain $x^p \in Conv(I)(II)(III)$ and $x^p \in Conv(I)(III)$ where $UB \leq c^T \cdot x^p$ holds with UB as the best known upper bound to Z. Let $\eta_{(II)}$ be the best known lower bound to $Z_{(II)}$. The cutting planes (6.46) are valid to $Z_{(II)}$ and derived from the dual cutting planes (3.14) of the dual master problem of (6.45).

$$\left\{ \upsilon \geq 0 \,\middle|\, Z_{(II)} \leq c^T \cdot x^p + (d - B \cdot x^p)^T \cdot \upsilon \,\forall p \in \bar{P} \right\}$$
$$\subseteq \left\{ \upsilon \geq 0 \,\middle|\, \eta_{(II)} \leq c^T \cdot x^p + (d - B \cdot x^p)^T \cdot \upsilon \,\forall p \in F_{(II)} \right\} \tag{6.46}$$

Second, assume that $Conv(II)(III)$ is a polytope. Formulate (I) in Z as a soft constraint and introduce a slack variable $y \in \mathbb{R}_+^m$ that takes the penalized violation of (I). Let $f \in \mathbb{R}^m$ be a vector of big numbers. Using Geoffrion's Theorem 3.11, $Z_{(I)}$ is

$$Z_{(I)} = \min \left\{ c^T \cdot x \,\middle|\, x \in H(I) \cap Conv(II)(III) \right\} \tag{6.47}$$
$$= \min \left\{ c^T \cdot x + f^T \cdot y \,\middle|\, A \cdot x + y \geq b \wedge x \in Conv(II)(III) \wedge y \geq 0 \right\}. \tag{6.48}$$

Keeping Theorem 3.19 in mind, the Benders' reformulation yields

Lemma 6.11

$$Z_{(I)} = \min_{x \in Conv(II)(III)} \max_{p \in P} \left\{ c^T \cdot x + (b - A \cdot x)^T \cdot \mu^p \,\middle|\, (b - A \cdot x)^T \cdot \mu^r \leq 0 \,\forall r \in R \right\} \tag{6.49}$$

with P (R) as the index set of all extreme points μ^p (extreme rays μ^r) to $\{\mu \geq 0 | \mu \leq f\}$.

Proof Let O (S) be the index set of all extreme points x^o (extreme rays x^s) of $Conv(II)(III)$. Use Minkowski's Theorem 3.2 to represent $Conv(II)(III)$, linearize

6.3 A Lagrangean hybrid heuristic

and rearrange (6.49) to yield

min η subject to

$$\sum_{o \in O} \lambda^o = 1 \qquad \tilde{\eta}$$

$$\eta + \left(A^T \cdot \mu^p - c\right)^T \cdot \left(\sum_{o \in O} x^o \cdot \lambda^o + \sum_{s \in S} x^s \cdot \lambda^s\right) \geq b^T \cdot \mu^p \qquad \forall p \in P \qquad \tilde{\lambda}^p$$

$$\left(A^T \cdot \mu^r\right)^T \cdot \left(\sum_{o \in O} x^o \cdot \lambda^o + \sum_{s \in S} x^s \cdot \lambda^s\right) \geq b^T \cdot \mu^r \qquad \forall r \in R \qquad \tilde{\lambda}^r$$

$$\lambda^o \geq 0 \ \forall o \in O \wedge \lambda^s \geq 0 \ \forall s \in S \wedge \eta \in \mathbb{R}.$$

This LP is dualized and rearranged.

$$\max \tilde{\eta} + b^T \cdot \left(\sum_{p \in P} \mu^p \cdot \tilde{\lambda}^p + \sum_{r \in R} \mu^r \cdot \tilde{\lambda}^r\right) \qquad \text{subject to}$$

$$\sum_{p \in P} \tilde{\lambda}^p = 1 \qquad\qquad \eta \qquad (6.50)$$

$$\tilde{\eta} + (A \cdot x^o)^T \cdot \left(\sum_{p \in P} \mu^p \cdot \tilde{\lambda}^p + \sum_{r \in R} \mu^r \cdot \tilde{\lambda}^r\right) \leq c^T \cdot x^o \cdot \sum_{p \in P} \tilde{\lambda}^p \qquad \forall o \in O \qquad \lambda^o \qquad (6.51)$$

$$(A \cdot x^s)^T \cdot \left(\sum_{p \in P} \mu^p \cdot \tilde{\lambda}^p + \sum_{r \in R} \mu^r \cdot \tilde{\lambda}^r\right) \leq c^T \cdot x^s \cdot \sum_{p \in P} \tilde{\lambda}^p \qquad \forall s \in S \qquad \lambda^s \qquad (6.52)$$

$$\tilde{\eta} \in \mathbb{R} \wedge \tilde{\lambda}^p \geq 0 \ \forall p \in P \wedge \tilde{\lambda}^r \geq 0 \ \forall r \in R$$

Simplify (6.51) and (6.52) with (6.50), substitute with

$$\tilde{\eta} = \eta - b^T \cdot \left(\sum_{p \in P} \mu^p \cdot \tilde{\lambda}^p + \sum_{r \in R} \mu^r \cdot \tilde{\lambda}^r\right)$$

and reformulate $\{\mu \geq 0 | \mu \leq f\}$ with Minkowski's Theorem 3.2 to obtain

max η	subject to	(6.53)
$\eta \leq c^T \cdot x^o + (b - A \cdot x^o)^T \cdot \mu$	$\forall o \in O$	(6.54)
$(A \cdot x^s)^T \cdot \mu \leq c^T \cdot x^s$	$\forall s \in S$	(6.55)
$\eta \in \mathbb{R} \wedge 0 \leq \mu \leq f.$		(6.56)

Using Minkowski's Theorem 3.2 for a given μ, a convexification approach states

$$ZL_{(I)}(\mu) \leq \begin{cases} -\infty & \text{if } \exists s \in S : (c - A^T \cdot \mu)^T \cdot x^s < 0 \\ c^T \cdot x^o + (b - A \cdot x^o)^T \cdot \mu \ \forall o \in O & \text{otherwise} \end{cases}$$

from which follows that (6.54) and (6.55) are reformulated as

$$\eta \leq \min_{o \in O} \left\{ c^T \cdot x^o + (b - A \cdot x^o)^T \cdot \mu \,\middle|\, (c - A^T \cdot \mu)^T \cdot x^s \geq 0 \ \forall s \in S \right\}$$
$$= \min \left\{ c^T \cdot x + (b - A \cdot x)^T \cdot \mu \,\middle|\, x \in Conv(II)(III) \right\}$$
$$= ZL_{(I)}(\mu).$$

Since f are big numbers, $\mu \leq f$ is redundant. (6.53)–(6.56) becomes $\max_{\mu \geq 0} ZL_{(I)}(\mu) = Z_{(I)}$, which is the definition (3.11) of the Lagrangean dual.

An intuitive approach to Lemma 6.11 is as follows. From (3.36) for a given $x \in Conv(II)(III)$, the Benders' subproblem of (6.48) is

$$D(x) = c^T \cdot x + \max\left\{ (b - A \cdot x)^T \cdot \mu \,\middle|\, \mu \leq f \wedge \mu \geq 0 \right\}$$

with the trivial optimal solution $\mu_i = f_i$ if $b_i - A_i \cdot x > 0$ (otherwise, 0) for all $i = 1, \ldots, m$. An infeasible solution $x \notin H(I) \cap Conv(II)(III)$ yields an upper bound $D(x)$ to $Z_{(I)}$ because f contains big numbers. The minimization $\min_{x \in X} D(x)$ intuitively yields $x \in H(I) \cap Conv(II)(III)$, which coincides with $Z_{(I)}$ from (6.47).

As $Conv(II)(III)$ is bounded, assume $R = \emptyset$. Let $P_{(I)}$ with $P_{(I)} \subseteq P$ and $P_{(I)} \neq \emptyset$ be the index set of *some* Lagrangean multiplier μ^p to $ZL_{(I)}$ that are obtained from an algorithm that solves $Z_{(I)}$. Linearize (6.49) and either fully describe or approximate $Conv(II)(III)$ by the LP relaxation with l valid inequalities (IV) where $C \in \mathbb{R}^{l \times n}$ and $e \in \mathbb{R}^l$ (Nemhauser and Wolsey 1988, p. 106).

$B_{(I)} = \min \eta$ subject to

$$\eta \geq c^T \cdot x + (b - A \cdot x)^T \cdot \mu^p \qquad \forall p \in P_{(I)} \qquad \delta_p \qquad (6.57)$$
$$B \cdot x \geq d \qquad\qquad\qquad\qquad\qquad\qquad\qquad\quad\ \upsilon \qquad (II)$$
$$C \cdot x \geq e \qquad\qquad\qquad\qquad\qquad\qquad\qquad\quad\ u \qquad (IV)$$
$$\eta \in \mathbb{R} \wedge x \geq 0$$

6.3 A Lagrangean hybrid heuristic

Dualize this LP and add (6.61) from (6.46) to obtain the auxiliary LP

$$DL_{(I)} = \max b^T \cdot \sum_{p \in P_{(I)}} \mu^p \cdot \delta_p + d^T \cdot \upsilon + e \cdot u \quad \text{subject to} \tag{6.58}$$

$$\sum_{p \in P_{(I)}} \delta_p = 1 \quad\quad \eta \tag{6.59}$$

$$A^T \cdot \sum_{p \in P_{(I)}} \mu^p \cdot \delta_p + B^T \cdot \upsilon + C^T \cdot u \leq c \quad\quad x \tag{6.60}$$

$$(B \cdot x^p - d)^T \cdot \upsilon \leq c^T \cdot x^p - \eta_{(II)} \quad\quad \forall p \in F_{(II)} \tag{6.61}$$

$$\delta_p \geq 0 \ \forall p \in P_{(I)} \land \upsilon, u \geq 0. \tag{6.62}$$

This auxiliary problem $DL_{(I)}$ estimates new initial Lagrangean multiplier υ for $ZL_{(II)}$. Note that dual variables of the LP relaxation provide an estimate of the Lagrangean multipliers (Klose 2001, p. 94). The information transfer between $ZL_{(I)}$ and $ZL_{(II)}$ is established from the following straightforward interpretation of $DL_{(I)}$. In (6.60), the dualization of the Benders' cuts (6.57) yields a convex combination μ of all Lagrangean multiplier μ^p. Let a variable x_j be more *advantageous* in $ZL_{(I)}$ if its objective function coefficient $c_j - (A^j)^T \cdot \mu$ is small and if the coefficient B_i^j of the ith constraint (II) is large. Consequently, the more advantageous x_j is, the less the ith constraint (II) is a bottleneck for x_j and the opportunity costs υ_i are low because υ_i is more bounded in the ith constraint (6.60). Furthermore, the resulting objective function coefficients $c_j - (B^j)^T \cdot \upsilon$ of x_j in $ZL_{(II)}$ have a lower bound that is $(A^j)^T \cdot \mu$. The cutting planes (6.61) contain feasible solutions x^p to Z and to $ZL_{(II)}$. The larger the slack $B_i \cdot x^p - d_i > 0$ is, the more the ith constraint (II) is satisfied, the less it is a bottleneck constraint and the opportunity costs υ_i are low because υ_i is more bounded in the constraint (6.61). The other way around, if the ith constraint (II) is violated for all x^p with $p \in F_{(II)}$, the opportunity costs υ_i increase because υ is maximized in (6.58) and (6.61) is not restrictive since $B_i \cdot x^p - d_i \leq 0$ and $c^T \cdot x^p - \eta_{(II)} \geq 0$ holds.

Denote $Z^{(III)+(IV)}$ ($Z^{(III)}$) as the optimal objective function value of the LP relaxation of Z with (without) (IV). Let $DL_{(I)}^{(IV)}$ be $DL_{(I)}$ without (IV) added to $B_{(I)}$.

Lemma 6.12 $DL_{(I)}$ is solvable. $DL_{(I)} \leq Z^{(III)+(IV)}$ and $DL_{(I)}^{(IV)} \leq Z^{(III)}$ holds.

Proof Since $Conv(II)(III)$ is bounded, there exists a simple upper bound constraint $x \leq s$ for $B_{(I)}$ with $C = -E_n$, $e = -s$ and $s \in \mathbb{R}_+^n$ in (IV). Farkas' Lemma 3.4 states that the solution space of $DL_{(I)}$ is not empty if and only if the corresponding dual problem to

$$\max \left\{ 0^T \cdot \sum_{p \in P_{(I)}} \delta_p + 0^T \cdot \upsilon + 0^T \cdot u \ \middle| \ (6.59)\text{--}(6.62) \right\}$$

has an optimal objective function value that is zero. The corresponding dual problem is

$$\min \eta + c^T \cdot x + \sum_{p \in F_{(\text{II})}} \left(c^T \cdot x^p - \eta_{(\text{II})}\right)^T \cdot \gamma_p \quad \text{subject to}$$

$$\eta + \left(A^T \cdot \mu^p\right)^T \cdot x \geq 0 \qquad \forall p \in P_{(\text{I})} \qquad \delta_p \quad (6.63)$$

$$B \cdot x + \sum_{p \in F_{(\text{II})}} (B \cdot x^p - d) \cdot \gamma_p \geq 0 \qquad \qquad \upsilon$$

$$-x \geq 0 \qquad \qquad u \quad (6.64)$$

$$\eta \in \mathbb{R} \wedge x \geq 0 \wedge \gamma_p \geq 0 \ \forall p \in F_{(\text{II})}.$$

$x = 0$ follows from (6.64) and from the non-negativity of x. Thus, (6.63) becomes $\eta \geq 0$ and clearly, $\eta = 0$ holds for an optimal solution. The simplified problem is

$$W = \min \left\{ \sum_{p \in F_{(\text{II})}} \left(c^T \cdot x^p - \eta_{(\text{II})}\right)^T \cdot \gamma_p \,\middle|\, \sum_{p \in F_{(\text{II})}} (B \cdot x^p - d) \cdot \gamma_p \geq 0 \wedge \gamma_p \geq 0 \ \forall p \in F_{(\text{II})} \right\}$$

and because of $\eta_{(\text{II})} \leq c^T \cdot x^p \ \forall p \in F_{(\text{II})}$ and the non-negativity of γ, it follows $W \geq 0$ and the feasible solution $\gamma_p = 0 \ \forall p \in F_{(\text{II})}$ with $W = 0$ is optimal. Thus, $DL_{(\text{I})}$ is solvable.

For the second part, assume $F_{(\text{II})} = \emptyset$ to relax (6.61) completely. The dual problem is

$$Z^{(\text{III})+(\text{IV})}$$

$$= \min \left\{ c^T \cdot x \,\middle|\, A \cdot x \geq b \ (\text{I}) \wedge B \cdot x \geq d \ (\text{II}) \wedge C \cdot x \geq e \ (\text{IV}) \wedge x \geq 0 \right\}$$

$$= \max \left\{ b^T \cdot \mu + d^T \cdot \upsilon + e^T \cdot u \,\middle|\, A^T \cdot \mu + B^T \cdot \upsilon + C^T \cdot u \leq c \wedge \mu, \upsilon, u \geq 0 \right\}.$$

Observe that the dual problem and $DL_{(\text{I})}$ only differ in the definition of the variables μ.

$$\left\{ \mu = \sum_{p \in P_{(\text{I})}} \mu^p \cdot \delta_p \,\middle|\, \sum_{p \in P_{(\text{I})}} \delta_p = 1 \wedge \delta_p \geq 0 \ \forall p \in P_{(\text{I})} \right\} \subseteq \{\mu \geq 0\}$$

holds because $P_{(\text{I})} \subseteq P$. Consequently, $DL_{(\text{I})} \leq Z^{(\text{III})+(\text{IV})}$ that also holds if $F_{(\text{II})} \neq \emptyset$. $DL_{(\text{I})}^{(\text{IV})} \leq Z^{(\text{III})}$ follows from the same argument.

6.3 A Lagrangean hybrid heuristic

Corollary 6.13 $DL_{(I)} \leq Z_{(I)}$ and $DL_{(I)}^{(IV)} \leq Z_{(I)}, Z_{(II)}$ holds.

Proof Let (N) denote the non-negativity constraint $x \geq 0$. Since (IV) is a valid inequality to $Conv$(II)(III), Geoffrion's Theorem 3.11 states $H(I) \cap Conv$(II)(III) $=$ $H(I) \cap Conv$(II)(III)(IV) $\subseteq H(I)$(II)(IV)(N) and $Z^{(III)+(IV)} \leq Z_{(I)}$. Furthermore, $Z^{(III)} \leq Z_{(I)}, Z_{(II)}$ holds by the same theorem. Since Lemma 6.12 states $DL_{(I)} \leq Z^{(III)+(IV)}$ and $DL_{(I)}^{(IV)} \leq Z^{(III)}$, the result follows.

Remark 6.4 The continuous corresponding dual solution x to $DL_{(I)}$ can be used to determine valid inequalities (IV) by solving respective separation problems.

6.3.2 The Lagrangean Hybrid Heuristic

The Lagrangean hybrid approach is applied to the CPMP and to the Lagrangean duals $Z_{(C)}$ and $Z_{(P)}$. The *Lagrangean Hybrid Heuristic* alternates between two Lagrangean duals and approximates each with a Lagrangean heuristic. Upper and lower bounds to the CPMP are accessible in every iteration. Two alternative hybrid heuristics are obtained depending on the Lagrangean heuristic with which the Lagrangean Hybrid Heuristic starts. In what follows, the variant that commences with $Z_{(C)}$ is presented and outlined in Fig. 6.3. An iteration consists of four steps:

- $Z_{(C)}$ is approximated with a Lagrangean heuristic that commences with the initial Lagrangean multiplier μ^{init}.
- Solve $DL_{(C)}$ and obtain the dual variables υ that update υ^{init}.
- $Z_{(P)}$ is approximated with a Lagrangean heuristic that commences with the initial Lagrangean multiplier υ^{init}.
- Solve $DL_{(P)}$ and obtain the dual variables μ that update μ^{init}.

In order to yield a better approximation of the convex hull $Conv$(C)(X)(Y) in $B_{(P)}$, the valid inequalities of the Knapsack polytope (O), (L) and the variable upper bound (V) are used. The computational results from Sect. 7.2 show that

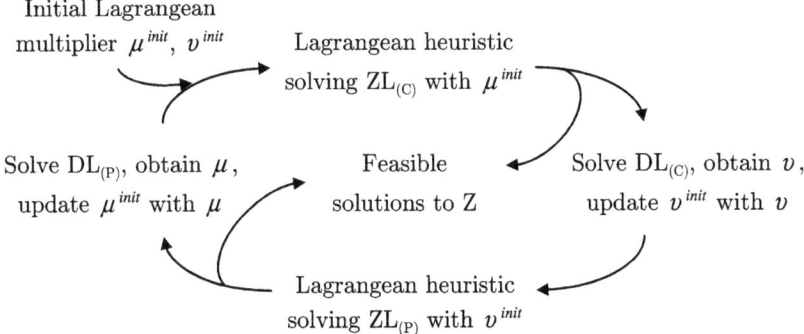

Fig. 6.3 The Lagrangean Hybrid Heuristic

6.9 Algorithm. Initial extended covers for the CPMP
  Input: An instance to the CPMP
  Output: A set $\mathcal{C}$ of extended covers
  Initialization: $\mathcal{C} \leftarrow \emptyset, t \leftarrow 0, a \leftarrow 0, b \leftarrow 0, c \leftarrow 0$
1 **for** $t = 1$ **to** T **do**
2   Sort all maintenance activities with respect to $r_{i_1} \geq \ldots \geq r_{i_n}$
3   Find the position a with $r_{i_{a-1}} > \bar{r}_t \wedge r_{i_a} \leq \bar{r}_t$
4   Find the position b with $\sum_{h=a}^{b} r_{i_h} \leq \bar{r}_t \wedge \sum_{h=a}^{b+1} r_{i_h} > \bar{r}_t$
5   **if** $a \neq b$ **then**
6         $c \leftarrow b + 1$
7   **while** $c \leq n \wedge \sum_{h=a}^{b} r_{i_h} + r_{i_c} > \bar{r}_t$ **do**
8   Add the extended cover $\mathcal{C} \leftarrow \mathcal{C} \cup \{i_1, \ldots, i_b, i_c\}$
9             $c \leftarrow c + 1$

the relative error of $Z^{(X)(Y)}$ to $Z_{(C)}$ is rather small. Hence, the approximation of $Conv(P)(V)(X)(Y)$ by $H(P)(V)(S)$ is considered as sufficient and no valid inequalities are added to $B_{(C)}$. The mathematical formulations of the auxiliary LPs $DL_{(C)}$ and $DL_{(P)}$ are presented in Sect. 6.3.3.

6.3.2.1 Initial Extended Cover Inequalities

Algorithm 6.9 finds initial extended covers for (L) from the instance data. Let $\mathcal{C}$ be the set of sets that contains the initial extended covers. Set $t = 1$. Sort all maintenance activities $r_{i_1} \geq \ldots \geq r_{i_n}$ according to non-increasing values. Find the position a with $r_{i_{a-1}} > \bar{r}_t$ and $r_{i_a} \leq \bar{r}_t$. Find the position b with $\sum_{h=a}^{b} r_{i_h} \leq \bar{r}_t$ and $\sum_{h=a}^{b+1} r_{i_h} > \bar{r}_t$.[25] If $a = b$, set $t = t + 1$ because the obtained cover inequality $x_{i_a,t} + x_{i_c,t} \leq y_t$ is a 2nd-degree clique inequality that has been added to $B_{(P)}$ as (L). Otherwise, the set $I = \{i_a, \ldots, i_b, i_c\}$ with $c = b + 1, \ldots, n$ is a cover if $\sum_{h=a}^{b} r_{i_h} + r_{i_c} > \bar{r}_t$ holds. Lift the cover with its extension to obtain $\{i_1, \ldots, i_{a-1}\} \cup I$ and insert it to $\mathcal{C}$. Each set $C \in \mathcal{C}$ yields an extended cover inequality (O). Choose the next period $t = t + 1$ and terminate if $t > T$.

6.3.2.2 A Hybrid Heuristic

Algorithm 6.10 presents the Lagrangean Hybrid Heuristic that starts with $Z_{(C)}$. Let (x^*, y^*) be the best known feasible solution to the CPMP with an objective function value UB^* and LB^* denotes the best known Lagrangean lower bound. Let $\eta_{(C)}$ and

---

[25]Implementation detail: To make the algorithm efficient, use introsort for sorting the items in the first step and apply a binary search for finding the items a and b. Both algorithms are outlined in Footnotes 33 and 32 on page 58.

6.3 A Lagrangean hybrid heuristic

6.10 Algorithm. The Lagrangean Hybrid Heuristic for the CPMP starting with $ZL_{(C)}$

Input: An instance to the CPMP, an initial feasible solution to the CPMP if one was found and initial Lagrangean multiplier μ^{init}, υ^{init}

Output: The best known lower bound LB^* and the best known feasible solution with UB^* if one was found

Initialization: $(x, y) \leftarrow (0, 0)$, $\mu_t^* \leftarrow 0\ \forall t$, $\bar{\mu}_t \leftarrow 0\ \forall t$, $\upsilon_{it}^* \leftarrow \upsilon_{it}^{init}\ \forall i;t$, $\bar{\upsilon}_{it} \leftarrow 0\ \forall i;t$, $P_{(C)} \leftarrow \emptyset$, $P_{(P)} \leftarrow \emptyset$, $F_{(C)} \leftarrow \emptyset$, $F_{(P)} \leftarrow \emptyset$, $LB \leftarrow 0$, $UB^* \leftarrow T+1$, $LB^* \leftarrow LB^{init}$, $\eta_{(C)} \leftarrow LB^{init}$, $\eta_{(P)} \leftarrow LB^{init}$

1 **if** *An initial feasible solution to the CPMP exists* **then** Update UB^*
2 In $DL_{(P)}$, insert extended clique inequalities (L) from Algorithm 3.5 and extended cover inequalities (O) from Algorithm 6.9
3 **repeat**
4   Solve $Z_{(C)}$ with a modified Algorithm 6.8. Update $P_{(C)}$, $F_{(C)}$, $F_{(P)}$, μ^*, $\eta_{(C)}$, $\eta_{(P)}$
5   Solve $DL_{(C)}$ (6.71)–(6.77) and obtain the dual variables $\bar{\upsilon}$. Update $\eta_{(C)}$, $\eta_{(P)}$
6      $\upsilon^{init} \leftarrow 0.25 \cdot \upsilon^* + 0.75 \cdot \bar{\upsilon}$
7   Solve $Z_{(P)}$ with a modified Algorithm 6.8. Update $P_{(P)}$, $F_{(P)}$, $F_{(C)}$, υ^*, $\eta_{(P)}$
8   Find extended cover inequalities (O) and insert them in $DL_{(P)}$
9   Solve $DL_{(P)}$ (6.79)–(6.85) and obtain the dual variables $\bar{\mu}$. Update $\eta_{(P)}$
10     $\mu^{init} \leftarrow 0.25 \cdot \mu^* + 0.75 \cdot \bar{\mu}$
11 **until** *A time limit is reached or Algorithm 6.8 found an optimal solution*
12 $LB^* \leftarrow \max\{\eta_{(C)}, \eta_{(P)}\}$

$\eta_{(P)}$ be the best known lower bounds to $Z_{(C)}$ and $Z_{(P)}$, respectively. In the Lagrangean Heuristic from Algorithm 6.8, denote (x, y) as a feasible solution to $ZL_{(C)}$ ($ZL_{(P)}$) and let μ^* (υ^*) be the best Lagrangean multiplier that yielded the best lower bound $LB_{(C)}$ ($LB_{(P)}$) during the current iteration of Algorithm 6.8. Note that every time the Lagrangean Heuristic is called $LB_{(C)}$ and $LB_{(P)}$ are determined anew.

In the initialization phase, valid inequalities are added to $DL_{(P)}$. Insert all extended clique inequalities (L) by applying Algorithm 3.5 to every period once.[26] Insert initial extended cover inequalities (O) from Algorithm 6.9.[27] Determine initial Lagrangean multiplier μ^{init} and υ^{init} (Lemma 6.9). Set $\upsilon^* = \upsilon^{init}$. Initialize $\eta_{(C)}$ and $\eta_{(P)}$ with the initial lower bound $LB^{init} = \max\{ZL^{init}, Z^{(C)}\}$ (Lemma 6.10).[28] If the First Fit Heuristic (Algorithm 6.1) finds an initial feasible solution, then update (x^*, y^*) and UB^*. Otherwise, set $UB^* = T + 1$.

In an iteration, approximate $Z_{(C)}$ with the Lagrangean Heuristic (Algorithm 6.8) that contains some modifications: An additional termination criterion is that the Lagrangean Heuristic terminates if either $LB_{(C)}$ or UB^* has not improved for 200 successive iterations or if half of the computational time has passed. Furthermore, if $LB_{(C)}$ is improved, update μ^* and insert μ in $P_{(C)}$. If $LB_{(C)}$ is improved and

---

[26]Implementation detail: Omit cliques with a cardinality of one because (V) has been already inserted.
[27]Some inequalities are redundant since (L) is tighter than (V).
[28]$ZL^{init} = \sum_{i=1}^{n}\sum_{t=1}^{T-\pi_i+1} \upsilon_{it}^{init}$ (Lemma 6.9) and $Z^{(C)} = \lfloor \frac{T}{\pi_1} \rfloor$ (Lemma 4.17).

$\sum_{t=1}^{T} y_t \geq UB^*$, insert (x, y) in $F_{(C)}$. If (x^*, y^*) is improved, then insert (x^*, y^*) in $F_{(C)}$ and $F_{(P)}$. In order to keep the auxiliary problems $DL_{(C)}$ and $DL_{(C)}$ small, no more than 200 Lagrangean multiplier and 200 feasible solutions can be inserted per iteration and such insertions are kept for 10 iterations before they are removed. The Lagrangean Heuristic returns μ^* and $LB_{(C)}$ that is used to update $\eta_{(C)}$ and $\eta_{(P)}$.

Solve $DL_{(C)}$ (6.71)–(6.77) and obtain the dual variables $\bar{\upsilon}$. Update $\eta_{(C)}$ as well as $\eta_{(P)}$ with $DL_{(C)}$ and calculate υ^{init} with the convex combination $\upsilon^{init} = 0.25 \cdot \upsilon^* + 0.75 \cdot \bar{\upsilon}$.

Approximate $Z_{(P)}$ with the modifications of the Lagrangean Heuristic (Algorithm 6.8) presented above but insert υ in $P_{(P)}$, insert the feasible solution (x, y) to $ZL_{(P)}$ in $F_{(P)}$. The Lagrangean Heuristic returns υ^* and $LB_{(P)}$ that updates $\eta_{(P)}$.

In the next step, extended cover inequalities (O) are added to $DL_{(P)}$.[29] These are obtained from the separation problem (3.56) where x is obtained from the continuous dual solutions to $DL_{(C)}$, to $DL_{(P)}$ (Remark 6.4) and from feasible solutions of the Continuous KPs, which are calculated in the Knapsack Heuristic. The separation problem (3.56) is heuristically solved with the methods from Sect. 3.6.2 because it is a binary $\mathcal{NP}$-hard KP.[30] Each of these generated covers is lifted by its extension to yield (O). The total number of such inserted inequalities (O) is limited to 1000.[31]

Solve $DL_{(P)}$ (6.79)–(6.85), obtain the dual variables $\bar{\mu}$, update $\eta_{(P)}$ with $DL_{(P)}$ and calculate μ^{init} with the convex combination $\mu^{init} = 0.25 \cdot \mu^* + 0.75 \cdot \bar{\mu}$.

The algorithm terminates if a time limit is reached or if one of the two Lagrangean heuristics has found the optimal solution. It returns $LB^* = \max \{\eta_{(C)}, \eta_{(P)}\}$ and (x^*, y^*) if a feasible solution was found.

The lower bound $\eta_{(C)}$ ($\eta_{(P)}$) is updated with $DL_{(C)}$ and $LB_{(C)}$ ($DL_{(P)}$, $DL_{(C)}$, $LB_{(P)}$ and $LB_{(C)}$). This follows from Corollary 6.13 and from $LB_{(C)} \leq Z_{(P)}$ because the presented heuristics to $ZL_{(C)}$ from the Sects. 6.2.1.1–6.2.1.5 solve $ZL_{(C)}^{(X)(Y)}$.[32] Neither $DL_{(P)}$ nor $LB_{(P)}$ are valid lower bounds to $Z_{(C)}$ because of the valid inequalities (O) and (L). Example 6.3 and Theorem 5.22 show that no dominance

---

[29] Implementation detail: Covers with a cardinality of two are not inserted because (L), which is tighter, has been added already.

[30] Implementation detail: In one iteration of the Knapsack Heuristic (Algorithm 6.7), T KPs are solved. When the tth KP is considered, obtain the continuous solution to the Continuous KP from the critical item and solve the separation problem (3.56) with the methods from Sect. 3.6.2. This generates at most T extended cover inequalities (O). Covers with one or two elements are omitted because they are already added since they are special cases of (V) or of 2nd-degree clique inequalities (L).

As a consequence of using a heuristic to solve the separation problem (3.56), not all existing cover inequalities are found and a cover is not necessarily minimal.

[31] Implementation detail: Use a triplet with checksums and a hash table as explained in Sect. 6.4 to prevent a multiple insertion of the same cutting plane or cover inequality.

[32] This follows because $LB_{(C)} \leq Z^{(X)(Y)}$ and $Z^{(X)(Y)} \leq Z_{(P)}$ hold by Geoffrion's Theorem 3.11.

6.3 A Lagrangean hybrid heuristic

between $Z_{(C)}$ and $Z_{(P)}$ exists.[33] This is advantageous since the Lagrangean Hybrid Heuristic accesses four different lower bounds.

Example 6.3 The Example 6.2 demonstrates $Z_{(C)} > Z_{(P)}$. The following instance has $Z_{(C)} < Z_{(P)}$. Let $n = 2$, $T = 3$, $\pi = (2, 2)^T$, $r = (4, 5)^T$ and $\bar{r} = (9, 5, 9)^T$. The optimal integer solution has $Z = 2$ with $x_{11} = x_{13} = x_{21} = x_{23} = 1$ and $y_1 = y_3 = 1$. The LP relaxation has $Z^{(X)(Y)} = 1\frac{8}{9}$ with $x = \begin{pmatrix} \frac{4}{9} & \frac{5}{9} & \frac{4}{9} \\ \frac{4}{9} & \frac{5}{9} & \frac{4}{9} \end{pmatrix}$ and $y = \left(\frac{4}{9}, 1, \frac{4}{9}\right)^T$. Note that $y_2 = 1$ because (C) provides a tight lower bound on this variable. $Z_{(C)} = 1\frac{8}{9}$ has

$$\lambda_1 = \tfrac{5}{9} \wedge x^1 = \begin{pmatrix} 0 & 1 & 0 \\ 0 & 1 & 0 \end{pmatrix} \wedge y^1 = \begin{pmatrix} 0 \\ 1 \\ 0 \end{pmatrix} \text{ and } \lambda_2 = \tfrac{4}{9} \wedge x^2 = \begin{pmatrix} 1 & 0 & 1 \\ 1 & 0 & 1 \end{pmatrix} \wedge y^2 = \begin{pmatrix} 1 \\ 1 \\ 1 \end{pmatrix}.$$

Observe that a variable fixation yields $y_2 = 1$ because of a negative objective function coefficient. $Z_{(P)} = 2$ has

$$\lambda_{11} = 1 \wedge x^1 = \begin{pmatrix} 1 \\ 1 \end{pmatrix} \text{ and } \lambda_{31} = 1 \wedge x^3 = \begin{pmatrix} 1 \\ 1 \end{pmatrix}.$$

The search process of the Lagrangean Hybrid Heuristic within both Lagrangean solution spaces is illustrated by the following two examples. In the first example, assume the following primal IP on the left and the corresponding dual problem of its LP relaxation on the right.

$Z = \min 12 \cdot x_1 + 15 \cdot x_2$		$\max 7.5 \cdot \mu + 7.5 \cdot \upsilon - 7 \cdot \alpha_1 - 8 \cdot \alpha_2$
subject to		subject to
$x_1 + 2 \cdot x_2 \geq 7.5$	(I)	$\mu \geq 0$
$2 \cdot x_1 + x_2 \geq 7.5$	(II)	$\upsilon \geq 0$
$x_1 \leq 7$	(IV^{x_1})	$\alpha_1 \geq 0$
$x_2 \leq 8$	(IV^{x_2})	$\alpha_2 \geq 0$
$x_1 \in \mathbb{N}$	(III^{x_1})	$\mu + 2 \cdot \upsilon - \alpha_1 \leq 12$
$x_2 \in \mathbb{N}$	(III^{x_2})	$2 \cdot \mu + \upsilon - \alpha_2 \leq 15$

Figure 6.4 shows the dual solution space on the left and the primal solution space on the right. The objective functions have an attached arrow that illustrates the direction of optimization. Lines drawn through stand for the polytopes of the dual and primal solution space. Integer solutions are represented by dots in the right figure. Arabic numbers represent the algorithmic steps. Assume that the Lagrangean multipliers are modified as depicted on the left of Fig. 6.4 each time the Lagrangean Heuristic

---

[33]This result corresponds with the analogous relaxations of the SSCFLP (Klose and Drexl 2005; Klose 2001, pp. 218–222).

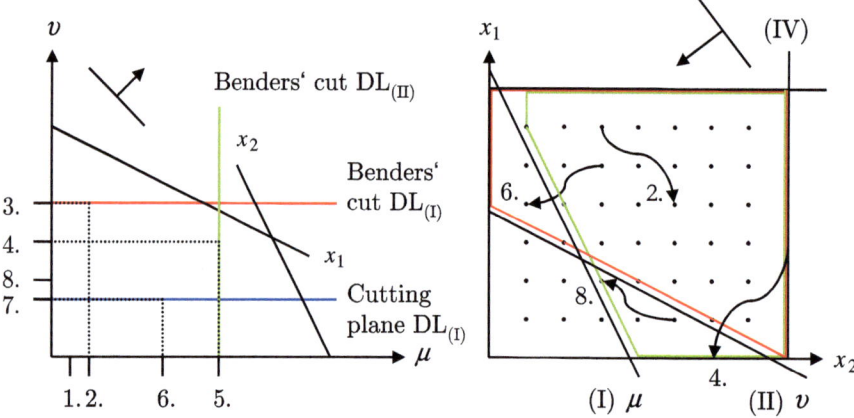

Fig. 6.4 An example to the Lagrangean Hybrid Heuristic[34]

is applied. On the right of Fig. 6.4, the search process of the Lagrangean heuristics in the primal solution space is sketched. Assume the following algorithmic steps:

1. Initialization with Lagrangean multipliers.
2. Calculate $ZL_{(I)}(\mu^{init})$ and obtain a feasible solution x (red polytope).
3. Solve $DL_{(I)}$ and let the dualized Benders' cut from step 2 (red halfspace) be restrictive. Obtain υ and update υ^{init}.
4. Calculate $ZL_{(II)}(\upsilon^{init})$ and obtain a feasible solution x (green polytope).
5. Solve $DL_{(II)}$ and let the dualized Benders' cut from step 4 (green halfspace) be restrictive. Obtain μ and update μ^{init}.
6. Calculate $ZL_{(I)}(\mu^{init})$ and obtain a feasible solution x (red polytope).
7. Solve $DL_{(I)}$ and let the cutting plane from step 2 (blue halfspace) be restrictive. Obtain υ and update υ^{init}.
8. Calculate $ZL_{(II)}(\upsilon^{init})$, obtain a feasible solution x (green polytope) and terminate.

In the second example, Algorithm 6.10 solves the same instance that is discussed in Fig. 6.1. Figure 6.5 shows the evolution of the upper and lower bounds over the iterations. The instance is solved to optimality because $\lceil LB^* \rceil = UB^*$ holds. LB^* is increased as the algorithm alternates between both Lagrangean heuristics. Furthermore, $LB_{(C)}$ and $LB_{(P)}$ slightly improve in every iteration. The lower bounds $DL_{(C)}$ and $DL_{(P)}$ increase but do not improve LB^*.

---

[34]Erratum: In the left of Fig. 6.4, the convexification of the variables is not correct. The Benders' cut $DL_{(I)}$ should be 2. as a vertical line, the Benders' cut $DL_{(II)}$ should be 4. as a horizontal line and the squared area between 2., 6. and the cutting plane is the third convexification.

6.3 A Lagrangean hybrid heuristic

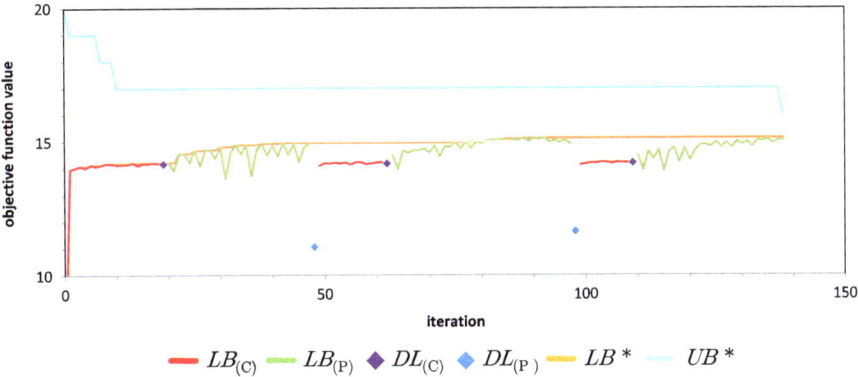

Fig. 6.5 Algorithm 6.10 solves one instance. $ZL_{(C)}$ and $ZL_{(P)}$ are solved to optimality. The lower bounds of $ZL_{(C)}$ and $ZL_{(P)}$ are denoted as $LB_{(C)}$ and $LB_{(P)}$, respectively. $DL_{(C)}$ and $DL_{(P)}$ are the lower bounds of the auxiliary LPs $DL_{(C)}$ and $DL_{(P)}$. LB^* and UB^* denote the best lower bound and the best upper bound to the CPMP

6.3.3 Mathematical Formulations of the Auxiliary LPs

Let the constraints (C) correspond to (I), the constraints (P), (V) to (II) and (Y), (X) to (III). (S) is considered as a valid inequality (IV). Following the hybrid approach, the two Lagrangean relaxations are $ZL_{(C)}$ and $ZL_{(P)(V)}$. The Lagrangean relaxation $ZL_{(P)}$ is used because $Z_{(P)} = Z_{(P)(V)}$ holds by Eq. (5.21).

6.3.3.1 The Auxiliary LP $DL_{(C)}$

Consider the formulation (5.51) for $ZL_{(C)}$ and apply Lemma (6.11) but omit the extreme rays because the polyhedron is bounded. The Benders' reformulation of $Z_{(C)}$ becomes

$Z_{(C)} = \min \eta$

subject to $(P) \wedge (V) \wedge (X) \wedge (Y)$

$$\eta \geq \sum_{t=1}^{T} \left(1 - \mu_t^p \cdot \bar{r}_t\right) \cdot y_t + \sum_{i=1}^{n} \sum_{t=1}^{T} r_i \cdot \mu_t^p \cdot x_{it} \qquad \forall p \in P$$

$\eta \in \mathbb{R}.$

The LP relaxation of the Benders' reformulation of $Z_{(C)}$ is

$B_{(C)} = \min \eta$ subject to

$$\eta + \sum_{t=1}^{T}(\mu_t^p \cdot \bar{r}_t - 1) \cdot y_t - \sum_{i=1}^{n}\sum_{t=1}^{T} r_i \cdot \mu_t^p \cdot x_{it} \geq 0 \quad \forall p \in P_{(C)} \quad \delta_p$$

$$\sum_{\tau=t}^{t+\pi_i-1} x_{i\tau} \geq 1 \quad \begin{array}{l}\forall i = 1,\ldots,n;\\ t = 1,\ldots,T-\pi_i+1\end{array} \quad v_{it} \quad \text{(P)}$$

$$y_t - x_{it} \geq 0 \quad \forall i = 1,\ldots,n; t = 1,\ldots,T \quad u_{it} \quad \text{(V)}$$

$$-x_{it} \geq -1 \quad \forall i = 1,\ldots,n; t = 1,\ldots,T \quad \alpha_{it} \quad (\text{S}^x)$$

$$-y_t \geq -1 \quad \forall t = 1,\ldots,T \quad \beta_t \quad (\text{S}^y)$$

$$\eta \in \mathbb{R} \wedge y_t \geq 0 \; \forall t = 1,\ldots,T \wedge x_{it} \geq 0 \; \forall i = 1,\ldots,n; t = 1,\ldots,T.$$

In order to yield $\text{DL}_{(C)}$, dualize $B_{(C)}$ and yield a LP

$$\max \sum_{i=1}^{n}\sum_{t=1}^{T-\pi_i+1} v_{it} - \sum_{i=1}^{n}\sum_{t=1}^{T}\alpha_{it} - \sum_{t=1}^{T}\beta_t \quad \text{subject to}$$

$$\sum_{p \in P_{(C)}} \delta_p = 1 \quad \eta \quad (6.65)$$

$$\sum_{\tau=\max\{1,t-\pi_i+1\}}^{\min\{t,T-\pi_i+1\}} v_{i\tau} - u_{it} - \sum_{p \in P_{(C)}} r_i \cdot \mu_t^p \cdot \delta_p - \alpha_{it} \leq 0 \quad \begin{array}{l}\forall i = 1,\ldots,n;\\ t = 1,\ldots,T\end{array} \quad x_{it} \quad (6.66)$$

$$\sum_{p \in P_{(C)}} (\mu_t^p \cdot \bar{r}_t - 1) \cdot \delta_p - \beta_t + \sum_{i=1}^{n} u_{it} \leq 0 \quad \forall t = 1,\ldots,T \quad y_t \quad (6.67)$$

$v_{it} \geq 0 \; \forall i = 1,\ldots,n; t = 1,\ldots,T-\pi_i+1 \wedge u_{it} \geq 0 \; \forall i = 1,\ldots,n; t = 1,\ldots,T$

$\delta_p \geq 0 \; \forall p \in P_{(C)} \wedge \alpha_{it} \geq 0 \; \forall i = 1,\ldots,n; t = 1,\ldots,T \wedge \beta_t \geq 0 \; \forall t = 1,\ldots,T.$

Let some variables $(\alpha_{i,t_1},\ldots,\alpha_{i,t_2})$ be sufficiently positive such that some $(v_{i,\tau_1},\ldots,v_{i,\tau_2})$ are raised through (6.66). Since the objective function value either remains the same or decreases, set $\alpha_{it} = 0 \; \forall i; t$. Simplify (6.67) with (6.65). The dual cut (6.75) is obtained from (5.41).[35]

---

[35] The Lagrangean dual $Z_{(P)} = \max_{v \geq 0} ZL_{(P)}(v)$ is obtained with (5.41) as given in (6.68)–(6.70) below. The extreme rays are omitted because the polyhedron of $ZL_{(P)}$ is bounded. (6.75) is obtained from (6.69).

6.3 A Lagrangean hybrid heuristic

$$DL_{(C)} = \max \sum_{i=1}^{n} \sum_{t=1}^{T-\pi_i+1} v_{it} - \sum_{t=1}^{T} \beta_t \qquad \text{subject to} \qquad (6.71)$$

$$\sum_{p \in P_{(C)}} \delta_p = 1 \qquad \eta \qquad (6.72)$$

$$\sum_{\tau=\max\{1,t-\pi_i+1\}}^{\min\{t,T-\pi_i+1\}} v_{i\tau} - u_{it} - \sum_{p \in P_{(C)}} r_i \cdot \mu_t^p \cdot \delta_p \leq 0 \qquad \begin{array}{l} \forall i = 1,\ldots,n; \\ t = 1,\ldots,T \end{array} \quad x_{it} \qquad (6.73)$$

$$\sum_{p \in P_{(C)}} \mu_t^p \cdot \bar{r}_t \cdot \delta_p - \beta_t + \sum_{i=1}^{n} u_{it} \leq 1 \qquad \forall t = 1,\ldots,T \qquad y_t \qquad (6.74)$$

$$\sum_{i=1}^{n} \sum_{t=1}^{T-\pi_i+1} \left(\sum_{\tau=t}^{t+\pi_i-1} x_{i\tau}^p - 1 \right) \cdot v_{it} \leq \sum_{t=1}^{T} y_t^p - \eta_{(P)} \qquad \forall p \in F_{(P)} \qquad (6.75)$$

$$v_{it} \geq 0 \; \forall i = 1,\ldots,n; t = 1,\ldots,T-\pi_i+1 \wedge$$
$$u_{it} \geq 0 \; \forall i = 1,\ldots,n; t = 1,\ldots,T \qquad (6.76)$$
$$\delta_p \geq 0 \; \forall p \in P_{(C)} \wedge \beta_t \geq 0 \; \forall t = 1,\ldots,T \qquad (6.77)$$

6.3.3.2 The Auxiliary LP $DL_{(P)}$

For the sake of simplicity, combine all three valid inequalities (O), (L), (V) to $ZL_{(P)}$ in

$$\sum_{i=1}^{n} c_{itc} \cdot x_{it} \leq C_{tc} \cdot y_t \qquad \forall t = 1,\ldots,T; c \in \bar{C}_t \qquad (6.78)$$

that is defined as follows. In order to formulate (V) for a period t, let $C_t''' = \{1,\ldots,n\}$ and $I_{tc}''' = \{c\} \; \forall c \in C_t'''$. Using the definitions for (O) and (L), define

$$\bar{C}_t = C_t' \cup C_t'' \cup C_t''' \qquad \forall t = 1,\ldots,T$$

$$Z_{(P)} = \max \eta \qquad \text{subject to} \qquad (6.68)$$

$$\eta \leq \sum_{t=1}^{T} y_t^p - \sum_{i=1}^{n} \sum_{t=1}^{T-\pi_i+1} \left(\sum_{\tau=t}^{t+\pi_i-1} x_{i\tau}^p - 1 \right) \cdot v_{it} \qquad \forall p \in P \qquad (6.69)$$

$$\eta \geq 0 \wedge v_{it} \geq 0 \; \forall i = 1,\ldots,n; t = 1,\ldots,T-\pi_i+1 \qquad (6.70)$$

$$c_{itc} = \begin{cases} 1 & \text{if } c \in C'_t \wedge i \in E'_{tc} \\ 1 & \text{if } c \in C''_t \wedge i \in E''_{tc} \\ 1 & \text{if } c \in C'''_t \wedge i \in I'''_{tc} \\ 0 & \text{otherwise} \end{cases} \quad \forall i = 1, \ldots, n; t = 1, \ldots, T; c \in \bar{C}_t$$

$$C_{tc} = \begin{cases} |I'_{tc}| - 1 & \text{if } c \in C'_t \\ 1 & \text{otherwise} \end{cases} \quad \forall t = 1, \ldots, T; c \in \bar{C}_t.$$

Consider $ZL_{(P)}$ as formulated in (5.42) and apply Lemma (6.11) but omit the extreme rays in the Benders' reformulation of $Z_{(P)}$ because the polyhedron is bounded.

$$Z_{(P)} = \min \eta$$

subject to $(6.78) \wedge (C) \wedge (X) \wedge (Y)$

$$\eta \geq \sum_{t=1}^{T} y_t - \sum_{i=1}^{n} \sum_{t=1}^{T} x_{it} \cdot \sum_{\tau=\max\{t-\pi_i+1,1\}}^{\min\{t,T-\pi_i+1\}} v_{i\tau}^p + \sum_{i=1}^{n} \sum_{t=1}^{T-\pi_i+1} v_{it}^p \quad \forall p \in P$$

$$\eta \in \mathbb{R}$$

The LP relaxation of the Benders' reformulation of $Z_{(P)}$ is

$$B_{(P)} = \min \eta \quad \text{subject to}$$

$$\eta - \sum_{t=1}^{T} y_t + \sum_{i=1}^{n} \sum_{t=1}^{T} x_{it} \cdot \sum_{\tau=\max\{t-\pi_i+1,1\}}^{\min\{t,T-\pi_i+1\}} v_{i\tau}^p \geq \sum_{i=1}^{n} \sum_{t=1}^{T-\pi_i+1} v_{it}^p$$

$$\forall p \in P_{(P)} \qquad \delta_p$$

$$\bar{r}_t \cdot y_t - \sum_{i=1}^{n} r_i \cdot x_{it} \geq 0 \qquad \forall t = 1, \ldots, T \qquad \mu_t \qquad (C)$$

$$C_{tc} \cdot y_t - \sum_{i=1}^{n} c_{itc} \cdot x_{it} \geq 0 \qquad \forall t = 1, \ldots, T; c \in \bar{C}_t \qquad u_{tc} \qquad (6.78)$$

$$-x_{it} \geq -1 \qquad \forall i = 1, \ldots, n; t = 1, \ldots, T \qquad \alpha_{it} \qquad (S^x)$$

$$-y_t \geq -1 \qquad \forall t = 1, \ldots, T \qquad \beta_t \qquad (S^y)$$

$$\eta \in \mathbb{R} \wedge y_t \geq 0 \; \forall t = 1, \ldots, T \wedge x_{it} \geq 0 \; \forall i = 1, \ldots, n; t = 1, \ldots, T.$$

A dualization of $B_{(P)}$ yields the auxiliary LP $DL_{(P)}$. (6.80) is used to yield the simplified constraint (6.82). The dual cut (6.83) is obtained from (5.81).

$$DL_{(P)} = \max \sum_{p \in P_{(P)}} \sum_{i=1}^{n} \sum_{t=1}^{T-\pi_i+1} v_{it}^p \cdot \delta_p - \sum_{i=1}^{n} \sum_{t=1}^{T} \alpha_{it} - \sum_{t=1}^{T} \beta_t \quad \text{subject to} \qquad (6.79)$$

$$\sum_{p \in P_{(P)}} \delta_p = 1 \qquad \eta \qquad (6.80)$$

$$\sum_{p \in P_{(P)}} \delta_p \cdot \sum_{\tau = \max\{t-\pi_i+1,1\}}^{\min\{t,T-\pi_i+1\}} v_{i\tau}^p - r_i \cdot \mu_t - \alpha_{it} - \sum_{c \in \bar{C}_t} c_{itc} \cdot u_{tc} \leq 0$$

$$\forall i = 1, \ldots, n; t = 1, \ldots, T \qquad x_{it} \qquad (6.81)$$

$$\bar{r}_t \cdot \mu_t - \beta_t + \sum_{c \in \bar{C}_t} C_{tc} \cdot u_{tc} \leq 1 \qquad \forall t = 1, \ldots, T \qquad y_t \qquad (6.82)$$

$$\sum_{t=1}^{T} \left(\bar{r}_t \cdot y_t^p - \sum_{i=1}^{n} r_i \cdot x_{it}^p \right) \cdot \mu_t \leq \sum_{t=1}^{T} y_t^p - \eta_{(C)} \qquad \forall p \in F_{(C)} \qquad (6.83)$$

$$\mu_t \geq 0 \ \forall t = 1, \ldots, T \land u_{tc} \geq 0 \ \forall t = 1, \ldots, T; c \in \bar{C}_t \qquad (6.84)$$

$$\delta_p \geq 0 \ \forall p \in P_{(P)} \land \alpha_{it} \geq 0 \ \forall i = 1, \ldots, n; t = 1, \ldots, T \land$$
$$\beta_t \geq 0 \ \forall t = 1, \ldots, T \qquad (6.85)$$

6.4 A Tabu Search Heuristic

Section 6.4.1 presents a tabu search heuristic to the CPMP that features a multi-state search. Section 6.4.2 shows how the tabu search objective function value can be efficiently calculated with bitshift operators.

6.4.1 The Tabu Search Heuristic

The CPMP objective function (Z) provides few information about the structure of a given solution because it only comprises the number of maintenance slots. The *Tabu Search Heuristic* has a tabu search objective function (6.86) with two auxiliary terms that characterize a solution better than (Z) solely does.

$$\min A \cdot \sum_{t=1}^{T} \max \left\{ 0, \sum_{i=1}^{n} r_i \cdot x_{it} - \bar{r}_t \cdot y_t \right\} + B \cdot \sum_{t=1}^{T} y_t - C \cdot \sum_{t=1}^{T} \max \left\{ 0, y_t - \sum_{i=1}^{n} \frac{r_i}{\bar{r}_t} \cdot x_{it} \right\}^2$$
$$(6.86)$$

Let $A, B, C \geq 0$ be goal weights. The first goal penalizes infeasibility of (C). The more the capacity is exceeded in a period, the higher the penalization. The second goal reflects (Z) and minimizes the number of maintenance slots. Subtracting the third goal and squaring the relative slack of (C), each period is given a bonus that becomes larger the emptier a period is. This goal guides the search towards a reduction of maintenance slots because planning out a maintenance activity from a relatively empty period creates a higher bonus than removing it from a relatively full period. The three goals are lexicographically optimized and the goal weights change in the ground state and in the final state.[36]

Given a solution (x, y) to the CPMP that may violate (C). The elementary operation *move* plans out a maintenance activity i from a period t^{out} and schedules it in t^{in}. The following procedure is applied if $t^{in} < t^{out}$: Remove all redundant tasks of i on the left of t^{in}. Set $t = t^{in}$. Let t' and t'' refer to the left and right neighbor of t. While t and the right neighbor do not overlap because $t + \pi_i < t''$ holds, (P) is violated. Hence, schedule an additional task in the period $\tau \in \{t' + \pi_i + 1, \ldots, t + \pi_i\} \setminus t^{out}$ that gives the smallest increase of (6.86). Ties are broken by the largest period. Because of $\tau \geq t' + \pi_i + 1$, the task in t does not become redundant. Remove redundant tasks on the right of τ. Set $t = \tau$ and repair all remaining violations of (P). If $t^{in} > t^{out}$, the aforementioned procedure works the other way around. The solution satisfies (P) and the worst case running time of a move is asymptotically upper bounded by $O(T)$.

Lemma 6.14 If the current solution (x, y) is feasible and $A \gg B, C$ holds that is the algorithm is not in the final state, then all moves that maintain feasibility dominate moves that violate (C).

Proof A move that leads to a violation of (C) increases the tabu objective function value (6.86) by at least A. If the current solution is already feasible, the change of (6.86) caused by a move that maintains feasibility is strictly smaller.

The neighborhood is constructed from a *scope* $\{t = \underline{s}, \ldots, \overline{s} \mid y_t = 1\}$ with a given period interval $[\underline{s}, \overline{s}]$. Consider all periods t^{out} of the scope, all tasks i scheduled in t^{out} and all periods $t^{in} \in \{t' + 1, \ldots, t'' - 1\} \setminus t^{out}$ in between the left and the right neighbor of t^{out}. Apply the move to the period t^{in} for a maintenance activity i if (x, y) is feasible and scheduling the maintenance activity i in t^{in} satisfies (C) or if the algorithm is in the final state. The periods that are not evaluated are pruned in the sense of Lemma 6.14. However, the preservation of (P) within a move can also cause a violation of (C), which is not taken into account in this pruning criterion. Having examined all periods, apply the move to all previously pruned periods where a maintenance activity i would violate (C) if (x, y) is feasible, neither a feasibility maintaining nor a non-tabu move has been found and the algorithm is not in the final state. The entire neighborhood is explored to find the best, non-

---

[36]Implementation detail: Preventing recalculations in (6.86), the slack of (C) and the tabu objective function value are updated per period whenever a maintenance activity is scheduled or planned out.

6.4 A Tabu Search Heuristic

tabu move that yields the solution with the lowest increase of (6.86) (steepest ascent mildest decent).[37]

The algorithm expands the search over multiple states that change the composition of the scope and the coefficients A, B, C in (6.86). In order to yield the highest intensification in the ground state, the scope has the smallest width $W = \text{WIDTH}$, the number of periods two scopes overlap is $L = 10$ and the violation of (C) is given the highest priority, followed by the number of open periods (Z) and then by the bonus (i.e., $C = 1$, $B = 1 + C \cdot \sum_{t=1}^{T} \max \left\{ 0, 1 - \frac{r^{min}}{\bar{r}_t} \right\}^2$ and $A = 1 + T \cdot B$). This setting only changes in the final state. In every state, the scope is shifted through the planning horizon from one side to the other in a given search direction. Figure 6.6 illustrates this approach. The first scope of a state is located either at the beginning or the end of the planning horizon. An iteration explores the neighborhood of the current scope for a certain number of iterations ITER and finds its best solution. When the best move closes open periods, the search is extended in this scope for additional ITER iterations. Subsequently, the next scope of the same width in the search direction is selected such that both scopes overlap for L periods. If the entire planning horizon has been explored, the state changes and the scope is shifted to the opposite direction. Specifically, if the best known feasible solution was improved or the first solution of the ground state became feasible, the next iteration starts from the ground state again with $W = \text{WIDTH}$ and $L = 10$. Otherwise, the algorithm is promoted to the next state with $W = 2 \cdot W$ and $L = 2 \cdot L$. This bidirectional search continues until the scope $\{t = 1, \ldots, T \mid y_t = 1\}$ was explored. The following final state strongly diversifies. The number of open periods (Z) is given the highest priority, followed by the bonus and by the violation of (C) (i.e., $A = 1$, $C = 1 + A \cdot \sum_{t=1}^{T} (r^{sum} - \bar{r}_t)$ and $B = 1 + C \cdot \sum_{t=1}^{T} \max \left\{ 0, 1 - \frac{r^{min}}{\bar{r}_t} \right\}^2$). Whereas the algorithm tried to find a feasible solution in the previous states, it now tends to close as many periods as possible. The neighborhood of the scope, which still covers the entire planning horizon, is explored for ITER^{final} iterations. Afterwards, the algorithm changes to the ground state where the first solution very likely violates (C). Hence, the algorithm commences again from the ground state if this first solution becomes feasible.

If the algorithm is not in the final state and the tabu objective function value is strictly smaller than A, the solution (x, y) is feasible. This verification in time $O(1)$ allows an efficient implementation of the dominance rule of Lemma 6.14.

The first tabu list is a short-term memory. Suppose that the best move schedules a maintenance activity i in period t^{in}. All inverse moves that try to reschedule this

---

[37] Implementation detail: It is efficient to store how the move changes the current solution instead of an entire solution. Such a *move object* contains a list that holds the periods in which tasks of a maintenance activity i are scheduled, a list where the tasks of i are planned out and a stack with the tabu search objective function values per period of (6.86). After every move this information is used to recover the solution from the beginning of the iteration. The scope is implemented as a list and is updated with the move object.

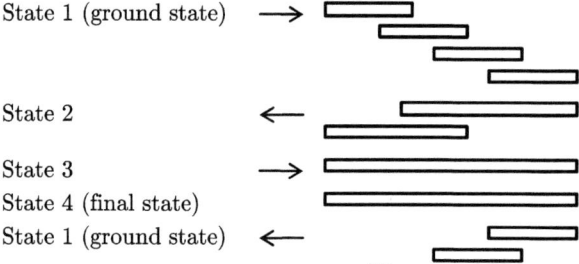

Fig. 6.6 Shifting the scope through the planning horizon in various states. An *arrow* indicates the search direction and a scope is represented by a *box*. In this example, the best known feasible solution could not have been improved at all

task to any other period are tabu for 5000 iterations.[38] However, such a move is never tabu if it improves the tabu objective function value of the best known feasible solution (aspiration level criterion). This tabu list is emptied whenever the number of open periods, the scope or the state changes. It is noteworthy to mention that as long as *no* feasible solution is found after a clearance of the tabu list, it is not active.

The second tabu list is a long-term memory and it should prevent cycling because it contains information about *all* visited solutions. Adapting an idea from Woodruff and Zemel (1993) and Chambers and Barnes (1998), a solution is encoded with checksums. The main solution characteristics are taken into account such that a fast evaluation is done with a *triplet* that consists of

- the number of open periods,
- an *Adler32* checksum (Deutsch and Gailly 1996) and[39]
- a *CRC32* checksum (Peterson and Brown 1961).[40]

---

[38] Implementation detail: Working on this tabu list requires $O(1)$. Let *iter* be an iteration of the tabu search algorithm and let $(a_{it})_{i=1,\ldots,n \wedge t=1,\ldots,T}$ be a matrix. A move that schedules the task of the maintenance activity i in period t^{in} is set tabu by $a_{i,t^{in}} = iter$. A move that tries to plan out this task from t^{out} is tabu if $iter < a_{i,t^{out}} + \text{SIZE}$ with SIZE as the length of the tabu list.

[39] Two 16-bit sums are calculated from a given message with the modulo of 65521. The Adler32 checksum is obtained from a concentration of both sums into a 32-bit integer by multiplying one sum with 2^{16} and adding the second sum. A short discussion of its errors is provided in Stone and Stewart (2002). The source code is available at Adler (1996) and it is programmed in C.

[40] The *Cyclic Redundancy Check (CRC)* check value is often referred to as CRC checksum although it is not a mathematical sum. The calculation is as follows: 32-bits are appended to a given message and this expression is divided by a given 33-bit polynomial. The remainder of the division is the CRC32 checksum. Tanenbaum (1981, pp. 128–132) provides a discussion of the errors of this checksum. This algorithm uses the well-known CRC-32 polynomial $x^{32} + x^{26} + x^{23} + x^{22} + x^{16} + x^{12} + x^{11} + x^{10} + x^8 + x^7 + x^5 + x^4 + x^2 + x^1 + 1$ (ANSI/IEEE Std 1985) that provides a sufficient error detection for this application. An exhaustive comparison of polynomials with a lower probability of undetected errors is given in Koopman (2002). The source code is available at Brown (1986) in C. The implementation uses look-up tables and the 'backwards' version.

6.4 A Tabu Search Heuristic

A comparison revealed that the computation of the Adler32 checksum is much faster than the CRC32 checksum but it has a higher probability of undetected errors (Deutsch and Gailly 1996; Thompson et al. 2004; Sheinwald et al. 2002). A similar behavior is reported for the 8-bit versions in Maxino (2006, p. 21) and Maxino and Koopman (2009). Thus, the algorithm uses an Adler32 and a CRC32 checksum to increase the precision. The solution is unambiguously encoded such that the checksums of two identical solutions are identical: Given a maintenance activity. List all periods with a task in an increasing order of indices and insert a separating character after the last period. Append the list of the next maintenance activity to this list.[41] A hash function 'total number of open periods + total slack of (C)' addresses a list of the presented triplets in a hash table (separate chaining strategy (Ottmann and Widmayer 2012, pp. 191–193)). If the triplet of the current solution coincides with one triplet of the list, the solution is tabu. Note that the tabu search objective function (6.86) provides neither an appropriate hash function nor a characteristic of a solution because the coefficients A, B, C are not constants.[42]

Algorithm 6.11 shows the Tabu Search Heuristic in full length. The algorithm is initialized with a feasible solution from the First Fit Heuristic (Algorithm 6.1). If no such solution was found, apply a modified First Fit Heuristic that satisfies (P) but may violate (C). Calculate (6.86), initialize the scope and start from the ground state. Find the best move among all non-tabu moves from the neighborhood of the scope. A solution is identified as feasible, if and only if the algorithm is not in the final state and its tabu objective function value is strictly smaller than A. If all moves are tabu, then proceed with the state change or scope change as explained below. Otherwise, obtain the solution from the best move, update the best known feasible solution and set the solution tabu in the long-term memory. If the number of open periods changed in the solution, clear the short-term memory. Otherwise, set the inverse move of the best move tabu in the short-term memory. If the algorithm is not in the final state and periods were closed in the solution, increase the number of iterations the algorithm explores this scope by ITER. The following four criteria induce a change of the current scope or state:

- The maximal number of iterations the algorithm evaluates the scope is met.
- The tabu objective function value (6.86) of a feasible (an infeasible) solution did not strictly improve after 150 (75) successive iterations. This allows an additional escape from a scope whose neighborhoods contain only 'poor' solutions. Note

---

[41] Implementation detail: Since both checksums require that the message consists of entries from 0 to 255, a representation of t with a precalculated tuple ($\lfloor \frac{t}{256} \rfloor, t \bmod 256$) is necessary if $t > 255$.

[42] Implementation detail: If the current solution has the same number of open periods as a triplet of the list, the solution is encoded and the Adler32 checksums are compared. Upon equivalence an additional comparison of the CRC32 checksum follows. Being computationally expensive, the Adler32 and the CRC32 checksums are calculated in that order, only if their values are necessary for a comparison and once calculated the values are stored in the aforementioned move object for further purposes (e.g., further scanning of the list and setting the solution tabu).

6.11 Algorithm. The Tabu Search Heuristic for the CPMP

Input: An instance as well as an initial solution to the CPMP that satisfies the period covering constraint (P) but may violate the capacity constraint (C)
Output: A lower bound $\lfloor \frac{T}{\pi_1} \rfloor$ and the best known feasible solution if one was found

1. Calculate the tabu objective function value and set the initial solution tabu
2. **while** *Termination criterion is not met* **do**
3.   Find the best, non-tabu move from the neighborhood of the scope
4.   **if** *A move was found that is not tabu* **then**
5.   Update the best known feasible solution and set the solution tabu
6.   **if** *The number of open periods changed* **then**
7.   Empty the short-term memory tabu list
8.   **else** Set the inverse move tabu in the short-term memory tabu list
9.   **if** *The algorithm is not in the final state and periods were closed* **then**
10.   Increase the number of iterations the scope is explored by ITER
11.   **if** *A criterion to change the scope or state is met* **then**
12.   Empty the short-term memory tabu list
13.   **if** *The algorithm is not in the final state* **then**
14.   **if** *The first solution of the ground state became feasible* **then**
15.   Reset scope, change search direction, choose first scope, <u>ground</u> state
16.   **else if** *The scope was not shifted through the planning horizon yet* **then**
17.   Choose the next scope
18.   **else if** *The best known feasible solution was improved in this state* **then**
19.   Reset scope, change search direction, choose first scope, <u>ground</u> state
20.   **else if** *The scope does not coincide with the planning horizon* **then**
21.   Double the scope width, change search direction, choose first scope
22.   **else**
23.   Adjust the goals, recalculate the tabu objective function value and set the <u>final</u> state
24.   **else**
25.   Adjust the goals, recalculate the tabu objective function value, reset the scope, change search direction, choose first scope and set the <u>ground</u> state

that a feasible solution is allowed a worsening of (6.86) by a small difference of 200.
- All moves of the neighborhood are tabu.
- The algorithm is in the final state and the number of open periods coincides with $\lfloor \frac{T}{\pi_1} \rfloor$. Note that it is not possible to have fewer open periods in a feasible solution than the UPMP (Lemma 4.17).

If a criterion is met that changes the current scope or state, clear the short-term memory. Finally, the next iteration commences with the exploration of a new scope. Note that the next higher state can only be attained if the entire planning horizon has been explored. The algorithm terminates if a time limit is reached or the optimal solution is found because the number of open periods of the best known feasible

6.4 A Tabu Search Heuristic

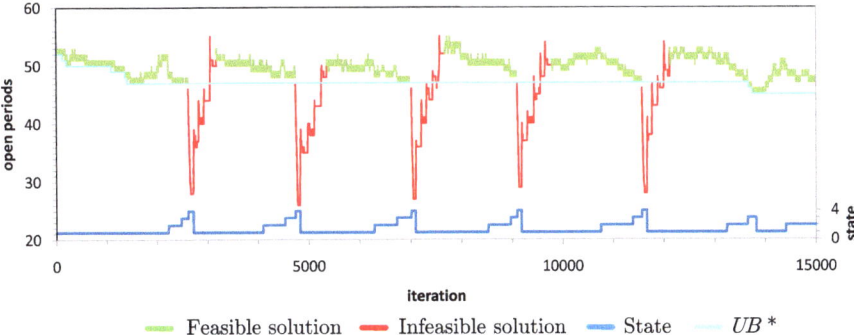

Fig. 6.7 Algorithm 6.11 solves one instance. The number of open periods is illustrated by a *green line* if the solution is feasible and in *red* if it is infeasible. The *blue line* illustrates the current state of the algorithm. UB^* is the best upper bound to the CPMP

solution coincides with $\lfloor \frac{T}{\pi_1} \rfloor$. It returns the UPMP lower bound as well as a feasible solution if one was found.

An implementation of all calculations of (6.86) with bit-shift operators decreases the computational time significantly. This is presented in Sect. 6.4.2. In order to add more flexibility, the parameters are random numbers within an interval: WIDTH $\in$ [14, 20], ITER $\in$ [100, 200] and ITER^{final} $\in$ [300, 500]. A random number is drawn *each time a parameter is used*. The pseudo-random generator Mersenne Twister (Matsumoto and Nishimura 1998) delivers respective integers with an uniform distribution.[43]

In Fig. 6.7, Algorithm 6.11 solves one instance. The evolution of objective function value in terms of open periods and of the best upper bound is shown over the iterations. Around iteration 2000, the objective function value increases and the algorithm changes from the ground state 1 to state 2. In the final state 4, infeasibility with respect to the capacity constraint is not penalized. Therefore, the algorithm focuses on closing periods and the number of open periods decreases. Afterwards, the algorithm changes to the ground state 1 and new periods are opened in order to establish feasibility. Since the algorithm shifts a scope through the planning horizon, one observes a stepwise increase of the number of open periods until the solution is feasible. Around iteration 14,000, the best feasible solution is improved in state 3. Thus, the algorithm commences from the ground state 1.

---

[43] Implementation detail: The C++ source code is available at Fog (2008).

Table 6.1 Bitshift implementation of $a \cdot c$

x	$a \cdot c$ Time s	$a \ll \log_2 c$ Time s
8	1.70	0.44
16	1.43	0.44
24	1.57	0.73
32	1.43	0.73
40	1.44	1.13
48	1.43	1.14
56	1.58	1.31

The computational time is given in seconds

6.4.2 Bitshifting in the Calculation of the Tabu Search Objective Function

The calculation of the tabu objective function value of (6.86) for a particular period is frequently used in the elementary operation move. Thus, an efficient implementation has a significant impact on the algorithm performance. Let the first two terms in (6.86) correspond to $a \cdot c$ and the last one to $\left(\frac{a}{b}\right)^2 \cdot c$. It holds that $a, b \in \mathbb{N}$ because of Remark 2.2 and $0 \leq \frac{a}{b} \leq 1$. The scaling factor c with $c \in \mathbb{N}$ and $c \gg a, b$ represents a goal weight. Two small computational studies were performed using the computational setting from Sect. 7.3. Using the following three approaches, the computation of the tabu objective function (6.86) is significantly sped up.

First, the data type of (6.86) is defined as an integer instead of a decimal fraction.

Second, a multiplication of the type $a \cdot c$ is implemented with a logical bitshift.[44] Calculate the goal weight c and then set $c = 2^{\lceil \log_2 c \rceil}$ to preserve the lexicographical optimization. With $\ll$ as the bitshift operator, the multiplication $a \cdot c$ with bitshifting becomes $a \ll \log_2 c$. A small computational study shows the effectiveness. $3 \cdot 10^7$ multiplicands a with $a \in [100, 1000]$ are generated and $c = 2^x$ is obtained from $x \in \{8, 16, \ldots, 56\}$. Table 6.1 shows the average computational times and reveals the superiority of bitshifting. Although 64-bit integer data types are used, the source code of the implementation of Algorithm 6.11 is based on a 32-bit implementation.[45]

Third, fractions of the type $\left(\frac{a}{b}\right)^2 \cdot c$ are calculated with bitshifts. In what follows, five alternatives are investigated to find the most effective calculation. Since $\frac{a}{b}, c \geq 0$, the conversion of a decimal to an integer via truncation of the fractional part is

---

[44] If and only if c is a power of two, a multiplication $a \cdot c$ is equivalent to a bitshift of the multiplicand a by $\log_2 c$ positions to the left in the CPU register.

[45] In a 32-bit CPU register, a signed non-zero integer a is shifted at most $30 - \lfloor \log_2 a \rfloor$ positions to the left with a single bitshift operation.

6.4 A Tabu Search Heuristic

equivalent to rounding down to the next integer. The naive implementation is

$$\left\lfloor \left(\frac{a}{b}\right)^2 \cdot c \right\rfloor. \tag{6.87}$$

Bitshifting yields two other arithmetically equivalent alternatives, namely (6.88) and (6.89). Note that b^2 is calculated upfront and used as a parameter in (6.88).

$$\left\lfloor \frac{a^2 \cdot c}{b^2} \right\rfloor = \left\lfloor \frac{a^2 \ll \log_2 c}{b^2} \right\rfloor \tag{6.88}$$

$$\left\lfloor \left(\frac{a \cdot \sqrt{c}}{b}\right)^2 \right\rfloor = \left\lfloor \left(\frac{a \ll \frac{1}{2} \cdot \log_2 c}{b}\right)^2 \right\rfloor \tag{6.89}$$

$$\left\lfloor \frac{a \cdot \sqrt{c}}{b} \right\rfloor^2 = \left\lfloor \frac{a \ll \frac{1}{2} \cdot \log_2 c}{b} \right\rfloor^2 \tag{6.90}$$

(6.87)–(6.90) yield lower bounds to $\left(\frac{a}{b}\right)^2 \cdot c$. It is well-known that the fractional part of a decimal fraction can be approximated by $\frac{1}{2}$. This idea cannot be applied because integers are used but a slight modification of (6.90) yields

$$\left\lfloor \frac{a \cdot \sqrt{c}}{b} \right\rfloor \cdot \left(1 + \left\lfloor \frac{a \cdot \sqrt{c}}{b} \right\rfloor\right) = \left\lfloor \frac{a \ll \frac{1}{2} \cdot \log_2 c}{b} \right\rfloor \cdot \left(1 + \left\lfloor \frac{a \ll \frac{1}{2} \cdot \log_2 c}{b} \right\rfloor\right). \tag{6.91}$$

Lemma 6.15 investigates the relative errors. Denote Δ as the fractional part of a decimal fraction. The relative error δ is defined as $\delta = \frac{|\epsilon|}{d^2}$ with the absolute error ϵ. Let $d = \frac{a \cdot \sqrt{c}}{b}$.

Lemma 6.15

$$\delta = \frac{\Delta}{d^2} \qquad \text{holds for (6.87), (6.88) and (6.89)} \tag{6.92}$$

$$\delta \approx \frac{2 \cdot \Delta}{\lfloor d \rfloor + 2 \cdot \Delta} \qquad \text{holds for (6.90)} \tag{6.93}$$

$$\delta \approx \frac{|2 \cdot \Delta - 1|}{\lfloor d \rfloor + 2 \cdot \Delta} \qquad \text{holds for (6.91)} \tag{6.94}$$

Proof Clearly, $d = \lfloor d \rfloor + \Delta$. Equations (6.87)–(6.89) are arithmetically equivalent. It follows $\epsilon = \Delta$. Since $\Delta \geq 0$, the relative error is $\frac{|\epsilon|}{d^2} = \frac{\Delta}{d^2}$. The absolute error and the relative error of (6.90) become

$$\epsilon = d^2 - \lfloor d \rfloor^2 = 2 \cdot \Delta \cdot \lfloor d \rfloor + \Delta^2$$

$$\delta = \frac{|\epsilon|}{d^2} = \frac{2 \cdot \Delta \cdot \lfloor d \rfloor + \Delta^2}{\lfloor d \rfloor^2 + 2 \cdot \lfloor d \rfloor \cdot \Delta + \Delta^2}$$

$$= \frac{\lfloor d \rfloor \cdot \left(2 \cdot \Delta + \frac{\Delta^2}{\lfloor d \rfloor}\right)}{\lfloor d \rfloor \cdot \left(\lfloor d \rfloor + 2 \cdot \Delta + \frac{\Delta^2}{\lfloor d \rfloor}\right)}$$

set $\frac{\Delta^2}{\lfloor d \rfloor} \approx 0$ because $\Delta \leq 1$ and d is a big number

$$\approx \frac{2 \cdot \Delta}{\lfloor d \rfloor + 2 \cdot \Delta}.$$

The absolute and relative error of (6.91) become

$$\epsilon = d^2 - \lfloor d \rfloor \cdot (1 + \lfloor d \rfloor) = 2 \cdot \Delta \cdot \lfloor d \rfloor - \lfloor d \rfloor + \Delta^2$$

$$\delta = \frac{|2 \cdot \Delta \cdot \lfloor d \rfloor - \lfloor d \rfloor + \Delta^2|}{\lfloor d \rfloor^2 + 2 \cdot \lfloor d \rfloor \cdot \Delta + \Delta^2}$$

$$= \frac{\lfloor d \rfloor \cdot \left|2 \cdot \Delta - 1 + \frac{\Delta^2}{\lfloor d \rfloor}\right|}{\lfloor d \rfloor \cdot \left(\lfloor d \rfloor + 2 \cdot \Delta + \frac{\Delta^2}{\lfloor d \rfloor}\right)}$$

set $\frac{\Delta^2}{\lfloor d \rfloor} \approx 0$ because $\Delta \leq 1$ and d is a big number

$$\approx \frac{|2 \cdot \Delta - 1|}{\lfloor d \rfloor + 2 \cdot \Delta}.$$

The relative error (6.92) decreases quadratically and is much smaller than the relative errors (6.93) and (6.94), which are approximately linear. Since $0 \leq \Delta < 1$, the subtraction of 1 in the enumerator of (6.94) provides a compensation of the rounding down. It is expected that (6.92) yields the smallest value, followed by (6.94) and then by (6.93). A computational study compares the approaches and yields information about the magnitude and the influence of c. The enumerator is $a \in [100, 1000]$ and the denominator is $b \in [a, 100 \cdot a]$. The multiplier $c = 2^x$ is obtained from $x \in \{14, 16, \ldots, 24\}$. $N = 3 \cdot 10^7$ fractions were generated and the average computational times are shown in Table 6.2. The last three columns show the *Mean Absolute Percentage Error (MAPE)* $N^{-1} \cdot \sum_{i=1}^{N} \delta_i$.

Table 6.2 Bitshift implementations of $\left(\frac{a}{b}\right)^2 \cdot c$

x	(6.87) Time s	(6.88) Time s	(6.89) Time s	(6.90) Time s	(6.91) Time s	(6.87)–(6.89) MAPE	(6.90) MAPE	(6.91) MAPE
14	2.15	2.55	2.13	0.98	1.00	10.96	33.57	16.26
16	2.15	2.55	2.14	0.97	1.00	2.62	19.01	10.35
18	2.15	2.55	2.14	0.98	1.01	0.64	9.21	4.76
20	2.15	2.55	2.13	0.98	1.01	0.16	4.84	2.41
22	2.15	2.55	2.13	0.98	1.01	0.04	2.46	1.23
24	2.15	2.55	2.14	0.97	1.01	0.01	1.22	0.62

The computational time is given in seconds

Considering the MAPE, the computational experiment shows the expected behavior of the relative errors. The equivalent approaches (6.87)–(6.89) yield the highest computational times. The MAPE of (6.90) is approximately as twice as large than the MAPE of (6.91) but the computational times are comparable. A multiplicand $c = 2^{18}$ in (6.89) yields a MAPE of 0.64 %, whereas (6.91) requires $c = 2^{24}$ to obtain a comparable MAPE of 0.62 %. Since the computational time of a given approximation is rather constant for all values of c, the choice of a sufficiently high c compensates the relative error. To sum up, the computational study reveals a trade-off between the computational time and the relative error. In the implementation of Algorithm 6.11, 64-bit integer data types are used and therefore, (6.91) with $c = 2^{24}$ is chosen because the relative error is sufficiently small. Note that the size of the CPU register and the values required for the lexicographical optimization in (6.86) set a physical upper bound on the magnitude of (6.89) and (6.91) for a given c. Preliminary computational studies showed that the number of iterations in the Tabu Search Heuristic (Algorithm 6.11) within a given computational time is one fifth to one fourth higher for (6.91) than for (6.87).

References

Adler, M. (1996). Implementation of the Adler32 checksum algorithm in C. Online. Accessed June 2010.

Ahuja, R. K., Magnanti, T. L., & Orlin, J. B. (1993). *Network flows: Theory, algorithms, and applications*. Upper Saddle River, NJ: Prentice-Hall.

ANSI/IEEE Std. (1985). *IEEE Standards for Local Area Networks: Carrier Sense Multiple Access with Collision Detection (CSMA/CD) Access Method and Physical Layer Specifications. ANSI/IEEE Std 802.3-1985* (p. 27).

Balakrishnan, A., Magnanti, T. L., & Wong, R. T. (1989). A dual-ascent procedure for large-scale uncapacitated network design. *Operations Research, 37*, 716–740.

Brown, G. S. (1986). Implementation of the CRC32 checksum algorithm in C. Online. Accessed June 2010.

Caprara, A., Fishetti, M., & Toth, P. (1999). A heuristic method for the set covering problem. *Operations Research, 47*, 730–743.

Chambers, J., & Barnes, J. W. (1998). *Reactive Search for Flexible Job Shop Scheduling*. ORP98-04, Graduate program in Operations Research and Industrial Engineering, The University of Texas at Austin.

Chvátal, V. (2002). *Linear programming*. New York: W.H. Freeman and Company.

Cormen, T. H., Leiserson, C. E., Rivest, R. L., & Stein, C. (2001). *Introduction to algorithms* (2nd ed.). Cambridge: The MIT Press.

Deutsch, P., & Gailly, J.-L. (1996). *ZLIB compressed data format specification version 3.3*. The Internet Engineering Task Force RFC 1950.

Erlenkotter, D. (1978). A dual-based procedure for uncapacitated facility location. *Operations Research, 26*, 992–1009.

Fog, A. (2008). Implementation of the Mersenne Twister algorithm in C++. Online. Accessed September 2010.

Gendron, B. (2002). A note on a dual-ascent approach to the fixed-charge capacitated network design problem. *European Journal of Operational Research, 138*, 671–675.

Guignard, M., & Spielberg, K. (1979). A direct dual method for the mixed plant location problem with some side constraints. *Mathematical Programming, 17*, 198–228.

Herrmann, J., Ioannou, G., Minis, I., & Proth, J. M. (1996). A dual ascent approach to the fixed-charge capacitated network design problem. *European Journal of Operational Research, 95*, 476–490.

Klose, A. (2001). *Standortplanung in distributiven Systemen*. Heidelberg: Physica.

Klose, A., & Drexl, A. (2005). Facility location models for distribution system design. *European Journal of Operational Research, 162*, 4–29.

Kolisch, R. (1996). Serial and parallel resource-constrained project scheduling methods revisited: Theory and computation. *European Journal of Operational Research, 90*, 320–333.

Koopman, P. J. (2002). 32-bit cyclic redundancy codes for internet applications. In *The International Conference on Dependable Systems and Networks (DSN 2002)* (pp. 459–468).

Matsumoto, M., & Nishimura, T. (1998). Mersenne Twister: A 623-dimensionally equidistributed uniform pseudo-random number generator. *ACM Transactions on Modeling and Computer Simulation, 8*, 3–30.

Maxino, T. C. (2006). *The Effectiveness of Checksums for Embedded Networks*. Master's thesis, Department of Electrical and Computer Engineering, Carnegie Mellon University, Pittsburgh.

Maxino, T. C., & Koopman, P. J. (2009). The effectiveness of checksums for embedded control networks. *IEEE Transactions on Dependable and Secure Computing, 6*, 59–72.

Nemhauser, G. L., & Wolsey, L. A. (1988). *Integer and combinatorial optimization*. New York: Wiley.

Ottmann, T., & Widmayer, P. (2012). *Algorithmen und Datenstrukturen* (5th ed.). Heidelberg: Spektrum Akademischer Verlag.

Papadimitriou, C. H., & Steiglitz, K. (1998). *Combinatorial optimization: Algorithms and complexity*. Mineola: Dover.

Peterson, W. W., & Brown, D. T. (1961). Cyclic codes for error detection. *Proceedings of the IRE, 49*, 228–235.

Sheinwald, D., Satran, J., Thaler, P., & Cavanna, V. (2002). *Internet protocol small computer system interface (iSCSI) cyclic redundancy check (CRC)/checksum considerations*. The Internet Engineering Task Force RFC 3385.

Stone, J., & Stewart, R. (2002). *Stream control transmission protocol (SCTP) checksum change*. The Internet Engineering Task Force RFC 3309.

Tanenbaum, A. S. (1981). *Computer networks*. New York: Prentice Hall.

Thompson, D. R., Maxwell, B., & Parkerson, J. P. (2004). Building the big message authentication code. In *Proceedings 8th World Multiconference on Systemics, Cybernetics and Informatics (SCI 2004)* (pp. 544–549).

Woodruff, D. L., & Zemel, E. (1993). Hashing vectors for tabu search. *Annals of Operations Research, 41*, 123–137.

Chapter 7
Computations for the Capacitated Planned Maintenance Problem

The main focus of this chapter is the performance of the heuristics from Chap. 6. Since the CPMP is a novel problem, Sect. 7.1 introduces an instance generation scheme and test-sets. In Sect. 7.2, the quality of all lower bounds from Theorem 5.22 of Sect. 5.3.2 is evaluated. All heuristics from Chap. 6 are evaluated in Sect. 7.3.

7.1 Instance Generation and Test-Sets

Preliminary computational studies revealed that the instance solvability and the performance of the heuristics significantly depend on n and T that determine the size of an instance, on the scarcity of $\bar{r}$ and on the variance of π and r. Let x_i be an instance parameter π_i, r_i or $\bar{r}_t$. Because of Remark 2.2, x_i is an integer and the random variable X is liable to a discrete uniform distribution of integers $X \sim U\left(x^{\varnothing} \cdot (1 - x^{\%}), x^{\varnothing} \cdot (1 + x^{\%})\right)$ where a scalar $0 \leq x^{\%} \leq 1$ determines a symmetric interval around the mean $x^{\varnothing}$. A choice of $x^{\varnothing}$ and $x^{\%}$ such that the interval bounds are integer prevents a distortion of the distribution. In real-world situations, production plans with periodically occurring maintenance slots would reasonably provide enough capacity such that every maintenance activity can be scheduled in every period. This yields

$$r^{\varnothing} \cdot (1 + r^{\%}) \leq \bar{r}^{\varnothing} \cdot (1 - \bar{r}^{\%}). \tag{7.1}$$

To increase the likelihood that a feasible solution exists, a value on $\bar{r}^{\varnothing}$ is derived from the following procedure. Note that a generated instance must satisfy Assumption 2.4.

7.1 Algorithm. A value for the mean $\bar{r}^\emptyset$

Input: $n \in \mathbb{N}$, $T \in \mathbb{N}$, instance generation parameter $\pi^\emptyset, \pi^\%, r^\emptyset, r^\%, \bar{r}^\%$
Output: R
Initialization: $(x, y) \leftarrow (0,0)$, $r_i \leftarrow 0 \; \forall i$, $\pi_i \leftarrow 0 \; \forall i$, $\bar{r}_t \leftarrow 0 \; \forall t$, $R \leftarrow 0$, $S \leftarrow \frac{n \cdot r^\emptyset}{\pi^\emptyset}$,
$\bar{r}_k^\emptyset \leftarrow 0 \; \forall k = 1, \ldots, 10000$, $k \leftarrow 0$

1 **do**
2     $R \leftarrow \lceil S \rceil$
3   **for** $k = 1$ **to** 10000 **do**
4   Draw n random numbers for π_i and r_i
5   Sort all pairs (π_i, r_i) according to non-decreasing values $\pi_1 \leq \ldots \leq \pi_n$
6        $\bar{r}_k^\emptyset \leftarrow R - 1$
7   **do**
8           $\bar{r}_k^\emptyset \leftarrow \bar{r}_k^\emptyset + 1$
9   Draw T random numbers $\bar{r}_t \in \left[\left\lceil \bar{r}^\emptyset \cdot (1 - \bar{r}^\%) \right\rceil, \left\lfloor \bar{r}^\emptyset \cdot (1 + \bar{r}^\%) \right\rfloor\right]$
10   Apply the Iterated Best-of-three Heuristic (Algorithm 6.3) to the generated instance and obtain a solution (x, y)
11   **while** (x, y) is infeasible or $r^\emptyset \cdot (1 + r^\%) > \left\lceil \bar{r}_k^\emptyset \cdot (1 - \bar{r}^\%) \right\rceil$
12   Calculate the sample mean $S \leftarrow \frac{1}{10000} \cdot \sum_{i=1}^{10000} \bar{r}_k^\emptyset$
13 **while** Percentage of instances where the Iterated Best-of-three Heuristic did not find a feasible solution is larger than $0.5\,\%$

7.1.1 A Value for the Mean $\bar{r}^\emptyset$

Algorithm 7.1 calculates a value for R such that the observed certainty to yield a feasible instance, which is generated with $\bar{r}^\emptyset = R$ and solved with the Iterated Best-of-three Heuristic, is larger than $99.5\,\%$. In each iteration, a value R is determined from a sample of 10,000 randomly generated instances.

Initialize with given instance generation parameter and $S = \frac{n \cdot r^\emptyset}{\pi^\emptyset}$. In an iteration, set $R = \lceil S \rceil$ and $k = 1$. Generate a partial instance with random values for π and r. Sort all pairs (π_i, r_i) according to non-decreasing values of π_i to satisfy Assumption (2.1). Set $\bar{r}_k^\emptyset = R$ and draw random values for $\bar{r}$ with $\bar{r}_k^\emptyset$ but round the lower (upper) interval bound to next integer. If r^{sum} is smaller than the upper interval bound, terminate the algorithm and adjust the instance generation parameter. While the Iterated Best-of-three Heuristic fails to find a feasible solution or constraint (7.1) is violated, increment $\bar{r}_k^\emptyset$ by 1, generate new random values for $\bar{r}$ and solve this instance as described above. Set $k = k + 1$, generate and solve the next instance until $k > 10000$. Set S to the sample mean of all $\bar{r}_k^\emptyset$. Until the percentage of instances where the Iterated Best-of-three Heuristic did not find a feasible solution is larger than a given tolerance of $0.5\,\%$, commence the next iteration. The algorithm terminates and returns an integer R.

Random numbers are obtained from the pseudo-random generator Mersenne Twister (Matsumoto and Nishimura 1998). Rounding up R at the beginning of every iteration ensures a monotonic increase. The algorithm commences with a value

7.2 Absolute Strength of the Lower Bounds

$S = \frac{n \cdot r^\emptyset}{\pi^\emptyset}$ that applies (C^y) as follows: To ensure feasibility, the expected minimal capacity of all periods $T \cdot R$ must suffice to schedule all maintenance activities if every maintenance activity i at least scheduled $\lfloor \frac{T}{\pi^\emptyset} \rfloor$ times (Lemma 4.17). From (C) follows $n \cdot r^\emptyset \cdot \lfloor \frac{T}{\pi^\emptyset} \rfloor \leq T \cdot R$ and an appropriate value is $\frac{n \cdot r^\emptyset}{\pi^\emptyset}$.

7.1.2 Test-Sets

The size of the instance is influenced by the instance generation parameter $n \in \{10, 15, 20, 25\}$ and $T \in \{20, 30, 40, 50, 100, 150, 200\}$. The scarcity of $\bar{r}$ is referred to as *capacity availability*. To each *pair* (n, T) three *test-sets* are obtained from three values for $\bar{r}^\emptyset = a \cdot n \cdot r^\emptyset$ with $a \in \{0.3, 0.5, 0.7\}$ and the capacity availability is referred to as *low*, *medium* and *high*. This gives 84 test-sets. The variances of π and r are considered as follows. Each test-set comprises nine *test-subsets* because of three values for $\pi^\% \in \{0.1, 0.3, 0.5\}$ and for $r^\% \in \{0.1, 0.3, 0.5\}$. Therefore, 756 test-subsets exist. A test-subset consists of 10 instances, a test-set of 90 instances and a pair (n, T) of 270 instances. Therefore, a solution approach to the CPMP can be evaluated against 7560 instances.

Other instance generation parameter are $r^\emptyset = 10$, $\pi^\emptyset = 10$ and $\bar{r}^\% = 0.2$. Note that $a = 1$ implies a period that is likely uncapacitated. The discrete uniform distribution is not distorted because the instance generation parameter yield integer interval bounds. Note that the values for $\bar{r}^\emptyset$ take a computational study with Algorithm 7.1 into account. The obtained values for R are shown in Table 7.1. One observes that R increases the larger $\pi^\emptyset$ or n becomes but R has a low sensitivity towards the other parameters. Appropriately, $\bar{r}^\emptyset$ for a given a increases linearly to n. Note that for a given pair (n, T) the lower interval bound of the symmetric interval around the mean $\bar{r}^\emptyset$ (e.g., $(1 - \bar{r}^\%) \cdot a \cdot n \cdot r^\emptyset$ with $\bar{r}^\% = 0.2$, $a = 0.3$, $r^\emptyset = 10$) is larger than the largest value for R from Table 7.1.

7.2 Absolute Strength of the Lower Bounds

Theorem 5.22 analyzes the relative strength of all relevant lower bounds from Sect. 5.3. In what follows, a computational study investigates the absolute strength of these 14 lower bounds.

The Lagrangean duals are solved as the RMP (3.23)–(3.26) from Sect. 3.3.1 with the column generation Algorithm 3.1 and initial columns are obtained from a modified First Fit Heuristic without the post optimization routines (Algorithm 6.1). $Z^{(C)}$ is calculated with Lemma 4.17. All computations of this section were performed on an Intel Pentium D 930 with 3.0 GHz, 3 GB memory, a Debian GNU Linux 2.6.8 operating system, ILOG CPLEX 8.0 and ILOG AMPL 8.0. CPLEX solved all instances with $n = 10$ and $T \in \{20, 30, 40\}$ to optimality. Tables 7.2 and 7.3 show

Table 7.1 Values R for the mean $\bar{r}^{\varnothing}$ for all test-subsets obtained with Algorithm 7.1

		T = 20			T = 30			T = 40			T = 50		
		$\pi^\%$	$\pi^\%$	$\pi^\%$	$\pi^\%$	$\pi^\%$	$\pi^\%$	$\pi^\%$	$\pi^\%$	$\pi^\%$	$\pi^\%$	$\pi^\%$	$\pi^\%$
n	$r^\%$	0.1	0.3	0.5	0.1	0.3	0.5	0.1	0.3	0.5	0.1	0.3	0.5
10	0.1	18	20	22	19	20	22	19	20	23	19	20	23
10	0.3	18	20	22	18	20	23	18	20	23	18	20	23
10	0.5	19	20	23	19	21	24	19	21	24	19	21	24
15	0.1	23	25	30	23	26	31	24	26	32	24	27	32
15	0.3	23	25	29	24	26	31	24	27	32	24	27	32
15	0.5	24	26	30	25	27	32	26	28	33	26	28	33
20	0.1	30	33	37	31	34	38	32	34	40	32	35	40
20	0.3	29	32	37	31	34	39	31	35	40	32	35	40
20	0.5	30	33	38	32	35	40	32	36	41	33	36	42
25	0.1	36	39	45	37	41	47	37	42	48	37	42	48
25	0.3	36	39	45	37	41	47	38	42	48	38	42	49
25	0.5	37	40	46	39	42	48	39	43	50	40	44	50

the relative errors of the lower bounds to the optimum of the CPMP. In Table 7.3, the lower bounds are rounded up to the next integer. The computational time to solve the lower bound to optimality is provided in Table 7.4. Blanks mean that the values were not available.

In Table 7.2, the relative errors follow the relative strengths from Theorem 5.22. Table 7.3 shows that rounding the lower bound up to the next integer is more effective the smaller T is. Note that this procedure distorts the comparison of the relative errors with each other because the absolute strength is determined by polyhedral properties. Nevertheless, the relative errors of Table 7.3 mostly evolve to the same tendency observed in Table 7.2. In Table 7.2, the Lagrangean lower bounds show the tendency to provide lower relative errors if the capacity availability is low.

The LP relaxation of (X) yields lower bounds that are in a comparable order of magnitude of the lower bounds where the integrality is kept but the quality of the lower bounds significantly decreases if (X) and (Y) are simultaneously relaxed. Specifically, this holds for the relative errors of the lower bounds $Z \geq Z^{(X)} \geq Z^{(X)(Y)}$, $Z^{(P)+(Q)}_{(R)} \geq Z^{(P)(X)+(Q)}_{(R)} \geq Z^{(P)(X)(Y)+(Q)}_{(R)} = Z^{(X)(Y)}$ and $Z^{(P)+(Q)} \geq Z^{(P)(X)+(Q)} \geq Z^{(P)(X)(Y)+(Q)}$.[1] Similar results are obtained for $Z^{(Y)}_{(V)} \geq Z^{(X)(Y)}$ and $Z_{(C)} = Z^{(X)}_{(C)} \geq Z^{(X)(Y)}$ where the LP relaxation of either (Y) or (X) yields relative errors comparable to $Z^{(X)(Y)}$.[2] However, this should be interpreted with caution because $Z^{(Y)}_{(V)}$ has $Conv(P)(C)(S)(X) \cap H(V)$ and $Z_{(C)}$ has $Conv(P)(V)(S)(Y) \cap H(C)$ (cf. Lemma 5.19 and Geoffrion's Theorem 3.11).

---

[1] $Z^{(P)(X)(Y)+(Q)}_{(R)} = Z^{(X)(Y)}$ holds because of the integrality property from Corollary 3.12.
[2] $Z_{(C)} = Z^{(X)}_{(C)}$ follows from equality (5.18).

7.2 Absolute Strength of the Lower Bounds

Table 7.2 Relative error in % of the lower bounds from Theorem 5.22 to the optimum of the CPMP generated by CPLEX

Test-set	$Z^{(X)}$	$Z_{(P)/(C)}$	$Z_{(P)}$	$Z_{(C)}$	$Z^{(P)+(Q)}_{(R)}$	$Z^{(P)(X)+(Q)}_{(R)}$	$Z^{(P)+(Q)}$	$Z^{(P)(X)+(Q)}$	$Z^{(Y)}_{(V)}$	$Z^{(V)(Y)}$	$Z^{(X)(Y)}$	$Z^{(V)(X)(Y)}$	$Z^{(P)(X)(Y)+(Q)}$	$Z^{(C)}$
10 n 20 T	1.87	12.03	12.39	15.43	6.39	8.79	15.88	16.56	15.46	25.27	15.71	25.39	29.63	41.69
Low	4.88	8.64	8.73	14.37	6.49	11.46	19.88	21.37	13.83	16.20	14.42	16.52	31.68	58.95
Medium	0.72	13.24	13.64	15.88	9.01	10.79	18.07	18.35	16.01	25.16	16.10	25.19	29.64	39.41
High	0.00	14.22	14.79	16.05	3.67	4.10	9.67	9.94	16.53	34.45	16.60	34.45	27.58	26.70
10 n 30 T	2.20		11.16	13.97	8.09	11.36	23.39	24.74		21.41	14.75	21.53	37.52	40.30
Low	5.20		7.62	13.69	6.07	11.93	23.00	26.12		14.91	13.79	15.20	35.00	58.67
Medium	1.04		11.38	13.31	8.81	11.59	26.79	27.41		19.50	14.42	19.55	36.74	38.25
High	0.37		14.47	14.90	9.41	10.54	20.38	20.70		29.81	16.04	29.83	40.82	23.97
10 n 40 T	2.43		11.97	13.99	8.63	11.56		20.89		21.79	15.30	21.93	34.83	42.61
Low	4.68		7.37	13.14	5.63	11.36		23.73		14.23	13.27	14.58	32.95	61.70
Medium	1.55		11.63	12.62	9.56	11.96		23.28		19.66	14.34	19.70	34.09	38.62
High	1.05		16.90	16.21	10.70	11.36		15.67		31.49	18.30	31.50	37.46	27.50

Table 7.3 Relative error in % of the lower bounds from Theorem 5.22, which are rounded up to the next integer, to the optimum of the CPMP generated by CPLEX

Test-set	$Z^{(X)}$	$Z_{(P)(C)}$	$Z_{(P)}$	$Z_{(C)}$	$Z^{(P+Q)}_{(R)}$	$Z^{(P)(X)+(Q)}_{(R)}$	$Z^{(P)+(Q)}$	$Z^{(P)(X)+(Q)}$	$Z^{(Y)}_{(V)}$	$Z^{(V)(Y)}$	$Z^{(X)(Y)}$	$Z^{(V)(X)(Y)}$	$Z^{(P)(X)(Y)+(Q)}$	$Z^{(C)}$
$10\,n\,20\,T$	1.87	1.60	1.84	4.29	0.17	2.29	15.88	16.56	4.44	12.29	4.57	12.34	18.59	41.69
Low	4.88	1.51	1.51	6.31	0.00	4.88	19.88	21.37	5.94	7.66	6.31	7.80	23.79	58.95
Medium	0.72	0.50	0.72	3.00	0.00	1.22	18.07	18.35	3.56	12.80	3.56	12.80	21.48	39.41
High	0.00	2.78	3.28	3.56	0.50	0.78	9.67	9.94	3.83	16.43	3.83	16.43	10.50	26.70
$10\,n\,30\,T$	2.20		4.55	7.06	1.80	4.42	23.39	24.74		14.13	7.65	14.24	31.34	40.30
Low	5.20		2.83	8.35	1.73	7.16	23.00	26.12		9.76	8.45	9.97	30.29	58.67
Medium	1.04		3.74	5.53	1.83	3.88	26.79	27.41		11.99	6.41	12.12	30.24	38.25
High	0.37		7.07	7.30	1.85	2.23	20.38	20.70		20.64	8.08	20.64	33.48	23.97
$10\,n\,40\,T$	2.43		7.24	8.85	4.12	7.26		20.89		16.12	10.31	16.24	29.98	42.61
Low	4.68		4.23	9.42	2.35	7.83		23.73		10.41	9.49	10.76	29.12	61.70
Medium	1.55		6.28	6.80	4.63	6.88		23.28		13.63	8.42	13.63	27.46	38.62
High	1.05		11.20	10.32	5.39	7.08		15.67		24.33	13.01	24.33	33.35	27.50

7.2 Absolute Strength of the Lower Bounds

Table 7.4 Computational time in seconds to obtain the values of Tables 7.2 and 7.3

Test-set	$Z^{(X)}$	$Z_{(P)(C)}$	$Z_{(P)}$	$Z_{(C)}$	$Z_{(R)}^{(P)+(Q)}$	$Z_{(R)}^{(P)(X)+(Q)}$	$Z^{(P)+(Q)}$	$Z^{(P)(X)+(Q)}$	$Z_{(V)}^{(Y)}$	$Z^{(V)(Y)}$	$Z^{(X)(Y)}$	$Z^{(V)(X)(Y)}$	$Z^{(P)(X)(Y)+(Q)}$	CPLEX
10 n 20 T	0.11	114.71	3.73	17.38	22.65	8.08	0.04	0.03	2574.40	0.02	0.04	0.02	0.02	0.30
Low	0.14	202.43	2.13	26.00	47.69	15.95	0.06	0.03	17.82	0.02	0.04	0.02	0.02	0.43
Medium	0.09	99.32	3.57	17.35	17.50	6.57	0.02	0.03	193.49	0.02	0.04	0.02	0.02	0.18
High	0.12	42.39	5.49	8.79	2.75	1.71	0.02	0.02	7511.89	0.02	0.03	0.02	0.02	0.31
10 n 30 T	1.09		5.72	375.23	691.71	160.40	75.68	0.11		0.03	0.09	0.04	0.03	322.45
Low	1.62		3.79	265.76	1568.78	322.70	226.93	0.19		0.06	0.09	0.04	0.03	959.94
Medium	0.83		6.08	249.65	435.19	124.71	0.06	0.07		0.03	0.09	0.04	0.03	3.52
High	0.83		7.29	610.29	71.15	33.80	0.04	0.08		0.02	0.09	0.04	0.03	3.89
10 n 40 T	20.48		9.23	4023.87	1517.18	1008.64		6.38		0.05	0.17	0.06	0.04	1470.23
Low	44.17		6.25	1564.51	3570.97	2582.63		16.67		0.10	0.16	0.06	0.04	1869.92
Medium	7.52		10.35	1793.63	863.43	382.86		0.91		0.03	0.18	0.06	0.04	2376.40
High	9.74		11.11	8713.46	117.15	60.44		1.56		0.03	0.17	0.07	0.04	164.35

The valid inequality (V) yields a significant improvement of the lower bound. This is reflected in the relative errors that follow the tendency $Z_{(V)}^{(Y)} \geq Z^{(V)(Y)}$ and $Z^{(X)(Y)} \geq Z^{(V)(X)(Y)}$. The relevance of (C) for the lower bounds is observable in the large differences between relative errors of Z, $Z_{(C)}$ and $Z^{(C)}$. Regarding the relevance of (P), the Lagrangean relaxation of (R) provides a significant improvement of the lower bound and the relative errors evolve to the tendency $Z \geq Z_{(R)}^{(P)+(Q)} \geq Z^{(P)+(Q)}$ and $Z \geq Z_{(R)}^{(P)+(Q)} \geq Z_{(P)}$. Analogous experiments for the uncapacitated variants are regarded as less insightful because the relative errors of $Z_{(C)}$ are already close to $Z^{(X)(Y)}$ and $Z^{(C)}$ coincides with $Z^{(C)(X)(Y)}$ (Lemma 4.17). Note that $Z_{(C)(R)}^{(P)+(Q)}$ and respectively $Z_{(R)}^{(C)(P)+(Q)}$ would provide relative errors in between.

Although there exists no dominance between $Z_{(P)}$ and $Z_{(C)}$, the relative errors of $Z_{(P)}$ are mostly lower than of $Z_{(C)}$. Furthermore, the Lagrangean decomposition $Z_{(P)/(C)}$ follows the tendency $Z_{(P)/(C)} \leq Z_{(P)}, Z_{(C)}$ but only slightly improves the relative errors from $Z_{(P)}$ and $Z_{(C)}$. In comparison with $Z^{(X)(Y)}$, the lower bounds $Z_{(C)}$, $Z^{(P)+(Q)}$, $Z^{(P)(X)+(Q)}$, $Z_{(V)}^{(Y)}$ and $Z^{(V)(Y)}$ provide rather high relative errors.

The high computational times of some Lagrangean relaxations is due to the tailing off effect. This effect massively occurs in $Z_{(P)/(C)}$, $Z_{(C)}$ and $Z_{(V)}^{(Y)}$. In $Z_{(P)/(C)}$, the tailing off effect is due to the UNDPs that require many iterations to be priced out whereas the KPs are priced out after relatively few iterations.

The comparably low relative errors of the relaxations $Z^{(X)}$, $Z_{(R)}^{(P)+(Q)}$ and $Z_{(P)}$ show that these relaxations appropriately reflect the structure of the CPMP. The computational time to solve the Lagrangean relaxation $ZL_{(R)}^{(P)+(Q)}$ is quite high, it has the same number of binary variables $n \cdot T + T$ as the CPMP has and is strongly $\mathcal{NP}$-hard (Table 5.10). In contrast to this, $Z^{(X)}$ provides a lower bound with a very good quality, T binary variables and a low computational time compared to CPLEX. However, $Z^{(X)}$ is an open problem (Table 5.10). The computational time of $Z_{(P)}$ is comparable to $Z^{(X)}$. The Lagrangean relaxation has $n \cdot T$ binary variables and a reasonable computational complexity, namely binary $\mathcal{NP}$-hard (Table 5.10). However, the lower bound provides neither the quality of $Z^{(X)}$ nor of $Z_{(R)}^{(P)+(Q)}$.

7.3 Performance of the Heuristics

All computations of this section were performed on an Intel Core i7-3770 with four 3.4 GHz cores as single thread applications with no hyperthreading, 32 GB memory, an Ubuntu 14.04 LTS operating system, ILOG CPLEX 12.5, programmed in C++ and complied with g++ 4.6.3. CPLEX solved all 4320 instances with $n \in \{10, 15, 20, 25\}$ and $T \in \{20, 30, 40, 50\}$ to optimality and no consumption of virtual memory was observed.

7.3.1 Heuristics to the Lagrangean Relaxations and Pseudo-Subgradient Optimization

Section 7.3.1.1 analyzes the performance of the heuristics that solve the Lagrangean relaxations. This particularly holds for the novel heuristics to $ZL_{(C)}$. The following two sections focus on subgradient and pseudo-subgradient optimization. In Sect. 7.3.1.2, different lower and upper bounds are applied in the Lagrangean heuristic to the CPMP. The approximation to the Lagrangean duals is evaluated in Sect. 7.3.1.3.

7.3.1.1 The Performance of the Heuristics

In what follows, the performance of the heuristics of the Lagrangean relaxations $ZL_{(P)}$ and $ZL_{(C)}$ are investigated. Furthermore, the benefits of considering potential integers and the complementary slackness criterion (6.37a) in the construction of a feasible solution to $ZL_{(C)}$ are analyzed.

For that purpose, the subgradients of the Lagrangean heuristic (Algorithm 6.8) are derived from an optimal solution to the Lagrangean relaxation. In every iteration, the relative error of all heuristics to the optimal objective function value of the Lagrangean relaxation is computed. Since the initial Lagrangean relaxation is solved to optimality (Lemma 6.2), the evaluation starts in the second iteration. Termination follows after maximal 100 iterations or proven optimality of the best feasible solution to the CPMP. Table 7.5 presents the applied algorithms. Tables 7.6 and 7.7 show the results for the lower and upper bounds as relative errors to the optimum of the Lagrangean relaxation.

Regarding the approximation of $ZL_{(P)}$, K mostly yields lower relative errors for the upper and lower bounds if the capacity availability evolves from high to low to medium. The upper bound improves the lower n and T are. The average relative error of the upper bound is 2.40 % whereas 4.62 % for the lower bound shows some potential of improvement.

The lower and upper bounds to $ZL_{(C)}$ provide lower relative errors when the capacity availability and T are low. UC alternates between the LP Lower Bound and the Dual Priority Rule Lower Bound and therefore, the relative errors lie in between the relative errors of ULP and UPR. Among the lower bounds, ULP provides the lowest average relative error of 1.50 %. The other heuristics provide lower bounds to $ZL_{(C)}^{(X)(Y)}$ and therefore, the values are higher, namely 3.41 % for UC, 5.50 % for UPR, 8.98 % for UPD and 13.58 % for USPP.

The lower bounds to $ZL_{(C)}$ provide a set of potential integers P and n sets of periods L_i to take the complementary slackness criterion (6.37a) into account. The influence of this dual information on the quality of a feasible solution is primarily investigated with ULP, ULP+Y+S, ULP+Y–S and ULP–Y+S because P and L_i are potentially non-empty and derived from the same optimal solution to $ZL_{(C)}^{(X)(Y)}$.

Table 7.5 Lower and upper bounds used in Tables 7.6 and 7.7

Algorithm	Description	Section
K	Knapsack Heuristic	Section 6.2.2
ULP	LP Lower Bound	Section 6.2.1.1
	Upper Bound Heuristic: best upper bound of ULP+Y+S, ULP+Y−S and ULP−Y+S	Section 6.2.1.6
ULP+Y+S	LP Lower Bound	Section 6.2.1.1
	Upper Bound Heuristic: with potential integers, with complementary slackness	Section 6.2.1.6
ULP+Y−S	LP Lower Bound	Section 6.2.1.1
	Upper Bound Heuristic: with potential integers, no complementary slackness	Section 6.2.1.6
ULP−Y+S	LP Lower Bound	Section 6.2.1.1
	Upper Bound Heuristic: no potential integers, with complementary slackness	Section 6.2.1.6
U−Y−S	Upper Bound Heuristic: no potential integers, no complementary slackness	Section 6.2.1.6
UC	Combined Lower Bound	Section 6.2.1.3
	Upper Bound Heuristic: no potential integers, with complementary slackness	Section 6.2.1.6
UPR	Dual Priority Rule Lower Bound	Section 6.2.1.2
	Upper Bound Heuristic: no potential integers, with complementary slackness	Section 6.2.1.6
UPD	Primal-dual Lower Bound	Section 6.2.1.4
	Upper Bound Heuristic: no potential integers, with complementary slackness	Section 6.2.1.6
USPP	Shortest Path Lower Bound	Section 6.2.1.5

Preliminary computational studies show that there exists no dominance between the upper bounds of ULP+Y+S, ULP+Y−S and ULP−Y+S. The average relative errors are 7.57, 6.82 and 9.89 %. All three upper bounds contribute to ULP and hence, taking the best upper bound yields a significant improvement such that ULP has the lowest average relative error of 4.64 %. Note that U−Y−S is not considered in ULP because it rarely provides an improvement that is in one out of approx. 1000 instances. The average relative errors of the upper bounds of UC, UPR, UPD and U−Y−S are 6.50, 8.36, 8.82 and 9.00 %.

The influence of the complementary slackness on the quality of the upper bound to $ZL_{(C)}$ is investigated from a comparison between UPR, UPD, ULP−Y+S and U−Y−S. The average relative errors of UPR, UPD and U−Y−S are 8.36, 8.82 and 9.00 % and ULP−Y+S provides 9.89 %. UPR, UPD and the Upper Bound Heuristic apply specific algorithms of the SWUPMP, namely the Algorithms 4.1 and 4.2, that exploit the same structure and construct a solution in the same direction that is from period 0 to period T. This might explain the low average relative errors of UPR and

UPD whereas the relative error of ULP–Y+S is rather high. The results for UPR, UPD and U–Y–S show that a consideration of the complementary slackness slightly improves the upper bound. Furthermore, there is a slight tendency that the quality of the dual information on complementary slackness changes for the better if the lower bound is better.

The average relative errors for ULP+Y–S, ULP+Y+S and U–Y–S are 6.82, 7.57 and 9.00 %. This indicates that taking potential integers into account significantly improves the obtained upper bound.

7.3.1.2 Applying Different Heuristics in the Lagrangean Heuristic

Important components of a subgradient algorithm are the lower bounds, the upper bounds and the subgradient. This section studies the initial lower bounds and the influence of different heuristics, which solve the Lagrangean relaxation, on the Lagrangean Heuristic.

The Lagrangean Heuristic is applied as presented in Algorithm 6.8 and the maximal computational time is set to 60 s. The heuristics and the lower bounds are provided in Table 7.8. Some abbreviations have been presented in Table 7.5. Considering the results of Sect. 7.3.1.1, ULP, UC, UPR and UPD are applied. Tables 7.9 and 7.10 present the results as relative errors of the lower and upper bounds to the optimum of the CPMP. The lower bounds are rounded up to the next integer. Table 7.11 shows the iterations per second.

In Table 7.9, the initial lower bound LB^{init} has an average relative error of 12.36 % followed by 15.04 % for ZL^{init} and 39.53 % for $Z^{(C)}$. This significant improvement of LB^{init} is due to the fact that there exists no dominance between ZL^{init} and $Z^{(C)}$ (Lemma 6.10). ZL^{init} improves, the lower the capacity availability is because the capacity $\bar{r}_t$ appears in (6.42) as a denominator of the objective function coefficients. $Z^{(C)}$ improves the higher the capacity availability is because the capacity constraint (C) becomes less restrictive.

In Table 7.9, the relative errors of the lower bounds of all Lagrangean heuristics improve if the capacity availability is low and T is small. Furthermore, Table 7.9 shows that subgradients from optimal solutions to the Lagrangean relaxation yield lower average relative errors than pseudo-subgradients from heuristic solutions. This result is expected because pseudo-subgradients roughly approximate the subdifferential. Hence, the increase of the Lagrangean lower bound after a subgradient step is less likely. Kh provides an average relative error of 8.32 % compared to 6.76 % from Ke. The average relative error of Ue is 6.67 % and the Lagrangean heuristics ULPh, UCh, UPRh, UPDh provide 8.32, 8.52, 10.93 and 11.79 %.

An important observation is that the lower bound from the Lagrangean Heuristic improves if a better heuristic to the Lagrangean relaxation is applied. In particular, the tendency of the average relative errors 8.32, 8.52, 10.93 and 11.79 % for ULPh, UCh, UPRh and UPDh from Table 7.9 follows the tendency of 1.50, 3.41, 5.50 and 8.98 % for ULP, UC, UPR and UPD from Table 7.6.

Table 7.6 Performance of the heuristics to $ZL_{(P)}$ and $ZL_{(C)}$ from Table 7.5

Test-set	K	ULP	UC	UPR	UPD	USPP
10 n 20 T	5.16	0.35	1.96	3.94	6.89	12.33
Low	6.09	0.11	1.02	2.04	3.74	5.84
Medium	4.12	0.56	2.74	5.20	9.45	17.55
High	5.28	0.37	2.12	4.57	7.47	13.61
10 n 30 T	11.93	0.91	2.72	4.64	7.94	12.02
Low	11.41	0.12	1.02	1.97	3.44	4.58
Medium	5.63	0.97	2.76	4.69	8.51	12.45
High	18.76	1.64	4.38	7.27	11.86	19.03
10 n 40 T	14.42	1.87	4.00	6.23	9.97	14.88
Low	12.57	0.24	1.17	2.14	3.77	4.82
Medium	6.54	2.10	4.54	7.10	11.44	16.08
High	24.16	3.26	6.28	9.46	14.71	23.74
10 n 50 T	11.88	2.48	4.59	6.79	10.86	14.83
Low	12.26	0.48	1.60	2.78	4.90	5.97
Medium	7.39	2.81	5.15	7.58	12.13	15.58
High	15.98	4.16	7.02	10.01	15.54	22.94
15 n 20 T	5.72	0.33	1.84	3.63	6.50	12.68
Low	3.03	0.06	0.64	1.33	2.83	4.53
Medium	2.21	0.40	2.62	5.16	9.18	17.86
High	11.91	0.53	2.27	4.40	7.48	15.66
15 n 30 T	3.45	1.19	3.18	5.30	8.68	13.59
Low	3.85	0.17	1.00	1.87	3.39	4.60
Medium	2.34	0.95	2.93	5.07	8.86	13.03
High	4.15	2.46	5.62	8.96	13.78	23.14
15 n 40 T	3.93	2.00	4.14	6.40	10.33	14.82
Low	4.68	0.37	1.49	2.67	4.69	5.68
Medium	2.63	2.16	4.57	7.11	11.49	15.89
High	4.48	3.46	6.36	9.42	14.82	22.88
15 n 50 T	4.88	2.77	4.97	7.27	11.50	15.70
Low	4.65	0.71	2.08	3.51	5.57	6.64
Medium	2.81	3.06	5.54	8.13	13.19	16.77
High	7.18	4.54	7.28	10.18	15.73	23.69
20 n 20 T	1.28	0.36	1.92	3.70	7.23	13.47
Low	1.48	0.03	0.69	1.47	2.58	4.20
Medium	1.45	0.48	2.50	4.65	9.87	18.36
High	0.90	0.56	2.56	4.98	9.25	17.86
20 n 30 T	1.44	0.90	2.71	4.72	7.65	11.80
Low	1.65	0.17	1.06	2.02	3.48	4.54
Medium	1.03	1.17	3.29	5.57	9.34	13.11
High	1.63	1.37	3.79	6.56	10.12	17.74

(contienued)

7.3 Performance of the Heuristics

Table 7.6 (continued)

Test-set	K	ULP	UC	UPR	UPD	USPP
20 n 40 T	2.66	1.83	3.86	6.07	9.70	13.63
Low	1.90	0.37	1.55	2.79	4.67	5.78
Medium	1.06	2.19	4.55	6.99	11.12	14.54
High	5.01	2.92	5.48	8.44	13.32	20.58
20 n 50 T	3.13	2.74	4.83	7.07	11.18	15.08
Low	1.99	0.73	2.10	3.53	5.68	6.68
Medium	1.01	3.21	5.61	8.12	13.09	16.41
High	6.40	4.29	6.79	9.57	14.78	22.14
25 n 20 T	0.94	0.19	1.39	3.11	5.71	10.58
Low	0.71	0.09	0.74	1.44	2.63	3.89
Medium	1.67	0.48	2.66	5.10	9.41	16.46
High	0.45	0.01	0.77	2.80	5.08	11.38
25 n 30 T	0.90	1.14	3.08	5.17	8.01	12.48
Low	0.88	0.17	1.02	1.95	3.22	4.20
Medium	0.68	1.11	3.21	5.49	8.69	12.80
High	1.15	2.14	5.02	8.08	12.13	20.45
25 n 40 T	1.07	1.93	4.07	6.38	9.80	13.72
Low	0.89	0.48	1.71	3.01	4.92	5.85
Medium	0.50	2.26	4.66	7.16	11.24	14.88
High	1.81	3.04	5.85	8.96	13.24	20.43
25 n 50 T	1.14	3.02	5.23	7.54	11.79	15.64
Low	0.98	0.79	2.22	3.69	5.94	6.97
Medium	0.58	3.43	5.89	8.44	13.62	16.50
High	1.85	4.85	7.57	10.48	15.80	23.44
Average	4.62	1.50	3.41	5.50	8.98	13.58

Relative error in % of the lower bounds to the optimum of the Lagrangean relaxation. The subgradient is calculated from an optimal solution to the Lagrangean relaxation

The upper bounds to the CPMP from Table 7.10 improve if the capacity availability is high and n as well as T are small. Ke and Kh provide comparably low average relative errors of 0.62 and 0.64 %. Furthermore, the values for the average number of iterations per second are comparable, which are $244.16\,s^{-1}$ for Ke and $251.57\,s^{-1}$ for Kh (see Table 7.11). Interestingly, Ue and ULPh yield average relative errors of 1.62 and 1.05 % that are higher than the average relative errors of UCh, UPRh and UPDh, which are 0.79, 0.79 and 0.89 %. The the average number of iterations $153.16\,s^{-1}$ is much smaller for Ue than for ULPh, UCh, UPRh and UPDh, which provide comparable values of 243.41, 248.86, 264.22 and $243.99\,s^{-1}$. Solving $ZL_{(C)}$ to optimality is very time consuming and this limits the possibilities to improve the best known feasible solution with the repair heuristic. However, Ue yields the best lower bound for $ZL_{(C)}$.

Altogether, UCh is the most preferable heuristic with pseudo-subgradient optimization for $ZL_{(C)}$ and combines the advantages of each applied lower bound. In

Table 7.7 Performance of the heuristics to $ZL_{(P)}$ and $ZL_{(C)}$ from Table 7.5

Test-set	K	ULP	ULP+Y+S	ULP+Y–S	ULP–Y+S	U–Y–S	UC	UPR	UPD
10 n 20 T	2.05	1.33	2.22	2.13	5.07	4.75	2.58	4.22	4.64
Low	1.72	0.31	0.57	0.42	1.96	1.25	0.81	1.34	1.15
Medium	1.95	2.55	3.67	4.09	8.07	7.30	4.06	5.75	6.63
High	2.47	1.14	2.41	1.87	5.18	5.70	2.87	5.56	6.15
10 n 30 T	2.67	2.67	4.37	4.11	5.62	5.43	3.82	5.09	5.67
Low	1.67	0.29	0.45	0.52	1.33	1.35	0.76	1.26	1.32
Medium	1.82	2.86	4.34	4.54	5.73	5.86	3.88	5.09	5.55
High	4.51	4.86	8.33	7.26	9.79	9.07	6.83	8.92	10.15
10 n 40 T	4.37	4.32	6.31	6.01	6.97	6.78	5.65	7.05	7.89
Low	1.52	0.59	0.82	0.77	1.43	1.41	1.05	1.53	1.61
Medium	1.55	4.88	6.64	6.99	7.46	8.00	6.21	7.66	8.60
High	10.03	7.50	11.47	10.27	12.02	10.92	9.69	11.96	13.47
10 n 50 T	2.67	4.62	6.24	6.26	6.75	6.85	5.87	7.19	8.04
Low	1.56	0.97	1.23	1.33	1.69	1.86	1.45	1.95	2.05
Medium	1.55	4.96	6.54	6.84	7.25	7.62	6.26	7.62	8.51
High	4.91	7.92	10.94	10.60	11.32	11.06	9.90	12.01	13.56
15 n 20 T	2.16	1.97	3.91	2.87	9.81	7.03	4.22	6.84	6.39
Low	1.60	0.19	0.30	0.35	1.95	2.14	0.80	1.46	1.20
Medium	1.42	1.96	3.70	3.61	15.79	9.76	5.34	9.29	7.85
High	3.45	3.75	7.74	4.66	11.69	9.20	6.52	9.77	10.11
15 n 30 T	1.82	4.57	7.58	6.87	9.00	8.65	6.30	8.13	8.51
Low	1.46	0.58	0.89	0.97	1.91	2.06	1.22	1.88	1.88
Medium	1.41	3.92	6.22	6.48	8.18	8.87	5.64	7.46	7.23
High	2.59	9.22	15.62	13.16	16.91	15.03	12.04	15.06	16.43
15 n 40 T	1.79	5.63	8.53	7.92	9.64	9.15	7.38	9.24	9.61
Low	1.38	1.10	1.56	1.53	2.49	2.57	1.76	2.45	2.60
Medium	1.35	6.18	8.59	9.20	10.16	10.69	7.97	9.88	9.96
High	2.64	9.60	15.45	13.04	16.28	14.20	12.41	15.40	16.27
15 n 50 T	2.95	7.00	9.63	9.82	10.39	10.66	8.61	10.34	11.25
Low	1.39	1.96	2.46	2.80	3.09	3.54	2.69	3.46	3.54
Medium	1.25	7.93	10.48	11.37	11.59	12.45	9.72	11.54	11.80
High	6.21	11.11	15.94	15.30	16.50	15.99	13.42	16.01	18.42
20 n 20 T	1.16	2.11	4.69	2.85	12.47	8.50	4.96	7.96	8.09
Low	1.19	0.18	0.28	0.25	2.46	2.31	0.78	1.49	1.51
Medium	1.22	2.54	5.46	3.36	19.02	12.24	6.79	11.06	10.38
High	1.07	3.60	8.33	4.94	15.94	10.94	7.32	11.32	12.38
20 n 30 T	1.32	3.89	7.34	6.29	9.52	8.97	5.66	7.68	8.04
Low	1.28	0.84	1.36	1.33	2.87	2.82	1.59	2.41	2.26
Medium	1.11	4.78	8.29	8.31	10.91	11.61	6.95	9.31	9.13
High	1.57	6.05	12.36	9.24	14.79	12.49	8.44	11.33	12.73

(continued)

7.3 Performance of the Heuristics

Table 7.7 (continued)

Test-set	K	ULP	ULP+Y+S	ULP+Y–S	ULP–Y+S	U–Y–S	UC	UPR	UPD
20 n 40 T	4.43	6.15	9.79	8.77	10.88	10.25	7.95	10.04	10.66
Low	1.18	1.55	2.17	2.39	3.21	3.67	2.44	3.36	3.29
Medium	1.04	7.11	10.66	10.63	12.03	12.32	9.15	11.38	11.54
High	11.06	9.80	16.53	13.29	17.41	14.75	12.25	15.39	17.16
20 n 50 T	4.00	7.49	11.19	10.70	11.87	11.55	9.28	11.20	12.02
Low	1.17	2.53	3.28	3.81	4.04	4.74	3.42	4.33	4.41
Medium	1.11	8.84	12.50	12.91	13.36	13.86	10.57	12.43	12.58
High	9.71	11.11	17.78	15.37	18.21	16.06	13.85	16.84	19.06
25 n 20 T	1.68	1.37	3.17	2.13	9.70	7.83	3.05	6.07	5.46
Low	0.85	0.44	0.79	0.75	2.57	2.33	1.10	1.82	1.96
Medium	3.08	3.56	8.57	5.46	20.50	12.81	6.53	9.99	9.67
High	1.12	0.11	0.15	0.18	6.03	8.36	1.51	6.39	4.74
25 n 30 T	1.40	5.52	10.95	9.07	12.81	11.55	7.32	9.47	9.77
Low	1.15	0.91	1.50	1.61	2.96	3.19	1.67	2.52	2.53
Medium	1.26	5.71	10.65	9.69	13.16	12.92	7.56	9.96	9.77
High	1.79	9.94	20.69	15.91	22.31	18.55	12.73	15.94	17.01
25 n 40 T	1.91	6.88	11.82	10.37	13.46	12.16	8.76	10.84	11.41
Low	1.11	2.01	2.91	3.23	4.07	4.48	3.02	4.08	4.05
Medium	0.97	7.85	12.67	12.21	15.21	14.78	10.03	12.35	12.65
High	3.64	10.77	19.88	15.67	21.09	17.21	13.23	16.08	17.54
25 n 50 T	2.08	8.65	13.42	12.92	14.31	13.88	10.49	12.46	13.58
Low	1.06	3.11	4.08	4.75	5.03	5.76	4.22	5.39	5.45
Medium	1.07	10.04	15.10	15.58	16.06	16.70	11.94	13.97	14.39
High	4.12	12.81	21.09	18.43	21.83	19.18	15.30	18.01	20.90
Average	2.40	4.64	7.57	6.82	9.89	9.00	6.50	8.36	8.82

Relative error in % of the upper bounds to the optimum of the Lagrangean relaxation. The subgradient is calculated from an optimal solution to the Lagrangean relaxation

UCh, the heuristic UC alternates between ULP and UPR but UCh yields a relative error of 8.52 % that is much closer to 8.32 % for ULPh than to 10.93 % for UPRh. Among the heuristics for $ZL_{(C)}$, UCh provides the best upper bound to the CPMP with an average relative error of 0.79 %.

7.3.1.3 The Approximation of the Lagrangean Duals

The effectiveness of subgradient and pseudo-subgradient optimization is evaluated by a comparison to the respective Lagrangean dual.

The Lagrangean Heuristic from Algorithm 6.8 is slightly modified to derive the largest possible Lagrangean lower bound as follows: Since there exists no dominance between ZL^{init} and $Z^{(C)}$ (Lemma 6.10), the lower bound $Z^{(C)}$ is never applied in the Lagrangean heuristic. The heuristic does not terminate if the lower

Table 7.8 Lagrangean heuristics and lower bounds used in Tables 7.9, 7.10, 7.11, 7.12, and 7.13

Algorithm	Description	Section
LB^{init}	Initial lower bound after the first iteration of the Lagrangean Heuristic	Section 6.2.3
ZL^{init}	Initial Lagrangean lower bound	Section 6.2.3
$Z^{(C)}$	Lower bound to the UPMP	Section 4.5
Ke	Lagrangean Heuristic, $ZL_{(P)}$ solved to optimality (Remark 6.3)	Section 6.2.3
Kh	Heuristic K in the Lagrangean Heuristic	Section 7.3.1.1
Ue	Lagrangean Heuristic, $ZL_{(C)}$ solved to optimality (Remark 6.3)	Section 6.2.3
ULPh	Heuristic ULP in the Lagrangean Heuristic	Section 7.3.1.1
UCh	Heuristic UC in the Lagrangean Heuristic	Section 7.3.1.1
UPRh	Heuristic UPR in the Lagrangean Heuristic	Section 7.3.1.1
UPDh	Heuristic UPD in the Lagrangean Heuristic	Section 7.3.1.1

bound, which is rounded up to the next integer, coincides with the upper bound. Instead, a termination follows after maximal 60 s of computational time or if convergence of the lower bound is observed. The latter is detected if the difference between the maximal and minimal lower bound of the last ten iterations is smaller than 0.001. The results are presented in the Table 7.12 as relative errors of the lower bound to the respective Lagrangean dual. Table 7.13 shows the relative errors of the lower bounds to the optimum of the CPMP and the relative errors of $Z_{(P)}$, $Z_{(C)}$ and $Z^{(X)(Y)}$ from Table 7.2 are provided. Note that the lower bounds are not rounded up to the next integer.

The average relative errors of LB^{init}, ZL^{init} and $Z^{(C)}$ to $Z_{(C)}$ are 6.46, 10.52 and 31.97 %, whereas the average relative errors of LB^{init}, ZL^{init} and $Z^{(C)}$ to $Z_{(P)}$ are slightly higher that is 9.01, 13.07 and 33.33 %. This is due to the fact that the relative errors to the optimum of the CPMP are lower for $Z_{(P)}$ than for $Z_{(C)}$ (see Table 7.13). The relative errors of LB^{init}, ZL^{init} and $Z^{(C)}$ mostly follow the tendencies that are observed in Table 7.9.

A subgradient algorithm provides an approximation of the Lagrangean dual if the Held-Karp step size is used. Note that this formula is slightly changed to adapt it to pseudo-subgradients. In Table 7.12, subgradients that are obtained from optimal solutions to the Lagrangean relaxation yield lower average relative errors than pseudo-subgradients, which are derived from heuristics. Specifically, the average relative errors of Ke and Ue are 5.59 and 3.91 % whereas Kh and UCh yield 8.28 and 4.88 %. As discussed in Sect. 7.3.1.2, this is an expected result since pseudo-subgradients roughly approximate the subdifferential and thus, an increase of the Lagrangean lower bound after a subgradient step is less likely.

In Table 7.13, the approximation of the Lagrangean duals $Z_{(P)}$ and $Z_{(C)}$ mostly improves for all four Lagrangean heuristics Ke, Kh, Ue and UCh if few capacity is available. The average relative errors of Ke, Kh, Ue and UCh are 14.99, 17.75, 16.42 and 17.45 % whereas the average relative error of $Z^{(X)(Y)}$ is 15.25 %. This shows that $Z^{(X)(Y)}$ provides lower average relative errors than the Lagrangean heuristics

Kh, Ue and UCh. In general, this is not favorable for many applications such as branch and bound approaches because Lagrangean relaxation is often applied to yield significantly better lower bounds than those provided by the LP relaxation. Therefore, the Lagrangean heuristics can be further improved.

The results show that pseudo-subgradient optimization is a promising approach. Kh and UCh improve the initial Lagrangean lower bound ZL^{init} and the relative errors of Kh and UCh are in the same order of magnitude than the values provided by the subgradient approaches for Ke and Ue.

7.3.2 Heuristics to the Planned Maintenance Problem

The performance of all heuristics that yield upper bounds to the CPMP are evaluated. In Sect. 7.3.2.1, the construction heuristics from Sect. 6.1 are investigated. Section 7.3.2.2 focuses on variants of the Lagrangean Hybrid Heuristic from Sect. 6.3. In Sect. 7.3.2.3, the Lagrangean heuristics from Sect. 6.2, a Lagrangean Hybrid Heuristic and the Tabu Search Heuristic from Sect. 6.4 are evaluated in greater detail.

Table 7.14 provides abbreviations and references of the applied heuristics and approaches. Some abbreviations have been presented in Tables 7.5 and 7.8. The findings of Sect. 7.3.1.2 are taken into account and therefore, UCh is applied.

7.3.2.1 The Construction Heuristics

The construction heuristics FF, OF and OS are applied as presented in the Algorithms 6.1 and 6.2. Table 7.15 shows the results as relative errors of the upper bounds to the optimum of the CPMP. If a heuristic terminates because the obtained solution is infeasible, the respective instance is not considered in the evaluation of the relative error. This causes a bias in Table 7.15; e.g., the relative error 18.24 % for FF with $n = 25, T = 50$ and a low capacity availability is smaller than the respective relative error 18.45 % for B3, which is theoretically not possible.

The number of infeasible solutions of all heuristics is 2, 2, 8 and 28 for $n = 10, 15, 20$ and 25. The average percentage of infeasible solutions from the total number of infeasible solutions is 0.00 % for IB3, B3, OS and 25.00 % for OF as well as 75.00 % for FF. Note that IB3 and B3 yield feasible solutions for every instance in every test-set of Sect. 7.1.

There is a tendency that the relative error improves if n and T are small. For OF and OS, a capacity availability from low to high to medium tends to improve the relative error. Considering FF, a capacity availability from medium to low to high tends to improve the relative error. FF has an average relative error of 15.60 % followed by 19.05 % for OF and 22.57 % for OS.

The average relative error 11.98 % for B3 is lower than for FF, OF and OS. However, FF yields the upper bound of B3 in 77.04 % of all instances; OF in 56.90 %

Table 7.9 Performance of the Lagrangean heuristics and lower bounds from Table 7.8

Test-set	LB^{init}	ZL^{init}	Z$^{(C)}$	Ke	Kh	Ue	ULPh	UCh	UPRh	UPDh
10 n 20 T	10.79	12.77	41.69	1.93	5.01	4.35	5.00	5.22	8.07	9.07
Low	9.08	9.08	58.95	1.51	6.95	6.50	6.64	6.64	8.33	8.36
Medium	11.41	12.80	39.41	0.72	4.24	3.00	4.52	5.19	11.41	11.41
High	11.89	16.43	26.70	3.56	3.83	3.56	3.83	3.83	4.48	7.44
10 n 30 T	13.20	14.76	40.30	4.62	8.41	7.06	8.72	8.95	11.82	12.94
Low	11.62	11.62	58.67	2.83	8.90	8.35	8.66	8.80	10.16	11.06
Medium	12.14	12.30	38.25	3.74	7.62	5.53	8.41	8.41	11.23	12.14
High	15.85	20.36	23.97	7.30	8.71	7.30	9.09	9.64	14.07	15.62
10 n 40 T	14.76	16.79	42.61	7.96	11.19	9.04	11.19	11.36	13.71	14.33
Low	11.72	11.72	61.70	4.31	10.33	9.49	9.94	10.08	10.76	11.11
Medium	13.91	14.17	38.62	6.52	9.08	6.90	9.85	10.21	12.87	13.75
High	18.64	24.49	27.50	13.06	14.15	10.74	13.78	13.78	17.50	18.14
10 n 50 T	16.02	18.69	41.23	9.62	13.08	10.75	13.07	13.18	15.23	15.82
Low	13.86	13.86	60.80	5.32	11.47	10.83	11.20	11.25	12.70	13.25
Medium	15.20	16.01	38.16	9.10	11.50	8.87	11.84	12.13	14.81	15.20
High	19.01	26.20	24.73	14.43	16.27	12.56	16.16	16.16	18.17	19.01
15 n 20 T	8.53	11.77	39.33	2.15	3.75	3.25	3.70	3.96	5.70	7.55
Low	7.66	7.66	56.71	1.85	4.81	4.59	4.77	5.00	6.03	6.38
Medium	9.88	10.99	36.57	2.10	3.94	2.94	3.83	4.38	7.83	9.33
High	8.06	16.67	24.72	2.50	2.50	2.22	2.50	2.50	3.24	6.94
15 n 30 T	12.49	15.33	38.05	7.11	8.47	6.70	8.48	8.77	11.28	11.95
Low	8.91	8.91	56.49	3.82	5.51	5.32	5.46	5.76	7.42	7.75
Medium	11.78	13.26	33.99	5.61	7.38	5.04	8.58	8.74	11.59	11.78
High	16.77	23.81	23.67	11.89	12.52	9.73	11.41	11.82	14.83	16.33
15 n 40 T	13.93	16.35	41.30	8.82	10.24	7.71	9.92	10.39	13.19	13.61
Low	10.10	10.10	59.55	5.72	7.64	7.01	7.55	7.72	9.53	9.60
Medium	13.47	14.49	38.21	8.44	9.64	6.78	9.77	10.41	13.10	13.47
High	18.22	24.47	26.15	12.31	13.45	9.33	12.44	13.05	16.94	17.75
15 n 50 T	14.94	17.45	40.44	10.06	11.84	9.40	11.88	12.01	14.15	14.64
Low	11.67	11.67	59.42	6.47	9.17	8.22	9.08	9.29	10.98	11.16
Medium	14.71	15.11	37.47	9.10	10.84	8.26	11.32	11.49	14.13	14.71
High	18.44	25.56	24.43	14.61	15.50	11.73	15.24	15.24	17.33	18.06
20 n 20 T	8.40	11.43	38.35	1.89	3.24	2.90	3.35	3.35	5.75	7.35
Low	7.06	7.06	55.96	1.78	4.05	3.71	3.87	3.87	5.77	6.32
Medium	10.56	11.94	35.65	2.50	4.28	3.61	4.78	4.78	9.44	10.00
High	7.59	15.28	23.43	1.39	1.39	1.39	1.39	1.39	2.04	5.74
20 n 30 T	11.47	14.32	36.38	7.01	7.74	5.89	7.82	8.07	10.72	11.20
Low	8.81	8.81	56.14	4.82	6.20	6.07	6.07	6.46	7.78	8.23
Medium	11.63	12.66	32.44	6.66	7.03	4.88	7.40	8.20	11.10	11.63
High	13.97	21.48	20.56	9.54	9.99	6.72	9.99	9.54	13.28	13.74

(continued)

7.3 Performance of the Heuristics

Table 7.9 (continued)

Test-set	LB^{init}	ZL^{init}	Z$^{(C)}$	Ke	Kh	Ue	ULPh	UCh	UPRh	UPDh
20 n 40 T	12.99	15.99	40.10	8.70	9.23	6.98	9.34	9.69	11.83	12.58
Low	8.88	8.88	58.80	5.54	6.54	5.97	6.66	6.75	8.28	8.62
Medium	13.05	13.43	37.22	8.71	8.97	6.08	9.21	10.02	12.49	13.05
High	17.04	25.65	24.29	11.86	12.17	8.89	12.15	12.31	14.71	16.06
20 n 50 T	13.25	16.56	38.79	10.32	10.86	8.72	10.82	10.91	12.68	13.08
Low	9.92	9.92	58.10	6.66	7.84	7.21	7.97	8.09	9.22	9.42
Medium	13.04	14.22	35.89	9.76	9.86	7.63	9.86	10.25	12.65	13.04
High	16.79	25.55	22.39	14.53	14.89	11.31	14.64	14.38	16.18	16.79
25 n 20 T	8.91	11.59	37.79	1.79	2.58	2.12	2.81	3.00	5.52	7.44
Low	6.96	6.96	55.79	2.24	3.11	3.11	3.64	3.64	6.09	6.27
Medium	10.96	11.24	35.17	2.85	4.35	2.96	4.52	5.07	10.19	10.96
High	8.80	16.57	22.41	0.28	0.28	0.28	0.28	0.28	0.28	5.09
25 n 30 T	11.42	13.90	37.19	7.03	7.51	5.44	7.23	7.56	10.30	10.92
Low	7.58	7.58	56.33	4.36	4.89	4.36	4.79	5.03	6.41	6.97
Medium	12.43	12.96	33.93	6.63	7.32	4.54	7.46	7.99	11.51	12.43
High	14.24	21.15	21.30	10.10	10.32	7.43	9.43	9.65	12.99	13.36
25 n 40 T	13.34	16.08	40.41	8.95	9.41	7.81	9.47	9.52	12.28	13.10
Low	8.65	8.65	58.68	5.59	6.68	6.09	6.68	6.84	7.90	8.40
Medium	13.51	14.25	36.83	8.85	9.13	6.80	9.51	9.65	12.82	13.51
High	17.86	25.33	25.73	12.42	12.42	10.55	12.23	12.07	16.12	17.39
25 n 50 T	13.28	16.82	38.44	10.25	10.49	8.53	10.28	10.39	12.71	13.07
Low	9.47	9.47	57.80	6.84	7.33	6.70	7.39	7.52	8.81	9.21
Medium	13.80	14.90	35.17	10.49	10.72	7.61	10.63	10.83	13.38	13.70
High	16.56	26.09	22.36	13.43	13.43	11.29	12.81	12.81	15.94	16.31
Average	12.36	15.04	39.53	6.76	8.32	6.67	8.32	8.52	10.93	11.79

Relative error in % of the lower bounds, which are rounded up to the next integer, to the optimum of the CPMP generated by CPLEX. Termination after maximal 60 s

and OS in 40.88 %. In 68.06 % of all instances FF yields an upper bound that is strictly smaller than the upper bound provided by OF and OS. Respective values are 23.70 % for OF and 8.24 % for OS. Consequently, all construction heuristics FF, OF and OS contribute to B3.

B3 is identical to the upper bound provided by IB3 in the first iteration. The iterated approach significantly improves B3 and therefore, IB3 yields the lowest average relative error among the construction heuristics that is 6.29 %.

7.3.2.2 Variants of the Lagrangean Hybrid Heuristic

One parameter that influences the Lagrangean Hybrid Heuristic is the linear coefficient in the calculation of the Lagrangean multiplier of the next iteration. In particular, the Lagrangean heuristics commence an iteration with the Lagrangean

Table 7.10 Performance of the Lagrangean heuristics from Table 7.8

Test-set	Ke	Kh	Ue	ULPh	UCh	UPRh	UPDh
10 n 20 T	0.00	0.00	0.82	0.48	0.00	0.25	0.28
Low	0.00	0.00	1.90	1.15	0.00	0.19	0.19
Medium	0.00	0.00	0.56	0.28	0.00	0.56	0.28
High	0.00	0.00	0.00	0.00	0.00	0.00	0.37
10 n 30 T	0.17	0.16	1.33	0.77	0.53	0.33	0.48
Low	0.51	0.49	2.47	1.92	1.40	0.61	0.84
Medium	0.00	0.00	0.90	0.19	0.00	0.19	0.19
High	0.00	0.00	0.63	0.19	0.19	0.19	0.41
10 n 40 T	0.27	0.46	2.08	0.89	0.60	0.58	0.60
Low	0.69	0.98	3.31	2.16	1.02	1.00	1.04
Medium	0.12	0.40	2.11	0.14	0.26	0.40	0.42
High	0.00	0.00	0.83	0.37	0.53	0.34	0.34
10 n 50 T	0.72	0.59	2.39	1.14	0.88	0.88	1.06
Low	1.42	1.45	3.53	1.98	1.42	1.31	1.91
Medium	0.35	0.19	2.08	0.65	0.55	0.71	0.27
High	0.39	0.12	1.57	0.79	0.66	0.61	1.01
15 n 20 T	0.00	0.00	0.71	0.68	0.20	0.09	0.34
Low	0.00	0.00	1.48	1.67	0.59	0.00	0.00
Medium	0.00	0.00	0.65	0.37	0.00	0.28	0.65
High	0.00	0.00	0.00	0.00	0.00	0.00	0.37
15 n 30 T	0.39	0.40	1.29	0.97	0.44	0.37	0.55
Low	1.00	1.05	3.27	2.31	1.17	0.96	1.06
Medium	0.16	0.16	0.60	0.38	0.16	0.16	0.60
High	0.00	0.00	0.00	0.22	0.00	0.00	0.00
15 n 40 T	0.75	0.70	1.80	1.03	0.87	0.92	0.90
Low	1.75	1.59	3.71	2.64	2.10	2.09	1.85
Medium	0.50	0.50	0.85	0.25	0.50	0.50	0.50
High	0.00	0.00	0.85	0.19	0.00	0.16	0.34
15 n 50 T	1.27	1.34	2.72	1.52	1.41	1.40	1.37
Low	2.43	2.56	3.82	3.06	2.69	2.70	2.65
Medium	0.97	1.04	2.74	0.95	1.13	1.22	1.07
High	0.42	0.42	1.61	0.56	0.42	0.28	0.40
20 n 20 T	0.07	0.07	1.12	0.95	0.30	0.60	0.75
Low	0.22	0.22	2.59	2.07	0.89	0.67	1.20
Medium	0.00	0.00	0.50	0.50	0.00	0.56	0.50
High	0.00	0.00	0.28	0.28	0.00	0.56	0.56
20 n 30 T	0.45	0.55	0.98	0.64	0.74	0.65	0.71
Low	1.16	1.28	1.67	1.40	1.42	1.42	1.42
Medium	0.19	0.37	1.26	0.53	0.53	0.53	0.72
High	0.00	0.00	0.00	0.00	0.28	0.00	0.00

(continued)

7.3 Performance of the Heuristics

Table 7.10 (continued)

Test-set	Ke	Kh	Ue	ULPh	UCh	UPRh	UPDh
20 n 40 T	0.96	0.89	1.75	1.05	0.87	0.90	0.90
Low	2.20	2.12	3.65	2.47	2.07	2.04	2.03
Medium	0.35	0.22	1.12	0.35	0.22	0.35	0.35
High	0.32	0.32	0.48	0.32	0.32	0.32	0.32
20 n 50 T	1.75	1.66	2.57	2.00	1.84	1.75	1.95
Low	3.88	3.80	4.87	4.20	3.97	3.86	4.06
Medium	1.23	1.03	1.91	1.12	1.02	1.11	1.24
High	0.14	0.14	0.93	0.68	0.54	0.28	0.54
25 n 20 T	0.12	0.20	0.41	0.35	0.26	0.26	0.26
Low	0.37	0.59	1.22	1.04	0.78	0.78	0.78
Medium	0.00	0.00	0.00	0.00	0.00	0.00	0.00
High	0.00	0.00	0.00	0.00	0.00	0.00	0.00
25 n 30 T	0.46	0.51	1.09	1.00	0.68	0.62	0.74
Low	1.20	1.20	2.58	2.34	1.57	1.32	1.56
Medium	0.19	0.34	0.69	0.37	0.19	0.53	0.37
High	0.00	0.00	0.00	0.28	0.28	0.00	0.28
25 n 40 T	0.78	0.93	1.87	1.14	1.10	0.91	1.12
Low	2.01	2.18	3.88	2.84	2.71	2.26	2.60
Medium	0.33	0.46	1.12	0.44	0.46	0.33	0.45
High	0.00	0.14	0.62	0.14	0.14	0.14	0.30
25 n 50 T	1.79	1.79	2.97	2.22	1.99	2.09	2.16
Low	3.95	4.00	4.90	4.41	4.08	4.07	4.34
Medium	0.89	0.87	2.01	1.18	1.09	1.29	1.09
High	0.54	0.51	1.99	1.07	0.79	0.90	1.05
Average	0.62	0.64	1.62	1.05	0.79	0.79	0.89

Relative error in % of the upper bounds to the optimum of the CPMP generated by CPLEX. Termination after maximal 60 s

multiplier $\upsilon^{init} = (1-\alpha) \cdot \upsilon^* + \alpha \cdot \bar{\upsilon}$ and $\mu^{init} = (1-\alpha) \cdot \mu^* + \alpha \cdot \bar{\mu}$ where $0 \leq \alpha \leq 1$ holds, μ^*, υ^* are the best Lagrangean multiplier and $\bar{\mu}, \bar{\upsilon}$ are the dual variables of $DL_{(P)}$ and $DL_{(C)}$.

The Lagrangean Hybrid Heuristic UChKh is investigated as presented in Algorithm 6.10 and the maximal computational time is set to 60 s. However, different values of α are applied and the results are presented in Table 7.16 as relative errors of the upper and lower bounds, which are rounded up to the next integer, to the optimum of the CPMP.

The relative error of the upper bound (lower bound) improves if the capacity availability is high (low) and n as well as T are small. Setting $\alpha = 0.00$, only the best Lagrangean multiplier of the last iteration is taken into account. This yields average relative errors of 0.73 % for the upper bound and 8.17 % for the lower bound. Considering the dual variables of the auxiliary LPs $DL_{(P)}$ and $DL_{(C)}$ by $\alpha = 0.25, 0.50, 0.75$ and 1.00 improves the results and yields lower average relative

Table 7.11 Performance of the Lagrangean heuristics from Table 7.8

Test-set	Ke	Kh	Ue	ULPh	UCh	UPRh	UPDh
10 n 20 T	726.02	748.05	562.56	732.39	743.82	789.97	744.56
10 n 30 T	384.59	401.04	200.43	391.11	402.93	430.63	394.76
10 n 40 T	259.62	272.31	77.16	240.42	251.76	270.11	234.87
10 n 50 T	195.51	204.47	51.80	186.09	194.66	208.72	173.67
15 n 20 T	516.32	524.98	442.08	520.95	519.18	545.75	528.59
15 n 30 T	266.33	272.39	146.23	263.68	271.07	285.68	266.09
15 n 40 T	161.31	169.15	60.33	158.66	165.57	176.29	156.03
15 n 50 T	121.02	124.52	36.72	115.39	121.30	130.30	110.41
20 n 20 T	350.84	356.53	320.11	369.19	368.85	385.82	370.69
20 n 30 T	189.24	193.87	110.48	184.08	189.24	200.41	186.54
20 n 40 T	114.53	120.03	49.50	115.96	120.87	128.87	113.85
20 n 50 T	87.53	89.18	33.96	85.05	88.84	96.61	82.48
25 n 20 T	252.83	258.43	230.40	265.24	269.31	278.82	271.54
25 n 30 T	128.30	132.26	71.86	125.04	128.64	137.22	128.69
25 n 40 T	84.50	86.34	32.11	79.24	83.42	90.72	79.81
25 n 50 T	67.99	71.52	24.77	62.08	62.31	71.63	61.20
Average	244.16	251.57	153.16	243.41	248.86	264.22	243.99

Number of iterations per second in s^{-1} to yield the results from Tables 7.9 and 7.10

Table 7.12 Approximation of the Lagrangean dual by the Lagrangean heuristics and lower bounds from Table 7.8

Test-set	LB^{init}	ZL^{init}	$Z^{(C)}$	Ue	UCh	Test-set	LB^{init}	ZL^{init}	$Z^{(C)}$	Ke	Kh
10 n 20 T	6.97	12.00	31.29	7.15	6.93	10 n 20 T	10.23	15.25	33.29	9.95	11.22
Low	4.28	4.28	52.15	1.07	1.38	Low	10.18	10.18	55.05	5.61	8.11
Medium	6.99	11.03	28.32	6.27	6.18	Medium	9.43	13.44	30.08	8.74	9.36
High	9.65	20.69	13.40	14.12	13.24	High	11.09	22.12	14.74	15.49	16.20
10 n 30 T	5.57	9.20	30.82	2.63	3.51	10 n 30 T	8.54	12.18	32.46	4.32	7.26
Low	3.32	3.32	52.32	0.28	0.88	Low	9.66	9.66	55.39	2.14	7.67
Medium	6.64	7.43	28.79	2.17	3.75	Medium	8.69	9.48	30.16	4.08	5.82
High	6.74	16.85	11.36	5.43	5.91	High	7.27	17.39	11.82	6.73	8.28
10 n 40 T	6.85	10.35	33.80	1.95	4.18	10 n 40 T	8.26	11.78	34.24	2.52	6.36
Low	5.36	5.36	57.33	2.53	3.22	Low	9.19	9.19	58.74	0.79	7.33
Medium	7.11	8.36	29.79	0.90	4.08	Medium	8.14	9.41	30.37	2.44	4.87
High	8.09	17.34	14.29	2.41	5.25	High	7.44	16.74	13.60	4.32	6.88
Average	6.46	10.52	31.97	3.91	4.88	Average	9.01	13.07	33.33	5.59	8.28

Relative error in % of the lower bounds to $Z_{(C)}$. Termination of the Lagrangean Heuristic after maximal 60 s. Left: $Z_{(C)}$. Right: $Z_{(P)}$

errors of 0.64, 0.64, 0.65 and 0.66 % for the upper bounds and 7.96, 7.93, 7.91 and 7.92 % for the lower bounds. The actual choice of $\alpha > 0.00$ should be interpreted with caution because the differences of the average relative errors of the upper

7.3 Performance of the Heuristics

Table 7.13 Performance of the Lagrangean heuristics from Table 7.8

Test-set	$Z_{(P)}$	Ke	Kh	$Z_{(C)}$	Ue	UCh	$Z^{(X)(Y)}$
10 n 20 T	12.39	21.45	22.52	15.43	21.87	21.67	15.71
Low	8.73	14.00	16.20	14.37	15.31	15.58	14.42
Medium	13.64	21.51	22.02	15.88	21.52	21.42	16.10
High	14.79	28.83	29.35	16.05	28.79	28.00	16.60
10 n 30 T	11.16	15.17	17.78	13.97	16.43	17.13	14.75
Low	7.62	9.64	14.73	13.69	13.94	14.46	13.79
Medium	11.38	15.11	16.61	13.31	15.27	16.61	14.42
High	14.47	20.77	21.99	14.90	20.09	20.31	16.04
10 n 40 T	11.97	14.31	17.73	13.99	15.10	16.99	15.30
Low	7.37	8.12	14.20	13.14	13.31	13.98	13.27
Medium	11.63	13.89	16.00	12.62	13.45	16.21	14.34
High	16.90	20.93	23.00	16.21	18.54	20.79	18.30
Average	11.84	14.99	17.75	14.46	16.42	17.45	15.25

Relative error in % of the lower bounds to the optimum of the CPMP generated by CPLEX. Termination of the Lagrangean Heuristic after maximal 60 s

Table 7.14 Construction heuristics, metaheuristics, lower and upper bounds used in Tables 7.15, 7.16, 7.17, 7.18, 7.19, and 7.20

Algorithm	Description	Section
IB3	Iterated Best-of-three Heuristic	Section 6.1.3
B3	Best upper bound of FF, OF and OS	
FF	First Fit Heuristic	Section 6.1.1
OF	Overlap Heuristic with the priority rule flexibility	Section 6.1.2
OS	Overlap Heuristic with the priority rule simplicity	Section 6.1.2
Tabu	Tabu Search Heuristic	Section 6.4
Ke	Lagrangean Heuristic, $ZL_{(P)}$ solved to optimality (Remark 6.3)	Section 7.3.1.2
Kh	Heuristic K in the Lagrangean Heuristic	Section 7.3.1.2
Ue	Lagrangean Heuristic, $ZL_{(C)}$ solved to optimality (Remark 6.3)	Section 7.3.1.2
UCh	Heuristic UC in the Lagrangean Heuristic	Section 7.3.1.2
KeUe	Lagrangean Hybrid Heuristic, start with Ke and change to Ue	Section 6.3
KhUCh	Lagrangean Hybrid Heuristic, start with Kh and change to UCh	Section 6.3
UeKe	Lagrangean Hybrid Heuristic, start with Ue and change to Ke	Section 6.3
UChKh	Lagrangean Hybrid Heuristic, start with UCh and change to Kh	Section 6.3
B2e	Best bounds of Ue, Ke, each with halved computational time	
B2h	Best bounds of UCh, Kh, each with halved computational time	
BestLB	Best lower bound of Ke, Kh, Ue, UCh, UChKh, B2e, B2h, $Z^{(X)(Y)}$ that are rounded up to the next integer	

bounds are insignificant. To sum up, the value $\alpha = 0.75$ is chosen as presented in Algorithm 6.10.

Table 7.15 Performance of the construction heuristics from Table 7.14

Test-set	IB3	B3	FF	OF	OS
10 n 20 T	4.37	6.71	11.89	13.20	17.99
Low	6.28	8.88	13.30	15.22	19.22
Medium	3.87	5.31	12.70	10.57	15.76
High	2.96	5.93	9.67	13.80	18.98
10 n 30 T	6.77	12.28	17.01	17.33	20.63
Low	8.62	13.30	16.92	18.33	22.14
Medium	6.88	11.61	19.11	16.37	19.17
High	4.82	11.94	14.99	17.29	20.59
10 n 40 T	7.48	13.21	17.77	19.32	21.29
Low	9.55	15.29	19.34	20.23	21.90
Medium	8.06	12.60	19.64	18.85	20.50
High	4.82	11.73	14.33	18.87	21.48
10 n 50 T	8.33	14.51	18.70	19.98	22.39
Low	10.02	16.37	19.33	20.23	21.99
Medium	8.34	13.26	20.53	17.92	20.98
High	6.64	13.91	16.23	21.80	24.19
15 n 20 T	4.50	6.92	9.64	14.45	21.09
Low	7.08	9.84	11.91	17.71	19.59
Medium	3.54	5.63	9.50	12.31	20.53
High	2.87	5.28	7.50	13.33	23.15
15 n 30 T	5.29	11.08	15.51	17.34	21.16
Low	8.97	14.82	17.54	20.28	23.12
Medium	4.74	9.73	17.14	16.32	19.71
High	2.15	8.68	11.85	15.41	20.65
15 n 40 T	7.45	13.29	16.89	19.39	22.16
Low	10.01	15.54	17.49	21.24	23.96
Medium	6.63	11.63	18.38	16.98	20.42
High	5.72	12.70	14.80	19.94	22.09
15 n 50 T	8.09	15.32	19.26	22.25	24.02
Low	10.22	17.79	19.62	23.57	25.20
Medium	7.73	14.69	21.52	21.08	22.74
High	6.32	13.49	16.63	22.10	24.11
20 n 20 T	3.53	6.49	9.14	15.71	20.27
Low	5.52	8.75	10.64	18.66	20.38
Medium	3.41	5.81	9.57	13.46	17.57
High	1.67	4.91	7.22	15.00	22.87
20 n 30 T	5.78	12.38	15.78	19.63	22.45
Low	8.06	13.87	15.18	21.46	22.23
Medium	5.87	11.42	17.58	17.52	22.59
High	3.40	11.84	14.58	19.90	22.54

(continued)

7.3 Performance of the Heuristics

Table 7.15 (continued)

Test-set	IB3	B3	FF	OF	OS
20 n 40 T	6.97	14.32	17.75	20.98	24.28
Low	8.64	15.54	17.16	22.43	24.59
Medium	6.96	13.94	19.22	18.83	22.48
High	5.31	13.49	16.87	21.68	25.78
20 n 50 T	7.56	16.13	18.96	23.64	26.50
Low	9.79	17.31	18.16	24.98	26.55
Medium	7.67	15.00	20.23	20.47	24.07
High	5.21	16.08	18.48	25.48	28.87
25 n 20 T	3.08	5.53	9.38	14.42	21.41
Low	4.47	6.82	10.31	15.88	20.20
Medium	3.74	4.96	9.96	10.81	19.13
High	1.02	4.81	7.87	16.58	24.91
25 n 30 T	5.63	11.78	15.85	20.11	24.20
Low	8.60	14.56	16.51	22.38	24.75
Medium	5.13	9.79	17.77	17.57	21.78
High	3.17	11.00	13.26	20.37	26.06
25 n 40 T	7.14	15.01	16.89	22.24	24.39
Low	8.83	17.12	17.42	24.24	25.71
Medium	7.10	13.73	16.92	20.38	21.87
High	5.50	14.17	16.32	22.11	25.58
25 n 50 T	8.74	16.72	19.26	24.84	26.85
Low	10.89	18.45	18.24	26.40	26.85
Medium	8.41	16.17	21.79	23.72	24.72
High	6.92	15.53	17.74	24.41	28.98
Average	6.29	11.98	15.60	19.05	22.57

Relative error in % of the upper bounds to the optimum of the CPMP generated by CPLEX

In what follows, the application of different Lagrangean heuristics and the effectiveness of the Lagrangean hybrid approach is investigated. For this purpose, the Lagrangean Hybrid Heuristic is applied as presented in Algorithm 6.10 and the maximal computational time is 180 s. Table 7.17 presents the results as relative errors of the upper and lower bounds, which are rounded up to the next integer, to the optimum of the CPMP.

Regarding the influence of the capacity availability n and T, the bounds follow the tendency as described for UChKh above.

The Lagrangean Hybrid Heuristic changes latest from one Lagrangean Heuristic to the other one after half of the computational time that is 90 s has elapsed. Hence, a comparison of KhUCh, UChKh and KeUe, UeKe with B2h and B2e shows the effectiveness of the Lagrangean hybrid approach. Note that in B2h and B2e each Lagrangean Heuristic has a computational time of 90 s and the best lower bound is taken. Considering the upper bounds, the average relative errors of KeUe, UeKe and B2e are 0.65, 0.71 and 0.82 %, whereas the average relative error of 0.54 %

Table 7.16 Relative error in % of the results of UChKh from Table 7.14 to the optimum of the CPMP generated by CPLEX for different values of α

Test-set	$\alpha = 0.00$	$\alpha = 0.25$	$\alpha = 0.50$	$\alpha = 0.75$	$\alpha = 1.00$	Test-set	$\alpha = 0.00$	$\alpha = 0.25$	$\alpha = 0.50$	$\alpha = 0.75$	$\alpha = 1.00$
10 n 20 T	0.00	0.00	0.00	0.00	0.00	10 n 20 T	4.81	4.76	4.69	4.76	4.69
Low	0.00	0.00	0.00	0.00	0.00	Low	6.64	6.50	6.50	6.50	6.50
Medium	0.00	0.00	0.00	0.00	0.00	Medium	3.96	3.96	3.74	3.96	3.74
High	0.00	0.00	0.00	0.00	0.00	High	3.83	3.83	3.83	3.83	3.83
10 n 30 T	0.49	0.23	0.16	0.26	0.24	10 n 30 T	8.30	8.07	8.00	8.00	8.05
Low	1.29	0.70	0.49	0.79	0.71	Low	8.80	8.45	8.59	8.45	8.59
Medium	0.19	0.00	0.00	0.00	0.00	Medium	7.60	7.44	7.09	7.25	7.25
High	0.00	0.00	0.00	0.00	0.00	High	8.49	8.31	8.31	8.31	8.31
10 n 40 T	0.55	0.36	0.34	0.38	0.44	10 n 40 T	11.08	10.84	10.92	10.87	10.87
Low	1.38	0.83	0.77	0.75	1.06	Low	10.03	9.94	9.94	9.94	9.94
Medium	0.26	0.26	0.26	0.40	0.26	Medium	9.58	8.96	9.21	9.21	9.21
High	0.00	0.00	0.00	0.00	0.00	High	13.62	13.62	13.62	13.46	13.46
10 n 50 T	0.85	0.74	0.76	0.81	0.73	10 n 50 T	12.91	12.61	12.62	12.65	12.64
Low	1.78	1.47	1.47	1.66	1.51	Low	11.15	11.09	11.09	11.09	11.09
Medium	0.52	0.36	0.57	0.51	0.43	Medium	11.67	11.19	11.46	11.57	11.54
High	0.25	0.39	0.25	0.25	0.25	High	15.92	15.55	15.30	15.30	15.30
15 n 20 T	0.14	0.00	0.00	0.00	0.00	15 n 20 T	2.63	2.46	2.46	2.51	2.43
Low	0.41	0.00	0.00	0.00	0.00	Low	3.27	2.97	2.97	3.11	3.11
Medium	0.00	0.00	0.00	0.00	0.00	Medium	4.35	4.13	4.13	4.13	3.91
High	0.00	0.00	0.00	0.00	0.00	High	0.28	0.28	0.28	0.28	0.28

7.3 Performance of the Heuristics

15 n 30 T	0.48	0.37	0.40	0.40	0.40	15 n 30 T	7.22	6.87	6.83	6.76	6.89
Low	1.28	0.94	1.05	1.05	1.03	Low	4.91	4.69	4.79	4.79	4.79
Medium	0.16	0.16	0.16	0.16	0.16	Medium	7.32	6.71	6.50	6.50	6.66
High	0.00	0.00	0.00	0.00	0.00	High	9.43	9.21	9.21	8.99	9.21
15 n 40 T	0.90	0.77	0.80	0.71	0.81	15 n 40 T	9.46	9.07	9.07	9.07	9.13
Low	2.20	1.94	1.90	1.76	1.94	Low	6.76	6.43	6.43	6.42	6.50
Medium	0.50	0.37	0.50	0.37	0.50	Medium	9.53	8.87	8.87	8.87	8.99
High	0.00	0.00	0.00	0.00	0.00	High	12.10	11.91	11.91	11.91	11.91
15 n 50 T	1.32	1.37	1.27	1.27	1.38	15 n 50 T	10.26	10.17	10.10	10.04	10.05
Low	2.50	2.52	2.44	2.46	2.68	Low	7.34	7.27	7.27	7.27	7.34
Medium	1.05	1.04	0.94	1.06	1.04	Medium	10.63	10.54	10.35	10.15	10.26
High	0.42	0.54	0.42	0.28	0.42	High	12.81	12.69	12.69	12.69	12.56
20 n 20 T	0.21	0.07	0.07	0.07	0.15	20 n 20 T	3.18	3.18	3.18	3.11	3.18
Low	0.63	0.22	0.22	0.22	0.44	Low	3.87	3.87	3.87	3.87	3.87
Medium	0.00	0.00	0.00	0.00	0.00	Medium	4.28	4.28	4.28	4.06	4.28
High	0.00	0.00	0.00	0.00	0.00	High	1.39	1.39	1.39	1.39	1.39
20 n 30 T	0.60	0.45	0.55	0.51	0.49	20 n 30 T	7.54	7.47	7.32	7.18	7.32
Low	1.28	1.16	1.28	1.16	1.28	Low	6.07	6.07	6.07	6.07	6.07
Medium	0.53	0.19	0.37	0.37	0.19	Medium	7.22	7.03	7.03	7.03	7.03
High	0.00	0.00	0.00	0.00	0.00	High	9.32	9.32	8.87	8.43	8.87
20 n 40 T	0.91	0.86	0.89	0.94	0.89	20 n 40 T	9.16	8.91	8.99	9.14	9.05
Low	2.04	1.90	2.13	2.13	2.12	Low	6.43	6.28	6.37	6.45	6.53
Medium	0.36	0.35	0.22	0.36	0.22	Medium	9.21	8.97	8.97	8.97	8.97
High	0.32	0.32	0.32	0.32	0.32	High	11.83	11.49	11.64	11.99	11.64

(continued)

Table 7.16 (continued)

Test-set		$\alpha=0.00$	$\alpha=0.25$	$\alpha=0.50$	$\alpha=0.75$	$\alpha=1.00$	Test-set		$\alpha=0.00$	$\alpha=0.25$	$\alpha=0.50$	$\alpha=0.75$	$\alpha=1.00$
20 n 50 T		1.53	1.63	1.52	1.57	1.70	20 n 50 T		10.62	10.53	10.37	10.29	10.15
	Low	3.64	3.55	3.75	3.56	3.80		Low	7.72	7.66	7.66	7.66	7.60
	Medium	0.82	1.22	0.82	1.02	1.03		Medium	9.86	9.66	9.56	9.56	9.56
	High	0.14	0.12	0.00	0.14	0.26		High	14.28	14.28	13.89	13.65	13.28
25 n 20 T		0.20	0.12	0.12	0.20	0.12	25 n 20 T		3.73	3.51	3.51	3.51	3.49
	Low	0.59	0.37	0.37	0.59	0.37		Low	4.81	4.59	4.59	4.59	4.59
	Medium	0.00	0.00	0.00	0.00	0.00		Medium	3.88	3.44	3.44	3.44	3.38
	High	0.00	0.00	0.00	0.00	0.00		High	2.50	2.50	2.50	2.50	2.50
25 n 30 T		0.55	0.53	0.44	0.44	0.44	25 n 30 T		8.28	7.95	8.00	7.95	7.93
	Low	1.46	1.58	1.32	1.32	1.32		Low	5.32	5.32	5.32	5.32	5.46
	Medium	0.19	0.00	0.00	0.00	0.00		Medium	7.89	7.57	7.73	7.57	7.38
	High	0.00	0.00	0.00	0.00	0.00		High	11.63	10.96	10.96	10.96	10.96
25 n 40 T		1.03	0.90	0.85	1.02	0.89	25 n 40 T		9.86	9.69	9.66	9.63	9.73
	Low	2.61	2.26	2.09	2.60	2.33		Low	7.48	7.39	7.31	7.48	7.40
	Medium	0.33	0.44	0.33	0.33	0.33		Medium	9.66	9.39	9.39	9.12	9.52
	High	0.14	0.00	0.14	0.14	0.00		High	12.44	12.28	12.28	12.28	12.28
25 n 50 T		1.86	1.87	2.03	1.84	1.93	25 n 50 T		11.65	11.30	11.20	11.16	11.07
	Low	4.07	3.88	4.20	4.13	4.13		Low	8.87	8.82	8.82	8.87	8.82
	Medium	0.87	1.08	1.11	0.87	0.89		Medium	10.95	10.57	10.74	10.56	10.59
	High	0.65	0.65	0.79	0.51	0.77		High	15.14	14.52	14.04	14.05	13.81
Average		0.73	0.64	0.64	0.65	0.66	Average		8.17	7.96	7.93	7.91	7.92

Termination after maximal 60 s. Left: Upper bounds. Right: Lower bounds, which are rounded up to the next integer

7.3 Performance of the Heuristics

Table 7.17 Variants of the Lagrangean hybrid heuristics from Table 7.14

Test-set	KeUe	KhUCh	UeKe	UChKh	B2e	B2h	Test-set	KeUe	KhUCh	UeKe	UChKh	B2e	B2h
10 n 20 T	0.00	0.00	0.00	0.00	0.00	0.00	10 n 20 T	1.89	4.74	1.80	4.63	1.74	4.81
Low	0.00	0.00	0.00	0.00	0.00	0.00	Low	1.67	6.66	1.67	6.50	1.51	6.64
Medium	0.00	0.00	0.00	0.00	0.00	0.00	Medium	0.72	3.74	0.72	3.56	0.72	3.96
High	0.00	0.00	0.00	0.00	0.00	0.00	High	3.28	3.83	3.00	3.83	3.00	3.83
10 n 30 T	0.20	0.24	0.13	0.12	0.12	0.16	10 n 30 T	4.58	7.85	4.51	7.95	4.47	8.05
Low	0.61	0.71	0.39	0.35	0.37	0.49	Low	2.94	8.55	2.94	8.45	2.83	8.59
Medium	0.00	0.00	0.00	0.00	0.00	0.00	Medium	3.74	6.91	3.74	7.09	3.74	7.25
High	0.00	0.00	0.00	0.00	0.00	0.00	High	7.07	8.08	6.85	8.31	6.85	8.31
10 n 40 T	0.36	0.30	0.48	0.23	0.22	0.30	10 n 40 T	7.04	10.65	6.96	10.59	6.96	10.87
Low	0.72	0.64	0.96	0.56	0.53	0.64	Low	4.46	9.94	4.38	9.75	4.31	10.08
Medium	0.35	0.26	0.49	0.12	0.12	0.26	Medium	6.24	8.70	6.24	8.70	6.13	9.08
High	0.00	0.00	0.00	0.00	0.00	0.00	High	10.43	13.31	10.27	13.31	10.43	13.46
10 n 50 T	0.69	0.46	0.87	0.54	0.63	0.36	10 n 50 T	8.20	12.62	8.22	12.55	8.14	12.87
Low	1.33	1.10	1.73	1.03	1.37	0.86	Low	5.84	11.16	5.90	11.09	5.25	11.07
Medium	0.35	0.17	0.51	0.33	0.26	0.09	Medium	7.69	11.39	7.59	11.25	7.59	11.39
High	0.39	0.12	0.37	0.25	0.26	0.12	High	11.07	15.30	11.17	15.30	11.57	16.16
15 n 20 T	0.00	0.00	0.00	0.00	0.71	0.00	15 n 20 T	2.06	3.42	2.06	3.51	3.25	3.75
Low	0.00	0.00	0.00	0.00	1.48	0.00	Low	1.85	4.59	1.85	4.59	4.59	4.81
Medium	0.00	0.00	0.00	0.00	0.65	0.00	Medium	2.10	3.16	2.10	3.44	2.94	3.94
High	0.00	0.00	0.00	0.00	0.00	0.00	High	2.22	2.50	2.22	2.50	2.22	2.50
15 n 30 T	0.43	0.36	0.44	0.40	1.25	0.35	15 n 30 T	5.72	7.83	5.72	7.90	6.70	8.11
Low	1.14	0.91	1.17	1.05	3.16	1.05	Low	3.93	5.32	3.93	5.32	5.32	5.51
Medium	0.16	0.16	0.16	0.16	0.60	0.00	Medium	3.95	7.22	3.95	7.41	5.04	7.20
High	0.00	0.00	0.00	0.00	0.00	0.00	High	9.29	10.96	9.29	10.96	9.73	11.63

(continued)

Table 7.17 (continued)

Test-set	KeUe	KhUCh	UeKe	UChKh	B2e	B2h	Test-set	KeUe	KhUCh	UeKe	UChKh	B2e	B2h
15 n 40 T	0.87	0.64	0.92	0.68	1.77	0.65	15 n 40 T	6.88	9.57	6.92	9.60	7.49	9.72
Low	2.12	1.43	2.10	1.68	3.62	1.44	Low	5.72	7.32	5.72	7.40	7.01	7.48
Medium	0.50	0.50	0.50	0.37	0.85	0.50	Medium	6.38	9.12	6.52	9.12	6.78	9.39
High	0.00	0.00	0.16	0.00	0.85	0.00	High	8.53	12.28	8.53	12.28	8.69	12.28
15 n 50 T	1.37	1.06	1.44	1.05	2.58	1.09	15 n 50 T	8.12	10.99	8.33	11.00	9.29	11.48
Low	2.56	2.36	2.85	2.21	3.75	2.18	Low	6.51	8.93	6.55	8.82	8.17	8.94
Medium	1.14	0.55	1.05	0.67	2.65	0.66	Medium	7.33	10.38	7.42	10.49	8.09	10.63
High	0.42	0.28	0.42	0.28	1.35	0.42	High	10.53	13.67	11.01	13.68	11.61	14.87
20 n 20 T	0.17	0.07	0.07	0.07	0.07	0.07	20 n 20 T	1.92	3.18	1.80	3.11	1.82	3.18
Low	0.22	0.22	0.22	0.22	0.22	0.22	Low	2.31	3.87	1.96	3.87	1.78	3.87
Medium	0.28	0.00	0.00	0.00	0.00	0.00	Medium	2.06	4.28	2.06	4.06	2.28	4.28
High	0.00	0.00	0.00	0.00	0.00	0.00	High	1.39	1.39	1.39	1.39	1.39	1.39
20 n 30 T	0.54	0.45	0.54	0.51	0.41	0.55	20 n 30 T	5.57	7.10	5.57	7.10	5.41	7.34
Low	1.28	1.16	1.28	1.16	1.03	1.28	Low	5.28	6.07	5.28	6.07	4.82	6.07
Medium	0.34	0.19	0.34	0.37	0.19	0.37	Medium	4.70	7.03	4.70	7.03	4.70	6.85
High	0.00	0.00	0.00	0.00	0.00	0.00	High	6.72	8.21	6.72	8.21	6.72	9.10
20 n 40 T	0.90	0.78	0.94	0.78	0.81	0.78	20 n 40 T	6.67	8.94	6.57	8.97	6.58	9.02
Low	2.04	1.81	2.15	1.81	1.89	1.79	Low	5.62	6.37	5.45	6.45	5.37	6.45
Medium	0.35	0.22	0.35	0.22	0.22	0.22	Medium	5.83	8.97	5.69	8.97	5.80	8.97
High	0.32	0.32	0.32	0.32	0.32	0.32	High	8.57	11.49	8.57	11.49	8.57	11.64
20 n 50 T	1.65	1.47	1.81	1.34	1.53	1.42	20 n 50 T	8.11	10.13	8.19	10.18	8.32	10.62
Low	3.68	3.50	3.99	3.18	3.58	3.44	Low	6.46	7.72	6.59	7.66	6.33	7.71
Medium	1.13	0.92	1.02	0.83	1.02	0.82	Medium	6.85	9.39	6.95	9.47	7.45	9.76
High	0.14	0.00	0.42	0.00	0.00	0.00	High	11.03	13.29	11.03	13.40	11.17	14.38

7.3 Performance of the Heuristics

	Upper bounds						Lower bounds					
25 n 20 T	0.06	0.06	0.14	0.06	0.12	0.14	1.43	2.51	1.43	2.43	1.43	2.51
Low	0.19	0.19	0.41	0.19	0.37	0.41	2.10	3.11	2.10	3.11	2.10	3.11
Medium	0.00	0.00	0.00	0.00	0.00	0.00	1.91	4.13	1.91	3.91	1.91	4.13
High	0.00	0.00	0.00	0.00	0.00	0.00	0.28	0.28	0.28	0.28	0.28	0.28
25 n 30 T	0.46	0.50	0.59	0.41	0.46	0.42	5.30	6.89	5.30	6.76	5.30	7.08
Low	1.20	1.32	1.58	1.23	1.20	1.06	4.12	4.79	4.12	4.79	4.12	4.79
Medium	0.19	0.19	0.19	0.00	0.19	0.19	4.36	6.66	4.36	6.50	4.36	7.02
High	0.00	0.00	0.00	0.00	0.00	0.00	7.43	9.21	7.43	8.99	7.43	9.43
25 n 40 T	0.87	0.84	0.98	0.75	0.75	0.78	7.12	9.01	7.26	9.02	7.24	9.17
Low	2.10	2.18	2.43	1.92	1.92	2.01	5.42	6.42	5.43	6.42	5.11	6.60
Medium	0.33	0.33	0.33	0.33	0.33	0.33	6.45	8.87	6.56	8.73	6.69	8.99
High	0.19	0.00	0.19	0.00	0.00	0.00	9.48	11.75	9.78	11.91	9.92	11.91
25 n 50 T	1.81	1.72	1.94	1.69	1.65	1.65	7.92	10.01	8.17	9.83	8.22	10.26
Low	4.13	3.88	4.00	3.82	3.88	3.82	6.32	7.21	6.32	7.20	6.24	7.33
Medium	0.77	0.78	0.89	0.88	0.67	0.77	6.93	10.26	7.41	9.84	7.39	10.63
High	0.54	0.51	0.93	0.37	0.40	0.37	10.50	12.56	10.78	12.44	11.04	12.81
Average	0.65	0.56	0.71	0.54	0.82	0.54	5.53	7.84	5.55	7.82	5.77	8.05

Relative error in % of the results to the optimum of the CPMP generated by CPLEX. Termination after maximal 180 s. Left: Upper bounds. Right: Lower bounds rounded up to the next integer for UChKh is comparable to 0.54 % for B2h and 0.56 % for KhUCh. Regarding the lower bounds, KeUe, UeKe and B2e provide average relative errors 5.53, 5.55 and 5.77 % and UChKh, KhUCh and B2h yield 7.82, 7.84 and 8.05 %. Therefore, the Lagrangean hybrid approach slightly improves the upper and lower bounds of the individual Lagrangean heuristics that run with half of the computational time. The improvement is better if the Lagrangean relaxation is solved to optimality. UChKh yields the best upper bound and KeUe the best lower bound.

7.3.2.3 The Metaheuristics

The performance of the metaheuristics is evaluated for all 7560 instances presented in Sect. 7.1.

Tabu is applied as stated in Algorithm 6.11 and all other heuristics as presented in the previous sections. Since Tabu contains a stochastic element, the average of four computations is taken. Considering the findings of Sect. 7.3.2.2, the Lagrangean Hybrid Heuristic UChKh is evaluated. In Table 7.18, the results are given as the relative error of the upper and lower bounds, which are rounded up to the next integer, to the optimum of the CPMP. Table 7.18 shows the computational time in seconds, which is required to solve an instance to optimality with CPLEX. The computational time tends to drastically increase if the values for n, T are high and the capacity availability is low. Due to this fact, instances with $T > 50$ and $n > 25$ could not be solved to optimality. Consequently, the Tables 7.19 and 7.20 present the results as relative errors of the upper and lower bounds, which are rounded up to the next integer, to the best known lower bound BestLB. Among all lower bounds from Theorem 5.22 evaluated in Sect. 7.2, the computational time of the promising lower bound $Z^{(X)}$ was too high for the instances with large values of n and T. Therefore, BestLB comprises only the lower bounds from the heuristics and from $Z^{(X)(Y)}$.

In order to estimate the relative errors to the optimum of the CPMP for the periods $T \in \{100, 150, 200\}$, a linear regression is performed with the averages per test-set from Table 7.18 for the periods $T \in \{20, 30, 40, 50\}$ where the relative errors are calculated with the optimum. Figure 7.1 shows the linear regression graphs of the relative errors of the upper bounds provided by Ke, Kh, Ue, UCh, UChKh, Tabu and of BestLB from Table 7.18. Note that the linear regression graphs meet the origin. For every approach, the linear regression graphs show similar tendencies for different values of n. It is well-known that the relative errors of upper and lower bounds to the optimum increase with the instance size. However, Fig. 7.1 shows that the absolute value of the slope of BestLB is much larger than the value of slope of the upper bounds. Since Tables 7.19 and 7.20 evaluate against BestLB, the relatively large values in both tables are due to BestLB. Therefore, it is expected that an evaluation against the optimum would yield relative errors in the magnitude of at most 5 % for Ke, Kh, UCh, UChKh and 9 % for Tabu and Ue.

The relative errors of the upper and lower bounds mostly improve if n and T become small. The lowest average relative errors of the upper bounds are provided by the Lagrangean heuristics Ke, Kh, UChKh, UCh followed Tabu and Ue. These

7.3 Performance of the Heuristics

Table 7.18 Performance of the metaheuristics from Table 7.14

Test-set	Tabu	Ke	Kh	Ue	UCh	UChKh	CPLEX Time s	Test-set	Ke	Kh	Ue	UCh	UChKh	BestLB
10 n 20 T								10 n 20 T						
Low	0.00	0.00	0.00	0.82	0.00	0.00	0.06	Low	1.84	4.93	4.35	5.22	4.63	1.74
Medium	0.00	0.00	0.00	1.90	0.00	0.00	0.09	Medium	1.51	6.95	6.50	6.64	6.50	1.51
High	0.00	0.00	0.00	0.56	0.00	0.00	0.04	High	0.72	4.02	3.00	5.19	3.56	0.72
10 n 30 T	0.00	0.00	0.00	0.00	0.00	0.00	0.05	10 n 30 T	3.28	3.83	3.56	3.83	3.83	3.00
Low	0.15	0.08	0.07	1.33	0.46	0.12	1.38	Low	4.55	8.19	7.06	8.95	7.95	4.47
Medium	0.11	0.25	0.22	2.47	1.19	0.35	3.14	Medium	2.83	8.80	8.35	8.80	8.45	2.83
High	0.16	0.00	0.00	0.90	0.00	0.00	0.46	High	3.74	7.46	5.53	8.41	7.09	3.74
10 n 40 T	0.19	0.00	0.00	0.63	0.19	0.00	0.53	10 n 40 T	7.07	8.31	7.30	9.64	8.31	6.85
Low	0.29	0.16	0.33	1.89	0.44	0.23	32.28	Low	7.60	10.98	8.99	11.36	10.59	6.87
Medium	0.30	0.48	0.72	3.01	0.73	0.56	62.62	Medium	4.31	10.33	9.49	10.08	9.75	4.31
High	0.37	0.00	0.26	1.97	0.26	0.12	19.15	High	6.42	8.84	6.90	10.21	8.70	6.03
10 n 50 T	0.19	0.00	0.00	0.69	0.34	0.00	15.08	10 n 50 T	12.08	13.78	10.58	13.78	13.31	10.27
Low	0.88	0.52	0.44	2.12	0.58	0.54	471.47	Low	9.20	12.93	10.33	13.18	12.55	7.92
Medium	0.75	1.18	1.11	3.35	1.19	1.03	536.84	Medium	5.19	11.47	10.83	11.25	11.09	5.19
High	0.63	0.26	0.09	1.73	0.29	0.33	332.65	High	8.61	11.39	8.57	12.13	11.25	7.49
15 n 20 T	1.26	0.12	0.12	1.27	0.26	0.25	544.91	15 n 20 T	13.80	15.92	11.58	16.16	15.30	11.07
Low	0.00	0.00	0.00	0.64	0.12	0.00	0.10	Low	2.15	3.75	3.25	3.96	3.51	2.06
Medium	0.00	0.00	0.00	1.26	0.37	0.00	0.14	Medium	1.85	4.81	4.59	5.00	4.59	1.85
High	0.00	0.00	0.00	0.65	0.00	0.00	0.09	High	2.10	3.94	2.94	4.38	3.44	2.10
	0.00	0.00	0.00	0.00	0.00	0.00	0.06		2.50	2.50	2.22	2.50	2.50	2.22

(continued)

Table 7.18 (continued)

Test-set	Tabu	Ke	Kh	Ue	UCh	UChKh	CPLEX Time s	Test-set	Ke	Kh	Ue	UCh	UChKh	BestLB
15 n 30 T	0.21	0.36	0.32	1.17	0.40	0.40	4.95	15 n 30 T	6.69	8.26	6.70	8.77	7.90	3.82
Low	0.48	0.91	0.96	2.92	1.05	1.05	11.15	Low	3.82	5.51	5.32	5.76	5.32	6.32
Medium	0.16	0.16	0.00	0.60	0.16	0.16	2.88	Medium	5.47	7.20	5.04	8.74	7.41	3.95
High	0.00	0.00	0.00	0.00	0.00	0.00	0.81	High	10.78	12.07	9.73	11.82	10.96	9.29
15 n 40 T	0.81	0.62	0.61	1.66	0.76	0.68	108.26	15 n 40 T	8.35	9.73	7.45	10.39	9.60	5.72
Low	1.27	1.49	1.33	3.46	1.77	1.68	261.48	Low	5.72	7.48	7.01	7.72	7.40	7.64
Medium	0.30	0.37	0.50	0.85	0.50	0.37	19.77	Medium	7.92	9.12	6.78	10.41	9.12	6.38
High	0.85	0.00	0.00	0.66	0.00	0.00	43.52	High	11.42	12.60	8.57	13.05	12.28	8.41
15 n 50 T	2.34	1.10	1.06	2.50	1.21	1.05	2648.81	15 n 50 T	9.68	11.40	8.91	12.01	11.00	8.12
Low	2.17	2.22	2.01	3.68	2.24	2.21	7042.23	Low	6.47	9.10	8.10	9.29	8.82	6.33
Medium	2.34	0.67	0.74	2.46	0.96	0.67	431.49	Medium	8.71	10.35	7.66	11.49	10.49	7.22
High	2.52	0.42	0.42	1.35	0.42	0.28	472.71	High	13.85	14.74	10.96	15.24	13.68	10.82
20 n 20 T	0.00	0.07	0.07	0.93	0.22	0.07	0.13	20 n 20 T	1.82	3.24	2.90	3.35	3.11	1.75
Low	0.00	0.22	0.22	2.00	0.67	0.22	0.20	Low	1.78	4.05	3.71	3.87	3.87	1.78
Medium	0.00	0.00	0.00	0.50	0.00	0.00	0.12	Medium	2.28	4.28	3.61	4.78	4.06	2.08
High	0.00	0.00	0.00	0.28	0.00	0.00	0.08	High	1.39	1.39	1.39	1.39	1.39	1.39
20 n 30 T	0.45	0.37	0.55	0.98	0.71	0.51	5.46	20 n 30 T	6.87	7.46	5.89	8.07	7.10	5.41
Low	0.57	0.91	1.28	1.67	1.33	1.16	14.54	Low	4.82	6.20	6.07	6.46	6.07	4.82
Medium	0.44	0.19	0.37	1.26	0.53	0.37	1.11	Medium	6.48	6.85	4.88	8.20	7.03	4.70
High	0.35	0.00	0.00	0.00	0.28	0.00	0.73	High	9.32	9.32	6.72	9.54	8.21	6.72
20 n 40 T	1.12	0.78	0.74	1.48	0.82	0.78	137.18	20 n 40 T	8.51	9.08	6.74	9.69	8.97	6.46
Low	2.11	1.80	1.87	3.33	1.91	1.81	361.55	Low	5.45	6.45	5.97	6.75	6.45	5.29
Medium	0.53	0.22	0.22	0.63	0.22	0.22	32.49	Medium	8.57	8.97	5.69	10.02	8.97	5.69
High	0.72	0.32	0.14	0.48	0.32	0.32	17.51	High	11.51	11.83	8.57	12.31	11.49	8.41

7.3 Performance of the Heuristics

20 n 50 T	2.71	1.41	1.49	2.42	1.58	1.34	5718.31		20 n 50 T	10.00	10.55	8.32	10.91	10.18	8.02
Low	3.77	3.31	3.56	4.70	3.57	3.18	15816.43		Low	6.39	7.71	7.14	8.09	7.66	6.25
Medium	2.05	0.93	0.92	1.73	0.92	0.83	337.58		Medium	9.57	9.66	6.78	10.25	9.47	6.78
High	2.30	0.00	0.00	0.82	0.26	0.00	1000.91		High	14.03	14.28	11.03	14.38	13.40	11.03
25 n 20 T	0.08	0.12	0.14	0.35	0.12	0.06	0.16		25 n 20 T	1.72	2.43	2.07	3.00	2.43	1.43
Low	0.24	0.37	0.41	1.04	0.37	0.19	0.24		Low	2.24	3.11	2.97	3.64	3.11	2.10
Medium	0.00	0.00	0.00	0.00	0.00	0.00	0.16		Medium	2.63	3.91	2.96	5.07	3.91	1.91
High	0.00	0.00	0.00	0.00	0.00	0.00	0.09		High	0.28	0.28	0.28	0.28	0.28	0.28
25 n 30 T	0.54	0.46	0.42	0.97	0.55	0.41	4.73		25 n 30 T	6.44	7.03	5.44	7.56	6.76	5.26
Low	0.78	1.20	1.06	2.21	1.46	1.23	11.32		Low	4.12	4.89	4.36	5.03	4.79	4.00
Medium	0.51	0.19	0.19	0.69	0.19	0.00	1.88		Medium	5.99	6.55	4.54	7.99	6.50	4.36
High	0.32	0.00	0.00	0.00	0.00	0.00	0.99		High	9.21	9.65	7.43	9.65	8.99	7.43
25 n 40 T	1.53	0.72	0.78	1.67	0.85	0.75	1081.90		25 n 40 T	8.50	9.09	7.25	9.52	9.02	6.95
Low	2.49	1.84	2.01	3.71	2.09	1.92	3096.34		Low	5.27	6.68	6.01	6.84	6.42	5.11
Medium	0.85	0.33	0.33	0.98	0.33	0.33	59.27		Medium	8.14	8.49	6.45	9.65	8.73	6.45
High	1.26	0.00	0.00	0.32	0.14	0.00	90.09		High	12.10	12.10	9.30	12.07	11.91	9.30
25 n 50 T	3.15	1.58	1.61	2.86	1.82	1.69	17433.34		25 n 50 T	9.93	10.23	8.10	10.39	9.83	7.92
Low	4.63	3.82	3.70	4.71	3.70	3.82	49225.62		Low	6.64	7.21	6.51	7.52	7.20	6.19
Medium	2.46	0.67	0.77	2.01	0.98	0.88	624.50		Medium	10.09	10.41	7.16	10.83	9.84	6.93
High	2.36	0.26	0.37	1.86	0.79	0.37	2449.90		High	13.06	13.06	10.64	12.81	12.44	10.64
Average	0.89	0.52	0.54	1.49	0.67	0.54			Average	6.49	8.08	6.49	8.52	7.82	5.25

Relative error in % of the results to the optimum of the CPMP generated by CPLEX. Termination after maximal 180 s. Left: Upper bounds. Right: Lower bounds, which are rounded up to the next integer

Table 7.19 Performance of the metaheuristics from Table 7.14

Test-set	Ke	Kh	Ue	UCh	UChKh	Test-set	Ke	Kh	Ue	UCh	UChKh
10 n 20 T	0.09	3.19	2.61	3.47	2.88	20 n 20 T	0.15	1.57	1.24	1.68	1.44
Low	0.00	5.44	4.98	5.12	4.98	Low	0.00	2.28	1.93	2.09	2.09
Medium	0.00	3.30	2.28	4.46	2.83	Medium	0.44	2.44	1.78	2.94	2.22
High	0.28	0.83	0.56	0.83	0.83	High	0.00	0.00	0.00	0.00	0.00
10 n 30 T	0.07	3.75	2.61	4.51	3.51	20 n 30 T	1.46	2.04	0.48	2.66	1.69
Low	0.00	6.06	5.60	6.07	5.70	Low	0.00	1.37	1.25	1.63	1.25
Medium	0	3.73	1.80	4.68	3.36	Medium	1.78	2.15	0.19	3.51	2.34
High	0.22	1.46	0.44	2.79	1.46	High	2.60	2.60	0.00	2.83	1.49
10 n 40 T	0.74	4.27	2.20	4.68	3.86	20 n 40 T	2.10	2.70	0.30	3.34	2.58
Low	0.00	6.22	5.34	5.95	5.61	Low	0.17	1.20	0.70	1.50	1.20
Medium	0.39	2.90	0.89	4.34	2.76	Medium	2.89	3.29	0.00	4.36	3.29
High	1.84	3.69	0.37	3.75	3.22	High	3.23	3.60	0.19	4.15	3.25
10 n 50 T	1.34	5.31	2.53	5.59	4.90	20 n 50 T	2.10	2.69	0.31	3.08	2.28
Low	0	6.56	5.88	6.32	6.16	Low	0.14	1.53	0.92	1.92	1.47
Medium	1.19	4.12	1.09	4.93	3.98	Medium	2.91	3.00	0.00	3.65	2.80
High	2.83	5.25	0.62	5.51	4.55	High	3.25	3.55	0.00	3.67	2.57
10 n 100 T	0.79	4.67	2.60	4.64	4.09	20 n 100 T	1.26	1.90	0.51	2.12	1.43
Low	0.00	6.48	5.49	6.51	6.15	Low	0.17	1.22	0.85	1.72	1.18
Medium	0.59	3.58	1.32	4.10	3.39	Medium	1.52	2.16	0.16	2.48	1.53
High	1.79	3.96	0.99	3.31	2.72	High	2.09	2.31	0.52	2.17	1.59
10 n 150 T	0.11	3.85	2.54	3.72	3.37	20 n 150 T	1.14	1.61	0.73	1.50	1.15
Low	0.00	6.02	5.17	5.86	5.55	Low	0.44	1.22	0.79	1.44	1.13
Medium	0.00	3.10	1.65	3.48	2.90	Medium	0.84	1.16	0.44	1.32	0.93
High	0.32	2.43	0.81	1.82	1.66	High	2.14	2.45	0.97	1.74	1.39
10 n 200 T	0.25	3.94	2.94	3.82	3.44	20 n 200 T	1.06	1.59	0.76	1.35	0.99
Low	0.00	6.26	5.49	6.09	5.74	Low	0.40	1.22	0.72	1.45	0.87
Medium	0.06	2.99	2.09	3.29	2.78	Medium	1.09	1.64	0.50	1.48	0.95
High	0.70	2.57	1.25	2.08	1.79	High	1.69	1.92	1.05	1.12	1.16
15 n 20 T	0.09	1.69	1.19	1.90	1.45	25 n 20 T	0.29	1.00	0.64	1.57	1.00
Low	0.00	2.96	2.74	3.14	2.74	Low	0.14	1.01	0.87	1.54	1.01
Medium	0.00	1.83	0.83	2.28	1.33	Medium	0.72	2.00	1.06	3.17	2.00
High	0.28	0.28	0.00	0.28	0.28	High	0.00	0.00	0.00	0.00	0.00
15 n 30 T	1.00	2.59	1.01	3.11	2.22	25 n 30 T	1.18	1.77	0.18	2.29	1.50
Low	0.00	1.68	1.49	1.93	1.49	Low	0.12	0.89	0.36	1.03	0.79
Medium	1.52	3.25	1.09	4.79	3.46	Medium	1.63	2.19	0.19	3.63	2.14
High	1.49	2.83	0.44	2.60	1.71	High	1.78	2.22	0.00	2.22	1.56
15 n 40 T	1.52	2.97	0.64	3.70	2.85	25 n 40 T	1.58	2.18	0.30	2.64	2.11
Low	0.00	1.80	1.32	2.07	1.73	Low	0.16	1.59	0.91	1.76	1.33
Medium	1.54	2.74	0.40	4.09	2.74	Medium	1.73	2.09	0.00	3.29	2.33
High	3.03	4.37	0.19	4.95	4.08	High	2.86	2.86	0.00	2.86	2.67

(continued)

7.3 Performance of the Heuristics

Table 7.19 (continued)

Test-set	Ke	Kh	Ue	UCh	UChKh	Test-set	Ke	Kh	Ue	UCh	UChKh
15 n 50 T	1.66	3.50	0.82	4.17	3.04	25 n 50 T	2.14	2.45	0.18	2.61	2.01
Low	0.15	2.93	1.87	3.13	2.63	Low	0.46	1.06	0.33	1.38	1.05
Medium	1.56	3.26	0.44	4.50	3.41	Medium	3.34	3.67	0.22	4.12	3.05
High	3.28	4.30	0.16	4.87	3.09	High	2.61	2.61	0.00	2.33	1.92
15 n 100 T	1.01	2.36	0.83	2.55	1.98	25 n 100 T	1.46	1.61	0.40	1.63	1.03
Low	0.04	2.24	1.52	2.56	2.08	Low	0.20	0.53	0.23	0.97	0.46
Medium	0.98	2.22	0.27	2.76	2.06	Medium	1.48	1.59	0.10	1.70	1.07
High	2.00	2.61	0.71	2.33	1.81	High	2.70	2.70	0.86	2.22	1.55
15 n 150 T	0.69	1.87	0.92	1.95	1.42	25 n 150 T	1.19	1.47	0.57	1.33	0.88
Low	0.00	1.83	1.40	2.25	1.68	Low	0.52	1.10	0.34	1.24	0.70
Medium	0.57	1.79	0.66	2.19	1.32	Medium	1.10	1.25	0.43	1.34	0.79
High	1.49	1.99	0.71	1.40	1.25	High	1.96	2.05	0.93	1.41	1.16
15 n 200 T	0.71	1.94	1.16	1.81	1.44	25 n 200 T	1.29	1.56	0.78	1.18	0.87
Low	0.07	2.20	1.75	2.34	1.95	Low	0.59	1.13	0.57	1.15	0.81
Medium	0.60	1.66	0.76	1.82	1.28	Medium	1.40	1.56	0.78	1.35	0.94
High	1.47	1.96	0.98	1.27	1.08	High	1.88	2.00	0.99	1.04	0.85
						Average	1.02	2.57	1.14	2.81	2.19

Relative error in % of the lower bounds, which are rounded up to the next integer, to BestLB. Termination after maximal 180 s

heuristics yield average relative errors 0.52, 0.54, 0.54, 0.67, 0.89, 1.49 % in Table 7.18 and 12.59, 12.59, 12.62, 12.81, 14.23, 14.38 % in Table 7.20. Note that the large values are due to the relative error of BestLB. The most promising approaches are Ke, Kh and UChKh.

In what follows, specific attributes of the computational results are investigated. In Table 7.18, BestLB has an average relative error of 5.25 %. Table 7.19 shows that Ke and Ue strongly contribute to BestLB. There is a tendency that the relative errors of BestLB abruptly become better in Table 7.18 if the capacity availability is low or medium. Therefore, the relative errors of the upper bounds in Table 7.19 to the optimum are even more overestimated if the capacity availability is high.

As already discussed in Sect. 7.3.1.2, Ke and Ue solve the Lagrangean relaxation to optimality and therefore, yield better lower bounds than Kh and UCh. In Table 7.18, the average relative errors of Ke as well as Ue are 6.49 % whereas 8.08 and 8.52 % are obtained for Kh and UCh. Depending on the capacity availability Ke and Ue behave differently. If the capacity availability is low, Ke mostly provides lower relative errors than Ue. The other way around holds if the capacity availability is high. Kh and UCh clearly show this behavior for instances with large values of n and T.

Regarding the lower bounds, Tables 7.17 and 7.18 show that the average relative error 5.77 % for B2e is lower than the values for Ke and Ue that are both 6.49 %. Furthermore, the Lagrangean hybrid heuristics KeUe, UeKe provide lower average relative errors of 5.53 and 5.55 % than B2e, Ke, Ue, which yield 5.77, 6.49 and

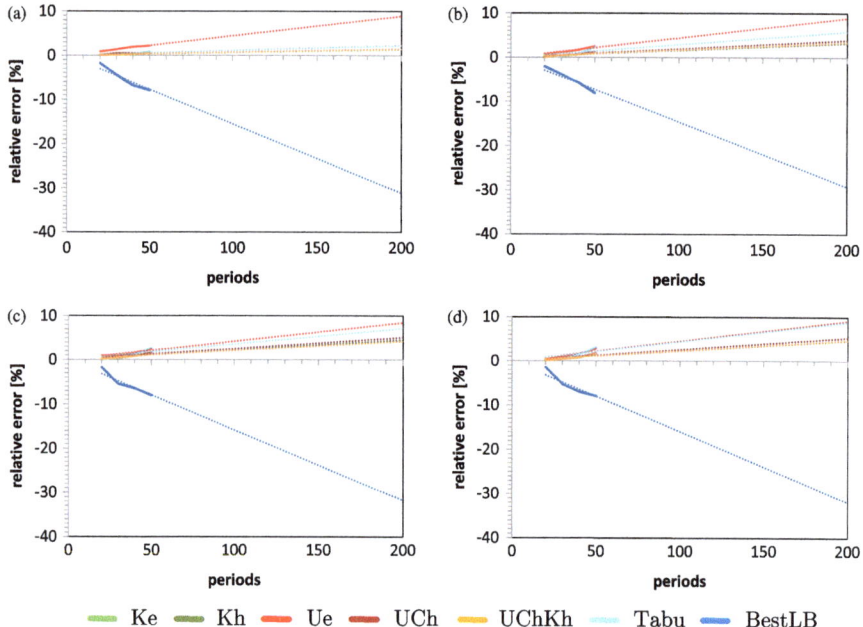

Fig. 7.1 Linear regression of the relative error of the upper bounds and of BestLB from Table 7.18 to the optimum of the CPMP generated by CPLEX. *Dotted lines* represent the linear regressions graphs of the approach given in the respective *color*. (**a**) $n = 10$, (**b**) $n = 15$, (**c**) $n = 20$, (**d**) $n = 25$

6.49 %. Similar observations are made for the pseudo-subgradient approaches. UChKh has a relative error of 7.82 %, KhUCh of 7.84 %, B2h of 8.05 % compared to Kh and UCh that provide 8.08 and 8.52 %. Therefore, similar results from Sect. 7.3.2.2 are obtained and the Lagrangean Hybrid Heuristics provide better lower bounds than the stand-alone approaches.

The Lagrangean heuristics Ke, Kh, Ue and UCh strongly evolve the upper and the lower bounds in the first seconds. For instance, Ke provides an average relative error of the upper bound 0.62 % and of the lower bound 6.76 % after 60 s in Tables 7.9 and 7.10. This is a significant improvement of the initial values of 12.36 % for LB^{init} from Table 7.9 and 6.29 % for IB3 from Table 7.15. In contrast to this, the relative improvement after 180 s is rather low because the respective average relative errors are 0.52 and 6.49 % in Table 7.18. Similar tendencies are observed for Kh, Ue and UCh. Note that Ke constantly evolves the upper bound and slightly outperforms KeUe and UeKe that yield average relative errors of 0.65 and 0.71 % after 180 s in Table 7.17.

Altogether, Ke, Kh, UChKh, B2h and KhUCh yield comparably low average relative errors of the upper bounds that are 0.52, 0.54, 0.54, 0.54, 0.56 % and UCh provides 0.64 %. Note that the values for B2h are taken from Table 7.17. The

7.3 Performance of the Heuristics

Table 7.20 Performance of the metaheuristics from Table 7.14

Test-set	Tabu	Ke	Kh	Ue	UCh	UChKh	Test-set	Tabu	Ke	Kh	Ue	UCh	UChKh
10 n 20 T	2.23	2.23	2.23	3.05	2.23	2.23	20 n 20 T	2.23	2.30	2.30	3.17	2.45	2.30
Low	1.79	1.79	1.79	3.69	1.79	1.79	Low	2.12	2.34	2.34	4.12	2.79	2.34
Medium	0.93	0.93	0.93	1.48	0.93	0.93	Medium	2.72	2.72	2.72	3.27	2.72	2.72
High	3.98	3.98	3.98	3.98	3.98	3.98	High	1.85	1.85	1.85	2.13	1.85	1.85
10 n 30 T	5.50	5.43	5.43	6.68	5.81	5.47	20 n 30 T	6.88	6.79	6.97	7.40	7.13	6.93
Low	3.25	3.38	3.36	5.60	4.32	3.48	Low	5.94	6.28	6.65	7.05	6.70	6.53
Medium	4.52	4.36	4.36	5.26	4.36	4.36	Medium	5.99	5.73	5.92	6.80	6.07	5.92
High	8.74	8.56	8.56	9.19	8.74	8.56	High	8.70	8.35	8.35	8.35	8.63	8.35
10 n 40 T	8.33	8.18	8.35	9.97	8.47	8.25	20 n 40 T	8.59	8.24	8.20	8.97	8.28	8.25
Low	5.00	5.17	5.41	7.79	5.43	5.25	Low	7.91	7.59	7.65	9.20	7.69	7.60
Medium	7.17	6.78	7.05	8.79	7.05	6.91	Medium	6.96	6.65	6.65	7.05	6.65	6.65
High	12.81	12.59	12.59	13.32	12.93	12.59	High	10.91	10.49	10.30	10.65	10.49	10.49
10 n 50 T	10.19	9.78	9.69	11.48	9.85	9.79	20 n 50 T	12.24	10.81	10.89	11.89	10.99	10.72
Low	6.36	6.81	6.72	9.09	6.83	6.65	Low	10.75	10.27	10.53	11.75	10.55	10.13
Medium	9.03	8.64	8.46	10.19	8.67	8.72	Medium	9.63	8.46	8.43	9.28	8.43	8.33
High	15.18	13.89	13.89	15.15	14.05	14.01	High	16.34	13.71	13.71	14.63	13.99	13.71
10 n 100 T	17.30	15.90	15.72	18.28	16.00	15.77	20 n 100 T	20.04	16.65	16.74	19.27	16.98	16.68
Low	10.84	13.56	13.31	15.67	13.57	13.45	Low	15.93	16.22	16.30	18.15	16.49	16.22
Medium	15.79	14.04	13.83	17.01	14.28	13.94	Medium	18.50	14.87	14.99	17.93	15.38	14.95
High	25.28	20.10	20.02	22.17	20.16	19.92	High	25.70	18.87	18.94	21.73	19.08	18.87

(continued)

Table 7.20 (continued)

Test-set	Tabu	Ke	Kh	Ue	UCh	UChKh	Test-set	Tabu	Ke	Kh	Ue	UCh	UChKh
10 n 150 T	20.78	19.12	19.01	22.15	19.30	19.01	20 n 150 T	25.87	21.67	21.59	24.65	21.95	21.75
Low	12.99	16.88	16.61	19.37	17.07	16.69	Low	19.20	20.99	20.96	23.12	21.29	20.90
Medium	19.06	17.33	17.15	21.05	17.65	17.36	Medium	24.46	19.97	19.80	22.96	20.37	20.22
High	30.30	23.16	23.27	26.04	23.17	22.97	High	33.94	24.04	24.02	27.88	24.20	24.14
10 n 200 T	23.62	21.92	21.94	25.11	22.19	22.05	20 n 200 T	27.65	23.14	23.19	26.14	23.65	23.33
Low	14.41	18.95	18.94	21.46	19.40	19.11	Low	20.41	22.86	22.97	24.67	23.32	22.95
Medium	21.37	19.98	19.93	23.22	20.23	19.97	Medium	25.89	21.30	21.35	24.74	21.74	21.40
High	35.09	26.84	26.95	30.66	26.95	27.08	High	36.65	25.25	25.25	29.01	25.90	25.63
15 n 20 T	2.61	2.61	2.61	3.24	2.74	2.61	25 n 20 T	1.85	1.89	1.90	2.11	1.89	1.83
Low	2.19	2.19	2.19	3.44	2.56	2.19	Low	2.73	2.86	2.90	3.53	2.86	2.68
Medium	2.69	2.69	2.69	3.33	2.69	2.69	Medium	2.44	2.44	2.44	2.44	2.44	2.44
High	2.96	2.96	2.96	2.96	2.96	2.96	High	0.37	0.37	0.37	0.37	0.37	0.37
15 n 30 T	7.00	7.14	7.10	7.97	7.19	7.19	25 n 30 T	6.80	6.73	6.68	7.23	6.81	6.67
Low	4.74	5.17	5.21	7.20	5.31	5.31	Low	5.23	5.65	5.51	6.66	5.91	5.68
Medium	4.81	4.81	4.65	5.26	4.81	4.81	Medium	5.65	5.33	5.33	5.83	5.33	5.14
High	11.44	11.44	11.44	11.44	11.44	11.44	High	9.53	9.20	9.20	9.20	9.20	9.20
15 n 40 T	8.79	8.58	8.57	9.66	8.72	8.64	25 n 40 T	9.68	8.84	8.90	9.84	8.97	8.87
Low	7.53	7.75	7.57	9.82	8.03	7.93	Low	8.15	7.46	7.63	9.43	7.71	7.54
Medium	7.55	7.62	7.75	8.11	7.75	7.62	Medium	8.13	7.59	7.59	8.25	7.59	7.59
High	11.28	10.38	10.38	11.04	10.38	10.38	High	12.77	11.47	11.47	11.83	11.61	11.47
15 n 50 T	11.79	10.45	10.39	11.94	10.55	10.38	25 n 50 T	12.48	10.81	10.84	12.17	11.06	10.92
Low	9.12	9.17	8.94	10.73	9.19	9.15	Low	11.59	10.72	10.59	11.67	10.59	10.72
Medium	10.46	8.69	8.76	10.56	8.98	8.68	Medium	10.27	8.39	8.50	9.79	8.72	8.59
High	15.80	13.48	13.48	14.54	13.48	13.32	High	15.59	13.31	13.42	15.04	13.88	13.44

7.3 Performance of the Heuristics

Test-set	Tabu	Ke	Kh	Ue	UCh	UChKh
15 n 100 T	19.53	16.80	16.70	19.40	17.01	16.86
Low	14.49	16.01	15.74	18.10	16.09	16.06
Medium	17.70	14.45	14.27	17.61	15.00	14.69
High	26.40	19.95	20.10	22.48	19.95	19.83
15 n 150 T	23.86	20.81	20.97	23.94	21.26	20.89
Low	17.22	20.31	20.41	22.52	20.62	20.17
Medium	21.78	18.68	18.87	22.19	19.35	18.74
High	32.58	23.43	23.64	27.11	23.81	23.76
15 n 200 T	25.76	22.95	22.84	25.99	23.36	22.96
Low	18.42	22.40	22.32	24.65	22.96	22.48
Medium	24.24	21.02	21.01	24.44	21.42	20.90
High	34.61	25.42	25.20	28.88	25.70	25.49

Test-set	Tabu	Ke	Kh	Ue	UCh	UChKh
25 n 100 T	21.98	17.73	17.60	20.09	18.04	17.64
Low	17.56	17.15	17.05	18.77	17.54	16.87
Medium	19.78	15.78	15.48	18.26	15.95	15.72
High	28.60	20.27	20.28	23.23	20.62	20.33
25 n 150 T	26.22	21.21	21.42	24.19	21.63	21.51
Low	20.48	20.63	21.05	23.29	21.32	21.06
Medium	24.87	19.73	19.72	22.86	20.13	19.96
High	33.32	23.27	23.49	26.43	23.43	23.50
25 n 200 T	28.75	23.77	23.71	26.66	24.21	23.82
Low	21.53	23.09	22.95	24.90	23.32	22.89
Medium	26.79	21.69	21.63	25.31	22.55	21.99
High	37.93	26.54	26.56	29.77	26.76	26.57
Average	14.23	12.59	12.59	14.38	12.81	12.62

Relative error in % of the upper bounds to BestLB. Termination after maximal 180 s

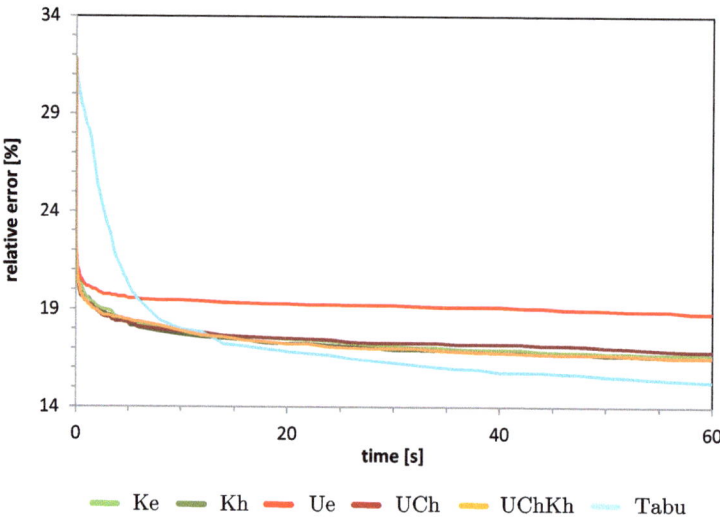

Fig. 7.2 Evolution of the relative error of the upper bounds of the CPMP from Table 7.14 to BestLB in the first 60 s for the results of the test-set $n = 15$ and $T = 100$ from Table 7.20 with low capacity availability

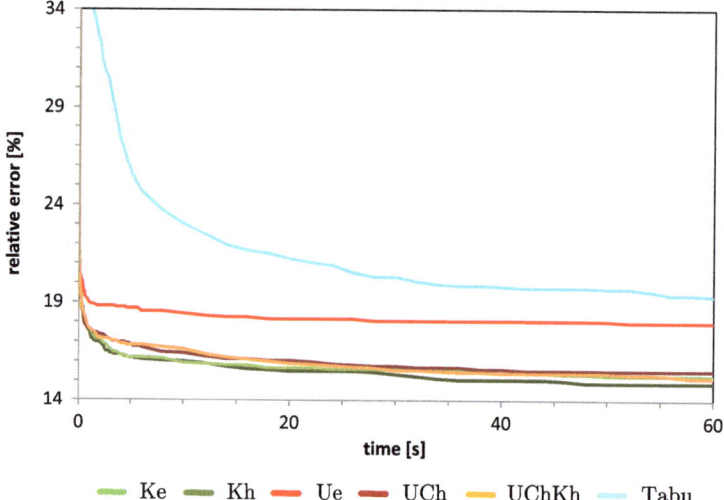

Fig. 7.3 Evolution of the relative error of the upper bounds of the CPMP from Table 7.14 to BestLB in the first 60 s for the results of the test-set $n = 15$ and $T = 100$ from Table 7.20 with medium capacity availability

Lagrangean Hybrid Heuristic UChKh yields an upper bound that is better or at least as good as the stand-alone approaches.

Figures 7.2, 7.3 and 7.4 exemplary depict the evolution of the upper bounds in the first 60 s for the test-set $n = 15$ and $T = 100$ from Table 7.20. Ke, Kh, Ue, UCh

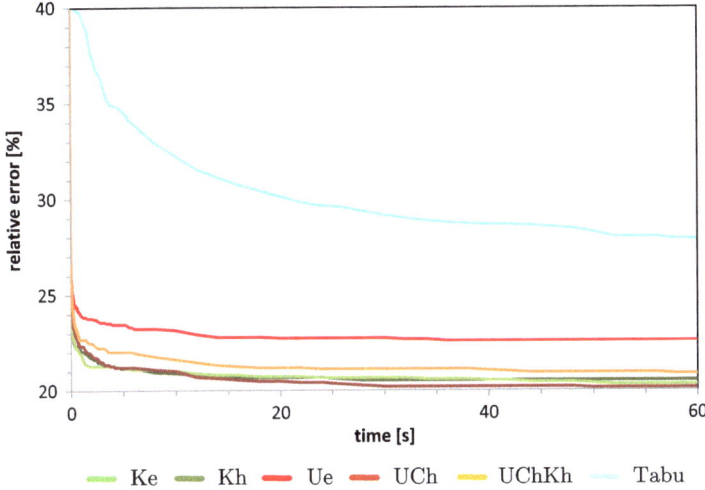

Fig. 7.4 Evolution of the relative error of the upper bounds of the CPMP from Table 7.14 to BestLB in the first 60 s for the results of the test-set $n = 15$ and $T = 100$ from Table 7.20 with high capacity availability

and UChKh quickly evolve the upper bounds in the first seconds. Tabu slowly but constantly improves the upper bound and yields the lowest average relative error in Fig. 7.2 where the capacity availability is low. In particular, Table 7.20 shows that Tabu outperforms the Lagrangean heuristics in many test-sets if the capacity availability is low. Note that this effect is well observable for large instances. This result indicates that the combination of the tabu search neighborhood and the guidance from the tabu search objective function (6.86) becomes more effective if the capacity availability is low. In contrast to this, the Lagrangean approaches strongly diversify via the guidance of the Lagrangean objective function and the construction heuristic IB3 and yield better results for the upper bounds if the capacity availability is high.

Reference

Matsumoto, M., & Nishimura, T. (1998). Mersenne twister: A 623-dimensionally equidistributed uniform pseudo-random number generator. *ACM Transactions on Modeling and Computer Simulation, 8*, 3–30.

Chapter 8
Final Remarks and Future Perspectives

Industrial production systems consist of components that gradually wear off and therefore, require maintenance. This thesis introduces the *Capacitated Planned Maintenance Problem (CPMP)* and the *Weighted Uncapacitated Planned Maintenance Problem (WUPMP)*. To the author's best knowledge both preventive maintenance problems have not been discussed in literature yet. Assume a single machine is maintained by a set of maintenance activities that have to be scheduled within a discretized planning horizon such that each maintenance activity is executed at least once within an individual, deterministic durability horizon. This so called period covering constraint ensures that no uncontrolled breakdowns occur. Every scheduled maintenance activity begins and ends in the same period and is executed by a single repairman. Scheduling at least one maintenance activity requires that a maintenance slot is established in the respective period.

The CPMP minimizes the total number of established maintenance slots such that a maximal available maintenance time, which is period dependent, is never exceeded (capacity constraint). Note that the maintenance time depends on the maintenance activity. The WUPMP minimizes the total sum of fixed and variable costs. Fixed costs are period dependent and caused by every established maintenance slot. Variable costs are period and maintenance activity dependent and charged whenever a maintenance activity is scheduled.

Both foresighted optimization approaches increase the reliability of the production system and consequently, a significantly higher overall efficiency is obtained. Appendix A.1 provides additional constraints for practical applications.

In what follows, the main contributions and results of the thesis are summarized.

The WUPMP is proven to be strongly $\mathcal{NP}$-hard even if each maintenance activity needs to be scheduled only once in the planning horizon and the fixed costs are identical. An exact algorithm is provided that solves the WUPMP in time $O(n \cdot T^{n+1} \cdot 2^n)$. Hence, the WUPMP is strongly polynomially solvable if the number of maintenance activities is a constant. Different problem variants are investigated and optimal, strongly polynomial algorithms are provided (cf. Table 4.4). The

WUPMP is a generalization of Uncapacitated Facility Location Problem but a special problem variant of the Uncapacitated Network Design Problem and of the Set Partitioning Problem. Consequently, the WUPMP is solvable by any algorithm that solves the Uncapacitated Network Design Problem or the Set Partitioning Problem. An optimal solution to the LP relaxation of the WUPMP comprises many components that are integer. Examples with the smallest possible dimensions are provided such that the polytope of the WUPMP LP relaxation contains fractional extreme points. However, some problem variants have an integral polytope. It is polyhedrally proven that the WUPMP has the single assignment property. This property states that the continuous variables, which define if a maintenance activity is scheduled or not, are implicitly integer. Furthermore, the WUPMP has a quasi-integral polytope, which allows for the development of integral Simplex algorithms that pivot only on integer basic feasible solutions of the LP relaxation to find an optimal integer solution. It is proven that the LP relaxation of a generalized period covering constraint for a single maintenance activity has an integral polytope and therefore, it is shown that the polytope of the corresponding dual linear problem is integral if the instance data comprises only finite integer values. Both properties hold for the WUPMP with one maintenance activity that is referred to as the *Single Weighted Uncapacitated Planned Maintenance Problem (SWUPMP)*. The SWUPMP is strongly polynomially solvable. In particular, the SWUPMP is solved by a shortest path algorithm that exploits specific structural properties. Furthermore, an algorithm is proposed that solves the corresponding dual problem of the SWUPMP LP relaxation to optimality. Both algorithms are used in several proposed algorithms for the WUPMP and the CPMP. The so called *Uncapacitated Planned Maintenance Problem (UPMP)* is the uncapacitated problem variant of the CPMP and a WUPMP where the fixed costs are identical but the variable costs are zero. The UPMP is strongly polynomially solvable and provides a lower bound to the CPMP. It is shown that the optimal objective function values of the integer problem and of the UPMP LP relaxation coincide. A problem variant of the UPMP is presented where the smallest convex set of optimal solutions to the LP relaxation has only integer extreme points. Furthermore, integral polytopes are obtained by two sets of inequalities. A small computational study showed that the UPMP LP relaxation yielded only optimal integer extreme points.

It is proven that the CPMP is strongly $\mathcal{NP}$-hard even if each maintenance activity needs to be scheduled only once in the planning horizon and the maximal available maintenance times for all periods are identical. Furthermore, the CPMP is binary $\mathcal{NP}$-hard even if there are just two periods. An exact algorithm solves the CPMP in time $O(\min\{n \cdot \frac{\log T}{\sqrt{T}} \cdot \bar{r}^{max\,T} \cdot 4^T, n \cdot T^{n+1} \cdot 2^n\})$. Therefore, the CPMP is pseudo-polynomially solvable if the number of periods is a constant and strongly polynomially solvable if either the number of maintenance activities is a constant or if the number of periods and the maximal capacity over all periods are constants. Other optimal, strongly polynomial and pseudo-polynomial algorithms to different problem variants are provided (cf. Table 5.5). Examples with the smallest possible dimensions are given such that the polytope of the CPMP LP relaxation has

8 Final Remarks and Future Perspectives

fractional extreme points. Some problem variants provide integral polytopes and one problem variant is presented where the smallest convex set of optimal solutions to the CPMP LP relaxation has only integer extreme points.

All possible lower bounds that can be derived from neglecting a subset of constraints completely, a Lagrangean relaxation of a subset of constraints or combinations of both are evaluated and their computational complexity is investigated (cf. Table 5.7). It is shown that only 14 out of the obtained 99 lower bounds are relevant. The relative strength is investigated and the lower bounds can either be solved as special cases of network flow problems or as facility location problems. The most promising lower bounds with respect to computational time and quality of the lower bound are $Z^{(X)}$ that is a CPMP formulation with a LP relaxed integrality constraint of the binary variables that define if a maintenance activity is scheduled or not, $Z_{(R)}^{(P)+(Q)}$ that is the Lagrangean dual of a Lagrangean relaxation of a subset of period covering constraints, which yields non-overlapping durability horizons in the Lagrangean subproblem, and $Z_{(P)}$ that is the Lagrangean dual of the Lagrangean relaxation of the period covering constraint that yields Knapsack Problems as Lagrangean subproblems.

Three construction heuristics for the CPMP are developed. The First Fit Heuristic is a serial heuristic and plans maintenance activities such that the difference between scheduled maintenance activities is maximal. The Overlap Heuristic is a parallel heuristic and subsequently determines a plan for each period that contains scheduled maintenance activities. Two priority rules for the Overlap Heuristic are presented. The computational results show that the three heuristics contribute in an iterated improvement approach such that the obtained Iterated Best-of-three Heuristic yields significantly better upper bounds than the best feasible solution of the three heuristics.

Two Lagrangean heuristics based on subgradient optimization are developed where either the capacity constraint or the period covering constraint is Lagrangean relaxed. Note that a subgradient is accessible in every iteration because a feasible solution to both Lagrangean relaxations can be found in polynomial time. In particular, the Lagrangean relaxation of the period covering constraint $ZL_{(P)}$ is binary $\mathcal{NP}$-hard and decomposes into a set of Knapsack Problems that are solved with well-known solution approaches (e.g., the Combo Algorithm). The Lagrangean relaxation of the capacity constraint $ZL_{(C)}$ is a problem variant of the WUPMP. However, the computational complexity status of this Lagrangean relaxation is unresolved. To this problem, five novel lower bounds are presented that solve the LP relaxation of $ZL_{(C)}$ either to optimality as an LP or heuristically with a priority rule heuristic or with a variant of the Primal-dual Simplex algorithm. Subsequently, a primal heuristic constructs a feasible solution to $ZL_{(C)}$ that takes so called dual information provided by the lower bounds into account. Specifically, the primal heuristic considers two of four complementary slackness criteria and preferably establishes maintenance slots where the LP relaxation has integer components. Computational results show that this dual information improves the quality of the obtained feasible solution to $ZL_{(C)}$. Furthermore, all heuristics are evaluated with respect to the optimal solution of the respective Lagrangean relaxation. In the presented Lagrangean Heuristic for the CPMP, a variant of the well-known Held-Karp step size is used to calculate Lagrangean multiplier and the update method of the step size parameter dynamically reacts on the quality and on the direction the Lagrangean lower bound evolves to. Initial Lagrangean multiplier for both Lagrangean relaxations $ZL_{(C)}$ and $ZL_{(P)}$ are provided such that the lower bounds of each Lagrangean relaxation are identical after the first iteration of the Lagrangean Heuristic. Furthermore, the Lagrangean relaxation is heuristically solved and so called pseudo-subgradients are obtained. Note that most approaches in literature prefer an optimal solution since optimal solutions allow to describe the subdifferential. A computational study shows the expected result that the approximation of the Lagrangean dual corresponds to the quality of the feasible solution to the Lagrangean relaxation. However, the Lagrangean dual is sufficiently approximated by pseudo-subgradients. One of the most effective Lagrangean Heuristics denoted as Ke solves the applied Lagrangean relaxation $ZL_{(P)}$ to optimality.

A novel problem independent and general hybrid approach is presented that uses the duality between Lagrangean relaxation and Benders' decomposition to link two Lagrangean relaxations with each other via an auxiliary LP. The corresponding dual solution to the auxiliary LP estimates novel initial Lagrangean multiplier for the second Lagrangean heuristic. Information to both Lagrangean relaxations is taken into account. Specifically, Lagrangean multiplier of the first Lagrangean relaxation are included as Benders' cuts and feasible solutions to the second Lagrangean relaxation and to the original integer problem as dual cutting planes. The approach is applied to the CPMP and the obtained Lagrangean Hybrid Heuristic alternates between both Lagrangean heuristics. Regarding the upper bound to the CPMP, the most effective hybrid heuristic, namely UChKh, alternates between the Lagrangean heuristics that apply $ZL_{(C)}$ and $ZL_{(P)}$. The computational results show that the upper and lower bounds of both Lagrangean Heuristics are improved by the Lagrangean Hybrid Heuristic. Moreover, this result also applies to the best upper and lower bounds each single Lagrangean Heuristic provides after half of the computational time has elapsed. This particularly holds if the Lagrangean relaxations are solved to optimality. Considering the upper bounds, this Lagrangean Hybrid Heuristic UChKh is slightly outperformed by the Lagrangean Heuristic Ke.

The Tabu Search Heuristic is a multi-state search where the objective function comprises three goals. One goal is the minimization of the number of established maintenance slots that corresponds to the CPMP objective function. A higher prioritized goals minimizes the penalized violation of the capacity constraint and a lesser prioritized goal maximizes the squared slack of the capacity constraint. This goal additionally guides the search process because each period is given a bonus that becomes larger the emptier a period is. The goal prioritization varies such that infeasibility is allowed in the final state. The elementary operation move plans out a single maintenance activity from a period and schedules it in another period such that the period covering constraint holds. In every state, the elementary operation is applied to a subsequence of consecutive periods, the so called scope, that is shifted through the planning horizon in a given search direction. If the best known feasible solution cannot be improved, the search commences in the next state with a broadened scope. The algorithm commences from the ground state again if the best known feasible solution has been improved.

Computational studies show that all Lagrangean heuristics quickly evolve the upper bound to the CPMP, whereas the Tabu Search Heuristic slowly but constantly evolves the upper bound. The Tabu Search Heuristic outperforms the Lagrangean heuristics on the long run if the capacity constraint is very restrictive.

Future research could focus on the development of a branch and bound approach to optimally solve the WUPMP, the CPMP and specific problem variants. The most applicable problem variant for real-world applications restricts the CPMP to identical maintenance times for each maintenance activity. Note that the complexity status of this problem is unresolved. The enumeration schemes from Lemmata 4.7, 5.9 and 5.14 could be applied in a best-first branch and bound approach where specific dominance rules could help to effectively reduce the size of the enumeration tree. Other approaches could involve the use of valid inequalities, variable fixation and reduction procedures. Specifically, variable fixation could be done with logical tests and Lagrangean probing, a tighter model formulation would yield better lower bounds and related optimization problems could provide valid inequalities and separation problems. Potentially interesting lower bounds could be provided by surrogate relaxation. Furthermore, approximation algorithms for the WUPMP and the CPMP could be investigated. Since the LP relaxation of the WUPMP has a quasi-integral polytope, an integral Simplex algorithm could be developed. Moreover, three problem variants of the CPMP have an unresolved computational complexity status. Regarding algorithms for the CPMP, the construction heuristics could be improved with alternative priority rules and other construction heuristics could use partial enumeration. Heuristics to the Lagrangean relaxation of the period covering constraint could be derived from dual-ascent procedures, Bender's decomposition, resource-directive decomposition and by adapting heuristics of related optimization problems; e.g. add- and drop-heuristics of the Uncapacitated Facility Location Problem. The tabu search approach could be improved by a variable neighborhood search where neighborhoods are applied that have a better performance if the maximal available maintenance time per period is rather high. Pseudo-subgradient optimization and the Lagrangean hybrid approach could be applied to other optimization problems; e.g. the related Capacitated Facility Location Problem. In order to increase the applicability for real-world applications, the CPMP could be augmented by additional side constraints that include precedence constraints, multiple machines, multiple repairmen, an integration with lot-sizing decisions and a cost oriented objective function.

Appendix A
Additional Material for the Capacitated Planned Maintenance Problem

Section A.1 proposes problem extensions of the CPMP. Mathematical formulations are provided in Sect. A.2.

A.1 Problem Extensions

The following mathematical models extend the CPMP formulation from Sect. 2.3. Since the CPMP occurs as a subproblem, all mathematical models are strongly $\mathcal{NP}$-hard (Lemma 5.7). The notation and the abbreviations from Sect. 2.3 apply.

A.1.1 Capacitated Weighted Planned Maintenance Problems

Imposing the capacity constraint (C), the CPMP and WUPMP become special cases of the generalized mathematical model

$$\min \sum_{t=1}^{T} f_t \cdot y_t + \sum_{i=1}^{n} \sum_{t=1}^{T} c_{it} \cdot x_{it} \quad \text{subject to } (P) \wedge (C) \wedge (V) \wedge (X) \wedge (Y).$$

A variant consists of two objective functions, which are lexicographically optimized. Among all slot plans with the minimal number of maintenance slots, one with minimal variable costs is selected. Setting $f_t = 1 \ \forall t$ and $c_{it} = 1 \ \forall i; t$, a CPMP is derived in which the number of scheduled maintenance activities is reduced for ecological and economic reasons.

$$\min \begin{pmatrix} \sum_{t=1}^{T} f_t \cdot y_t \\ \sum_{i=1}^{n} \sum_{t=1}^{T} c_{it} \cdot x_{it} \end{pmatrix}_{lexopt} \quad \text{subject to } (P) \wedge (C) \wedge (V) \wedge (X) \wedge (Y)$$

A.1.2 Multiple Machines

Assume $G \in \mathbb{N}$ identical machines are simultaneously maintained. Production requires that no machine is maintained but all machines operate. Therefore, a maintenance slot is established if at least one machine is maintained. No maintenance interdependency exists between the machines and each machine is maintained by a single repairman. The binary variable z_{itg} equals 1 if a maintenance activity $i = 1, \ldots, n$ is scheduled in a period $t = 1, \ldots, T$ on machine $g = 1, \ldots, G$ (otherwise, 0).

$$\min \sum_{t=1}^{T} y_t \quad \text{subject to}$$

$$\sum_{\tau=t}^{t+\pi_i-1} z_{i\tau g} \geq 1 \qquad \forall i = 1, \ldots, n; t = 1, \ldots, T - \pi_i + 1; g = 1, \ldots, G$$

$$\sum_{i=1}^{n} r_i \cdot z_{itg} \leq \bar{r}_t \cdot y_t \qquad \forall t = 1, \ldots, T; g = 1, \ldots, G$$

$$z_{itg} \leq y_t \qquad \forall i = 1, \ldots, n; t = 1, \ldots, T; g = 1, \ldots, G$$

$$z_{itg} \in \{0, 1\} \qquad \forall i = 1, \ldots, n; t = 1, \ldots, T; g = 1, \ldots, G$$

$$y_t \in \{0, 1\} \qquad \forall t = 1, \ldots, T$$

A.1.3 Multiple Repairmen

Let a single machine be maintained by $G_t \in \mathbb{N}$ repairmen that are available in period $t = 1, \ldots, T$ and assume identical skills for each repairman. If a maintenance activity $i = 1, \ldots, n$ is scheduled in a period $t = 1, \ldots, T$ by the repairman $g = 1, \ldots, G_t$, then the binary variable z_{itg} equal 1 (otherwise, 0).

A.1 Problem Extensions

$$\min \sum_{t=1}^{T} y_t \qquad \text{subject to}$$

$$\sum_{\tau=t}^{t+\pi_i-1} \sum_{g=1}^{G_\tau} z_{i\tau g} \geq 1 \qquad \forall i = 1,\ldots,n; t = 1,\ldots,T-\pi_i+1 \qquad (\text{A.1})$$

$$\sum_{i=1}^{n} r_i \cdot z_{itg} \leq \bar{r}_t \cdot y_t \qquad \forall t = 1,\ldots,T; g = 1,\ldots,G_t$$

$$z_{itg} \leq y_t \qquad \forall i = 1,\ldots,n; t = 1,\ldots,T; g = 1,\ldots,G_t$$

$$z_{itg} \in \{0,1\} \qquad \forall i = 1,\ldots,n; t = 1,\ldots,T; g = 1,\ldots,G_t$$

$$y_t \in \{0,1\} \qquad \forall t = 1,\ldots,T.$$

Because of the objective function and constraint (A.1), it is not beneficial to have a maintenance activity scheduled by more than one repairman. Thus, the following logical constraint is valid.

$$\sum_{g=1}^{G_t} z_{itg} \leq y_t \qquad \forall i = 1,\ldots,n; t = 1,\ldots,T$$

A.1.4 Maintenance Precedence Constraints

A maintenance activity $i = 1,\ldots,n$ can require a completion of a set maintenance activities $J_i \subseteq \{1,\ldots,n\}$ within the same maintenance slot. The technological design of a production system typically implies such precedence constraints for component replacements.

$$\min \sum_{t=1}^{T} y_t \quad \text{subject to } (P)\wedge(C)\wedge(V)\wedge(X)\wedge(Y)$$

$$x_{it} \leq x_{jt} \qquad \forall i = 1,\ldots,n; t = 1,\ldots,T; j \in J_i$$

A.1.5 Minimal Time Between Maintenance Activities

Assume that at least $p_i \in \mathbb{N}$ with $p_i \leq \pi_i$ $\forall i$ periods must elapse between two tasks of the same maintenance activity $i = 1,\ldots,n$.

$$\min \sum_{t=1}^{T} y_t \quad \text{subject to } (P) \wedge (C) \wedge (V) \wedge (X) \wedge (Y)$$

$$x_{it} + x_{i,t+h} \leq 1 \qquad \forall i = 1, \ldots, n; t = 1, \ldots, T - p_i; h = 1, \ldots, p_i$$

A.1.6 Integration with Lot-Sizing

Two commonly applied concepts decouple the interdependency between production planning and maintenance planning. In the first concept, maintenance is carried out in reserved time slots in the master production schedule and production is planned around the maintenance decisions. The other way around, the lot-sizing decisions come first and maintenance is done whenever the production capacity allows it. Besides other aspects, both concepts can cause long lead times or backlogging if production peaks occur. Some integrated approaches for stochastic maintenance models are presented in Sect. 2.2.

The following integrated approach combines maintenance and lot-sizing decisions in a medium term planning horizon. Specifically, the CPMP is combined with the *Capacitated Lot-Sizing Problem (CLSP)*[1] that is restrictive but covers fundamental aspects of lot-sizing. Let a single machine manufacture $J \in \mathbb{N}$ products within a discrete planning horizon $1, \ldots, T$ and each product j has a processing time $p_j \geq 0$. Production and maintenance is begun as well as completed in the same period and both share the available time $C_t \geq 0$ of period t (big bucket). There is at most one maintenance slot per period and assume $\bar{r}_t \leq C_t \ \forall t$. The demand $d_{jt} \geq 0$ for product j in period t is deterministic and requested at the end of a period. Let the variable q_{jt} be the production quantity of product j in period t. A positive inventory I_{jt} of product j in period t causes holding costs $h_j \geq 0$. If a product $j = 1, \ldots, J$ is manufactured in period $t = 1, \ldots, T$, let the binary variable z_{jt} equal 1 (otherwise, 0) to cause setup costs $s_j \geq 0$ if the setup state changes. The setup state between two

---

[1] The CLSP minimizes the total setup and holding costs such that the demand is satisfied and production capacities are never exceeded (Karimi et al. 2003). Furthermore, the CLSP is binary $\mathcal{NP}$-hard even for one product (Bitran and Yanasse 1982; Wolsey 1988, pp. 86–87; Florian et al. 1980) and strongly $\mathcal{NP}$-hard with setup times (Maes et al. 1991).

A.1 Problem Extensions

consecutive periods is not preserved. The mathematical model is stated as

$$\min \sum_{t=1}^{T} f_t \cdot y_t + \sum_{i=1}^{n}\sum_{t=1}^{T} c_{it} \cdot x_{it} + \sum_{j=1}^{J}\sum_{t=1}^{T} s_j \cdot z_{jt} + \sum_{j=1}^{J}\sum_{t=1}^{T} h_j \cdot I_{jt} \quad \text{subject to}$$

$$I_{j,t-1} + q_{jt} = I_{jt} + d_{jt} \qquad \forall j = 1,\ldots,J; t = 1,\ldots,T \qquad (A.2)$$

$$q_{jt} \leq z_{jt} \cdot \sum_{\tau=t}^{T} d_{j\tau} \qquad \forall j = 1,\ldots,J; t = 1,\ldots,T \qquad (A.3)$$

$$\sum_{j=1}^{J} p_j \cdot q_{jt} + \sum_{i=1}^{n} r_i \cdot x_{it} \leq C_t \qquad \forall t = 1,\ldots,T \qquad (A.4)$$

$$\sum_{i=1}^{n} r_i \cdot x_{it} \leq \bar{r}_t \cdot y_t \qquad \forall t = 1,\ldots,T \qquad (C)$$

$$\sum_{\tau=t}^{t+\pi_i-1} x_{i\tau} \geq 1 \qquad \forall i = 1,\ldots,n; t = 1,\ldots,T-\pi_i+1 \qquad (P)$$

$$x_{it} \leq y_t \qquad \forall i = 1,\ldots,n; t = 1,\ldots,T \qquad (V)$$

$$x_{it} \in \{0,1\} \qquad \forall i = 1,\ldots,n; t = 1,\ldots,T \qquad (X)$$

$$y_t \in \{0,1\} \qquad \forall t = 1,\ldots,T \qquad (Y)$$

$$z_{jt} \in \{0,1\} \qquad \forall j = 1,\ldots,J; t = 1,\ldots,T$$

$$q_{jt} \geq 0 \qquad \forall j = 1,\ldots,J; t = 1,\ldots,T$$

$$I_{jt} \geq 0 \land I_{j0} = 0 \qquad \forall j = 1,\ldots,J; t = 0,\ldots,T.$$

The objective function minimizes the total maintenance, the total setup and the total holding costs. The inventory balance constraint (A.2) states that the demand d_{jt} must be satisfied either by the inventory of the proceeding period $I_{j,t-1}$ or by the produced quantity q_{jt}. The excess defines the inventory I_{jt}. Because of (A.3), a setup state is incurred whenever the respective product is manufactured. The capacity constraint (A.4) limits the total processing time and the total maintenance time to the available time per period. The constraints (P), (C) and (V) define the maintenance of the machine. Despite the continuous variables q_{jt} and I_{jt}, all variable domains are binary.

The approach to couple production and maintenance decisions through the available time is applicable to more short-term oriented, small bucket lot-sizing problems that may also incorporate scheduling decisions. An introduction to different classes of lot-sizing and scheduling problems is provided in Suerie (2005) and Drexl and Kimms (1997). A survey on sequence dependency is provided in Zhu and Wilhelm (2006).

A.2 Mathematical Formulations

The CPMP is formulated in Sect. 2.3 as

$$Z = \min \sum_{t=1}^{T} y_t \qquad \text{subject to} \qquad (Z)$$

$$\sum_{\tau=t}^{t+\pi_i-1} x_{i\tau} \geq 1 \qquad \forall i=1,\ldots,n; t=1,\ldots,T-\pi_i+1 \qquad (P)$$

$$\sum_{i=1}^{n} r_i \cdot x_{it} \leq \bar{r}_t \cdot y_t \qquad \forall t=1,\ldots,T \qquad (C)$$

$$x_{it} \leq y_t \qquad \forall i=1,\ldots,n; t=1,\ldots,T \qquad (V)$$

$$x_{it} \in \{0,1\} \qquad \forall i=1,\ldots,n; t=1,\ldots,T \qquad (X)$$

$$y_t \in \{0,1\} \qquad \forall t=1,\ldots,T \qquad (Y)$$

In the second formulation, let the continuous variable I_{it} represent the remaining coverage of a maintenance activity i that is the maintenance activity i covers I_{it} succeeding periods at the beginning of a period t. All other variables and parameters are identical to the CPMP. The mathematical formulation becomes

$$L = \min \sum_{t=1}^{T} y_t \qquad \text{subject to}$$

$$\sum_{i=1}^{n} r_i \cdot x_{it} \leq \bar{r}_t \cdot y_t \qquad \forall t=1,\ldots,T \qquad (C)$$

$$x_{it} \leq y_t \qquad \forall i=1,\ldots,n; t=1,\ldots,T \qquad (V)$$

$$1 \leq I_{it} \leq \pi_i \qquad \forall i=1,\ldots,n; t=1,\ldots,T \qquad (A.5)$$

$$I_{i,t-1} - 1 \geq I_{it} - \pi_i \cdot x_{it} \qquad \forall i=1,\ldots,n; t=1,\ldots,T \qquad (A.6)$$

$$I_{i,t-1} - 1 \leq I_{it} + \pi_i \cdot x_{it} \qquad \forall i=1,\ldots,n; t=1,\ldots,T \qquad (A.7)$$

$$\pi_i \cdot x_{it} \leq I_{it} \qquad \forall i=1,\ldots,n; t=1,\ldots,T \qquad (A.8)$$

$$I_{it} \geq 0 \land I_{i0} = \pi_i \qquad \forall i=1,\ldots,n; t=0,\ldots,T$$

$$x_{it} \in \{0,1\} \qquad \forall i=1,\ldots,n; t=1,\ldots,T$$

$$y_t \in \{0,1\} \qquad \forall t=1,\ldots,T.$$

A.2 Mathematical Formulations

The objective function value minimizes the number of maintenance slots in the planning horizon. The constraint (C) restricts the capacity per period. The lifetime I_{it} is restricted to the interval $I_{it} \in [1, \pi_i]$ because of (A.5). If a maintenance activity i is not scheduled in period t since $x_{it} = 0$, (A.6) and (A.7) state that the lifetime I_{it} equals the lifetime of the previous period $t-1$ decreased by one.[2] Otherwise, (A.6) and (A.7) are not restrictive but the lifetime I_{it} equals the coverage π_i because of (A.5) and (A.8). Assumption (2.6) sets $I_{i0} = \pi_i$ $\forall i$. The continuous variable I_{it} is implicitly integer. The variable domains of x and y are integer. Correctness of the MIP is shown with Lemma A.1.

Lemma A.1

$$\sum_{\tau=t}^{t+\pi_i-1} x_{i\tau} \geq \alpha_i \quad \forall i = 1, \ldots, n; t = 1, \ldots, T - \pi_i + 1 \tag{A.9}$$

with $\frac{1}{\pi_i} \leq \alpha_i \leq 1$ $\forall i$ holds for the constrains (A.5) and (A.6).

Proof Rearrange (A.6) to obtain

$$x_{it} \geq \frac{I_{it} - I_{i,t-1} + 1}{\pi_i}$$

and sum this inequality up over all periods $\tau = t, \ldots, t + \pi_i - 1$ to yield

$$\sum_{\tau=t}^{t+\pi_i-1} x_{i\tau} \geq \frac{I_{it} - I_{i,t-1} + 1}{\pi_i} + \ldots + \frac{I_{i,t+\pi_i-1} - I_{i,t+\pi_i-2} + 1}{\pi_i}$$

$$= \frac{I_{it} + \ldots + I_{i,t+\pi_i-1} - I_{i,t-1} - \ldots - I_{i,t+\pi_i-2} + \pi_i}{\pi_i}$$

$$= \frac{I_{i,t+\pi_i-1} - I_{i,t-1} + \pi_i}{\pi_i} = \alpha_i.$$

Note that $I_{i,t-1} \geq I_{i,t+\pi_i-1}$ because of (A.6). Hence, $I_{i,t+\pi_i-1} = I_{i,t-1}$ defines $\alpha_i \leq 1$. Using the largest difference via $I_{i,t+\pi_i-1} = 1$ and $I_{i,t-1} = \pi_i$ from (A.5), $\alpha_i \geq \frac{1}{\pi_i}$ holds.

The integrality of x forces the left-hand side of (A.9) to be at least 1 and the right-hand side of (A.9) can be feasibly rounded up. Thus, (A.9) becomes (P) and the constraints (A.7) and (A.8) are redundant. Most importantly, the lemma implies

---

[2] The two constraints (A.6) and (A.7) are inspired by the inventory balance constraints (A.2), which are found in mathematical formulations of lot-sizing problems such as the CLSP (Karimi et al. 2003).

that the LP relaxation $L^{(X)(Y)}$ is less tight than $Z^{(X)(Y)}$ from Sect. 2.3 that is $L^{(X)(Y)} \leq Z^{(X)(Y)}$ holds. That is why this formulation of the CPMP is not investigated further.

In the third formulation of the CPMP, the capacity constraint (C) is added to the formulation of the WUPMP as an extended UFLP from Sect. 4.2. Let $f_t = 1 \ \forall t$ and $c_{it} = 0 \ \forall i; t$.

$$\min \sum_{t=1}^{T} f_t \cdot y_t + \sum_{i=1}^{n}\sum_{t=1}^{T} c_{it} \cdot x_{it} + \sum_{j=1}^{J}\sum_{t=0}^{T} h_{jt} \cdot z_{jt} \quad \text{subject to} \qquad (4.14)$$

$$\sum_{i=1}^{n} r_i \cdot x_{it} \leq \bar{r}_t \cdot y_t \qquad \forall t = 1,\ldots,T \qquad (C)$$

$$\sum_{t=0}^{T} z_{jt} = 1 \qquad \forall j = 1,\ldots,J \qquad (4.15)$$

$$z_{jt} \leq x_{it} \qquad \forall i = 1,\ldots,n; j \in I_i; t = 1,\ldots,T \qquad (4.16)$$

$$x_{it} \leq y_t \qquad \forall i = 1,\ldots,n; t = 1,\ldots,T \qquad (4.17)$$

$$x_{it} \in \{0,1\} \qquad \forall i = 1,\ldots,n; t = 1,\ldots,T \qquad (4.21)$$

$$y_t \in \{0,1\} \qquad \forall t = 1,\ldots,T \qquad (4.22)$$

$$z_{jt} \in \{0,1\} \qquad \forall j = 1,\ldots,J; t = 0,\ldots,T \qquad (4.23)$$

The fourth formulation states the CPMP as an extended SetPP based on the formulation presented in Sect. 4.2 where (C) is added. Let $f_t = 1 \ \forall t$ and $c_{it} = 0 \ \forall i; t$.

$$\min \sum_{t=1}^{T} f_t \cdot y_t + \sum_{i=1}^{n}\sum_{t=1}^{T} c_{it} \cdot x_{it} + \sum_{j=1}^{J}\sum_{t=0}^{T} h_{jt} \cdot z_{jt} \quad \text{subject to} \qquad (4.26)$$

$$\sum_{i=1}^{n} r_i \cdot x_{it} \leq \bar{r}_t \cdot y_t \qquad \forall t = 1,\ldots,T \qquad (C)$$

$$\sum_{t=0}^{T} z_{jt} = 1 \qquad \forall j = 1,\ldots,J \qquad (4.27)$$

$$z_{jt} + x_{it}^c + s'_{ijt} = 1 \qquad \forall i = 1,\ldots,n; j \in I_i; t = 1,\ldots,T \qquad (4.28)$$

$$x_{it} + y_t^c + s''_{it} = 1 \qquad \forall i = 1,\ldots,n; t = 1,\ldots,T \qquad (4.29)$$

$$x_{it} + x_{it}^c = 1 \qquad \forall i = 1,\ldots,n; t = 1,\ldots,T \qquad (4.30)$$

$$y_t + y_t^c = 1 \qquad \forall t = 1,\ldots,T \qquad (4.31)$$

$$s'_{ijt} \in \{0,1\} \qquad \forall i = 1,\ldots,n; j \in I_i; t = 1,\ldots,T \qquad (4.32)$$

A.2 Mathematical Formulations

$$s''_{it} \in \{0,1\} \qquad \forall i=1,\ldots,n; t=1,\ldots,T \qquad (4.33)$$

$$x_{it}, x^c_{it} \in \{0,1\} \qquad \forall i=1,\ldots,n; t=1,\ldots,T \qquad (4.34)$$

$$y_t, y^c_t \in \{0,1\} \qquad \forall t=1,\ldots,T \qquad (4.35)$$

$$z_{jt} \in \{0,1\} \qquad \forall j=1,\ldots,J; t=0,\ldots,T. \qquad (4.36)$$

The fifth formulation from Sect. 5.2 uses an exponential number of variables and states the CPMP as

$$\min \sum_{p \in P_0} f_p \cdot z_{0p} \qquad \text{subject to} \qquad (5.5)$$

$$\sum_{p \in P_i} z_{ip} = 1 \qquad \forall i=0,\ldots,n \qquad (5.6)$$

$$\sum_{i=1}^{n} \sum_{p \in P_i} r_i \cdot \hat{x}^p_{it} \cdot z_{ip} \leq \bar{r}_t \cdot \sum_{p \in P_0} \hat{y}^p_t \cdot z_{0p} \qquad \forall t=1,\ldots,T \qquad (5.7)$$

$$z_{ip} \in \{0,1\} \qquad \forall i=0,\ldots,n; p \in P_i. \qquad (5.8)$$

In the sixth formulation of the CPMP, the capacity constraint (A.10) is added to the network-transformation that is presented in Sect. 5.3.3.1. Using the parameters and variables defined in that section, define arc capacities as $\bar{R}_{ab} = \bar{r}_{\frac{b}{2}}$ if $(a,b) \in E'$ (otherwise, ∞) $\forall (a,b) \in E$. Let $F_{ab} = 1$ arise if $(a,b) \in E'$ (otherwise, 0) $\forall (a,b) \in E$ and let $C_{iab} = 0 \; \forall i; (a,b) \in E$.

$$\min \sum_{(a,b) \in E} F_{ab} \cdot \zeta_{ab} + \sum_{i=1}^{n} \sum_{(a,b) \in E} C_{iab} \cdot z_{iab} \quad \text{subject to} \qquad (5.52)$$

$$\sum_{i=1}^{n} r_i \cdot z_{iab} \leq \bar{R}_{ab} \cdot \zeta_{ab} \quad \forall (a,b) \in E \qquad (A.10)$$

$$\sum_{(a,b) \in E} z_{iab} - \sum_{(b,a) \in E} z_{iba} = \begin{cases} 1 & \forall i=1,\ldots,n; a=0 \\ 0 & \forall i=1,\ldots,n; a=1,\ldots,2\cdot T \\ -1 & \forall i=1,\ldots,n; a=2\cdot T+1 \end{cases} \qquad (5.53)$$

$$z_{iab} \leq \zeta_{ab} \quad \forall i=1,\ldots,n; (a,b) \in E \qquad (5.54)$$

$$z_{iab} \geq 0 \; \forall i=1,\ldots,n; (a,b) \in E \wedge \zeta_{ab} \in \{0,1\} \; \forall (a,b) \in E \qquad (5.55)$$

References

Bitran, G. R., & Yanasse, H. H. (1982). Computational complexity of the capacitated lot size problem. *Management Science, 28*, 1174–1186.

Drexl, A., & Kimms, A. (1997). Lot sizing and scheduling - Survey and extensions. *European Journal of Operational Research, 99*, 221–235.

Florian, M., Lenstra, J. K., & Rinnooy Kan, A. H. G. (1980). Deterministic production planning: Algorithms and complexity. *Management Science, 26*, 669–679.

Karimi, B., Fatemi Ghomia, S. M. T., & Wilson, J. M. (2003). The capacitated lot sizing problem: A review of models and algorithms. *Omega, 31*, 365–378.

Maes, J., McClain, J. O., & Van Wassenhove, L. N. (1991). Multilevel capacitated lotsizing complexity and LP-based heuristics. *European Journal of Operational Research, 53*, 131–148.

Suerie, C. (2005). *Time continuity in discrete time models*. Lecture notes in economics and mathematical systems (Vol. 552). Berlin: Springer.

Wolsey, L. A. (1988). *Integer programming*. New York: Wiley.

Zhu, X., & Wilhelm, W. E. (2006). Scheduling and lot sizing with sequence-dependent setup: A literature review. *IIE Transactions, 38*, 987–1007.

Index

3-DIMENSIONAL MATCHING, 144
3-PARTITION, 114

Adler32, 214
Aggregated covering constraint, 105
Aspiration level criterion, 53
Assumption, 8, 17, 56
Auxiliary LP, 195, 207
    $DL_{(C)}$, 207
    $DL_{(P)}$, 209

Benders'
    cuts, 50
    master problem, 50
    reformulation, 50
    subproblem, 49
Big O-notation, 25
Big number, 41
BIN PACKING, 150
Binary search, 58
Binomial coefficient, 84
    Upper bound, 114
Bitshift, 218
Bound, 31
    Combined Lower, 176
    Dantzig, 59
    Dual Priority Rule Lower, 174
    Initial Lagrangean lower, 189
    Initial lower, 192
    LP Lower, 173
    Müller-Meerbach, 60
    Martello-Toth, 60
    Primal-dual Lower, 176
    Shortest Path Lower, 183

Capacity, 7
    availability, 225
    residual, 18, 59
Capacity constraint, 16
Certificate
    concise, 26
Chvátal-Gomory inequality, 58
Clique, 56
    singleton, 58
Clique inequality, 56
    2nd-degree, 56
    extended, 57, 103
Column generation, 41
Combo Algorithm, 63
Convex hull, 31
Convexification, 37
Core, 62
Cover, 56
    minimal, 56
Cover inequality, 56
    extended, 57, 103
    initial extended, 202
Coverage, 7
CRC32, 214
Critical item, 59
Customer dummy, regular, 147

Dual cutting plane, 37
Dual decomposition, 39
Dynamic programming, 63

Efficiency of item, 56
EXACT COVER BY 3-SETS, 33
Exponential time algorithm, 26
Extension, 56
Extreme point, 28
Extreme ray, 28

Face, 28
    integral, 28
Facet, 28
Facility dummy, regular, 147
Farkas' Lemma, 30
Flexibility, 168
Function
    cost, 63
    non-differentiable, 43
    objective, 27

Halfspace, 28
Heap, 169
Heapsort, 58
Heuristic
    dual-ascent, 38
    First Fit, 166
    hill-climbing, 53
    Iterated Best-of-three, 170
    Knapsack, 186
    Lagrangean, 47, 191
    Lagrangean Hybrid, 201
    local search, 52
    Overlap, 167
    repair, 47
    Tabu Search, 211
    tabu search, 53
    Upper Bound, 183
Hyperplane, 28

Identity matrix, 27
Input length, 25
Integer
    potential, 174
Integer linearization property, 137
Integer Program, 30
Integrality constraint, 31
Integrality property, 38
Introsort, 58

KNAPSACK, 138

Lagrangean
    dual, 37

heuristic, 47
relaxation, 36
Lagrangean dual
    $Z_{(V)}^{(Y)}$, 161
    $Z_{(R)}^{(P)(X)+(Q)}$, 150
    $Z_{(R)}^{(P)+(Q)}$, 150
    $Z_{(C)}$, 156
    $Z_{(P)/(C)}$, 157
    $Z_{(P)}$, 154
Lagrangean hybrid approach, 201
Lagrangean multiplier, 36
    overestimation, 47
Least common denominator, 74
Least common multiple, 74
Linear constraint, 27
Linear Program, 27
    infeasible, 28
    solvable, 28
List, 19
LP relaxation, 31

Maintenance, 1
    Condition-Based, 2
    Corrective, 1
    Design Out, 2
    periodic, 10
    Planned, 2
    Preventive, 1
    Reliability Centered, 2
    Repair and Overhaul, 2
    scheduling, 2
    Time-Based, 3
    Total Productive, 2
Maintenance activity, 1
    labeled, 168
Maintenance slot, 7
    established, 7
Mean Absolute Percentage Error, 220
Memory management, 19
Minimal capacity constraint, 109
Mixed Integer Program, 30
Move, 52, 212
    inverse, 53
    object, 213
    tabu, 53
Multi-state search, 54

Neighbor left, right, 18, 168
Neighborhood, 52
Non-coupling capacity constraint, 108
$\mathcal{NP}$, 26

Index 285

$\mathcal{NP}$-complete, 26
  binary, 26
  strongly, 26

Objective function, 27
Objective function coefficient, 27

$\mathcal{P}$, 26
Pair, 225
PARTITION, 119
Period
  affected, 168
  closed, 18
  open, 18
Period covering constraint, 15
Phase
  diversification, 53
  intensification, 53
Plan
  maintenance, 18
  slot, 18
  task, 18
Polyhedron
  bounded and non-empty, 28
  integral, 28
  unbounded and non-empty, 28
Polynomial time algorithm, 26
Polytope, 28
Post optimization, 167
Pricing out, 34, 42
Primal recovery, 48
Principle of optimality, 64
Priority queue, 169
  of a fixed size, 188
Priority rule, 168, 175
  flexibility, 168
  simplicity, 168
Problem
  Capacitated Facility Location, 144
  Capacitated Lot-Sizing, 276
  Capacitated Planned Maintenance, 7, 15
  Continuous Knapsack, 57, 60
  core, 62
  decision, 25
  dual (linear), 29
  dual master, 37
  Fixed-Charge Capacitated Network Design, 146
  Generalized Bin Packing, 150
  Knapsack, 55
  LP relaxation of the Knapsack, 57
  Minimization Knapsack, 55
  open, 27
  optimization, 25
  Origin-Destination Integer Multi-Commodity Network Flow, 146
  primal (linear), 29
  primal master, 37
  Reduced Dual, 35
  Reduced Primal, 35
  Relaxed Reduced Primal, 177
  Restricted Master, 41
  Restricted Reduced Dual, 177
  separation, 57
  Set Partitioning, 32
  Shortest Path, 84
  Single Source Capacitated Facility Location, 150
  Single Weighted Uncapacitated Planned Maintenance, 89
  Subset Sum, 55
  Uncapacitated Facility Location, 32
  Uncapacitated Network Design Problem, 32
  Uncapacitated Planned Maintenance, 94
  Weighted Uncapacitated Planned Maintenance, 71
Pseudo-polynomial time algorithm, 26

Quant, 74
Quasi-integral, 32
Quicksort, 58

Ray, 28
  extreme, 28
Reaching algorithm, 90
Rearrangement inequality, 137
Reduced costs, 34
Reduction
  polynomial, 26
Right-hand side, 27

Scope, 212
Score, 169
Section, 104
Section covering constraint, 104
SET COVER, 83
Simple upper bound constraint, 16
Simplex algorithm, 34
  Primal-dual, 34
Single-assignment property, 77

Size of the instance, 25
Skill Level Upgrade, 2
Slot covering constraint, 104
Solution, 18, 27
 backward greedy, 61
 feasible, 27
 forward greedy, 60
 greedy, 61
 infeasible, 27
 reference, 165
 single item, 61
Solution space, 27
 empty, 28
Stack, 19
Stairway structure, 72
State, 63
 function, 63
Steepest-ascent mildest-decent strategy, 53
STEINER TREE IN GRAPHS, 32
Step size, 45
 Held-Karp, 47
 parameter, 47
Step size strategy
 constant, 47
 dynamic, 188
Subgradient, 43
 algorithm, 47
 optimization, 43
 pseudo, 187
 zig-zagging, 49
SUBSET SUM, 141
System, 1

Tabu list, 53
 long-term memory, 53
 short-term memory, 53
Tailing off effect, 42
Task, 7
 cover, 18
 redundant, 18
Task horizon, 7, 18
 complete, 18
 consecutive, 18
 covered, 18
 overlap, 18
Task redundancy constraint, 104
Task upper bound constraint, 106
Temporary schedule, 168
Test-set, 225
Test-subset, 225
Theorem
 Geoffrion's, 38
 Minkowski's, 29
Tight formulation, 97
Total dual integrality, 31
Total unimodularity, 31
Transformation
 facility, 147
 network, 145
Triplet, 214

Valid inequality, 28
Variable
 advantageous, 199
 decision, 63
 dual, 29
 fixation, 171
 potential primal basic, 35
 primal, 29
 slack, 27
 structure, 27
Variable upper bound constraint, 16
Vector, 27
VERTEX COVER, 32, 144

Worst case performance ratio, 61
Worst case running time, 25

MIX
Papier aus verantwortungsvollen Quellen
Paper from responsible sources
FSC® C105338

If you have any concerns about our products,
you can contact us on
ProductSafety@springernature.com

In case Publisher is established outside the EU,
the EU authorized representative is:
Springer Nature Customer Service Center GmbH
Europaplatz 3, 69115 Heidelberg, Germany

Printed by Libri Plureos GmbH
in Hamburg, Germany